Nanomaterials and Nanocomposites

Synthesis, Properties, Characterization Techniques, and Applications

T0174374

Nanomaterials and Nanocomposites

Synthesis, Properties, Characterization Techniques, and Applications

Rajendra Kumar Goyal

CRC Press
Taylor & Francis Group
Boca Raton London New York

CRC Press is an imprint of the
Taylor & Francis Group, an **informa** business

CRC Press
Taylor & Francis Group
6000 Broken Sound Parkway NW, Suite 300
Boca Raton, FL 33487-2742

First issued in paperback 2020

© 2018 by Taylor & Francis Group, LLC
CRC Press is an imprint of Taylor & Francis Group, an Informa business

No claim to original U.S. Government works

ISBN-13: 978-0-367-57278-5 (pbk)
ISBN-13: 978-1-4987-6166-6 (hbk)

Library of Congress Cataloging-in-Publication Data

Names: Goyal, Rajendra Kumar, author.
Title: Nanomaterials and nanocomposites : synthesis, properties, characterization techniques, and applications / Rajendra Kumar Goyal.
Description: Boca Raton : Taylor & Francis, CRC Press, 2018. | Includes bibliographical references and index.
Identifiers: LCCN 2017018880| ISBN 9781498761666 (hardback : alk. paper) | ISBN 9781315153285 (ebook)
Subjects: LCSH: Nanocomposites (Materials)
Classification: LCC TA418.9.N35 G685 2018 | DDC 620.1/15--dc23
LC record available at https://lccn.loc.gov/2017018880

Visit the Taylor & Francis Web site at
http://www.taylorandfrancis.com

and the CRC Press Web site at
http://www.crcpress.com

Dedicated to my wife, Manju; my lovely daughter, Muskan; and my

son, Prasun, for their loving support and patience. It is also dedicated

to my mother, Savitri, and my father, Gangasharan.

"If You Want to Shine Like a Sun, First Burn Like a Sun."

Bharat Ratna Dr. A P J Abdul Kalam

Contents

Preface xv
Author xvii

1. Introduction to Nanomaterials and Nanotechnology 1
   1.1   Introduction 1
   1.2   Nanotechnology in Past 2
   1.3   Nanotechnology in the Twenty-First Century 4
   1.4   Classification of Nanomaterials 4
         1.4.1   0-D Nanomaterials 5
         1.4.2   1-D Nanomaterials 5
         1.4.3   2-D Nanomaterials 6
         1.4.4   3-D Nanomaterials 6
   1.5   Size Effect on GBs 7
   References 9

2. Length Scale and Calculations 11
   2.1   Size Effect on Surface Area of Cubical Particles 11
   2.2   Size Effect on Surface Atoms of Cubical Particles 14
   2.3   Size Effect on Surface Atoms of Spherical Particles 15
   References 20
   Further Reading 20

3. Effect of Particle Sizes on Properties of Nanomaterials 21
   3.1   Thermal Properties 21
         3.1.1   Melting Point 21
         3.1.2   Heat Capacity 25
         3.1.3   Curie Temperature 25
         3.1.4   Coefficient of Thermal Expansion 27
   3.2   Electrical Properties 29
   3.3   Lattice Constant 31
   3.4   Phase Transformation 33
   3.5   Mechanical Properties 33
         3.5.1   Elastic Modulus 33
         3.5.2   Hardness and Strength 34
         3.5.3   Toughness 36
         3.5.4   Fatigue and Creep 39
   3.6   Magnetic Properties 41
   3.7   Optical Properties 43
   3.8   Wear Resistance 45
   3.9   Chemical Sensitivity 46
   3.10 Dielectric Constant 49
   References 50

4. What Have We Learnt From the Nature and History? 53
   4.1   Nanotechnology of Nature 53

4.1.1 Lotus Leaf 53
4.1.2 Morpho Butterfly 55
4.1.3 Peacock Feather.. 55
4.1.4 Cuttlefish... 56
4.1.5 Water Strider.. 57
4.1.6 Spider Silks 57
4.1.7 Beetles... 59
4.1.8 Glass Sponges (Euplectella).. 59
4.1.9 Gecko 60
4.1.10 Chameleon 61
4.1.11 Abalone Nacre 61
4.1.12 Magnetotactic Bacteria... 63
4.1.13 Antireflective Nanostructures of Moth Eyes......... 64
4.1.14 Natural Nanocomposite 64
4.1.15 Dentin... 65
4.1.16 Natural Solar Cell 66
4.2 Nanotechnology in Ancient History? 66
4.2.1 Damascus Sword 66
4.2.2 Stained Glass.. 68
4.2.3 Maya Blue Paint 70
4.2.4 Age-hardened Aluminum Alloys 70
References 71

5. Synthesis of Nanomaterials..73
5.1 Top-Down Approaches 73
5.1.1 Mechanical Alloying... 73
5.1.2 Severe Plastic Deformation 78
5.1.3 Lithography 81
5.2 Bottom-Up Approaches.. 88
5.2.1 Physical Vapor Deposition (PVD)............................. 88
5.2.2 Molecular-Beam Epitaxy 89
5.2.3 Chemical Vapor Deposition 90
5.2.4 Colloidal or Wet Chemical Route 91
5.2.5 Reverse Micelle Method 93
5.2.6 Green Chemistry Route 94
  5.2.6.1 Synthesis of Metallic NPs............................. 96
  5.2.6.2 Synthesis of Oxide NPs................................. 98
  5.2.6.3 Factors Affecting Size and Morphology of NPs 99
5.2.7 Sol-gel Method 99
5.2.8 Combustion Method 101
5.2.9 Atomic Layer Deposition... 102
References 104

6. Synthesis, Properties, and Applications of Carbon Nanotubes 107
6.1 Introduction 107
6.2 Synthesis of CNTs 107
6.2.1 Arc Discharge Method.. 108
6.2.2 Laser Ablation 110
6.2.3 Chemical Vapor Decomposition............................... 110

6.3 Characterization Techniques to Analyze the Purity 112
6.4 Purification of CNTs 113
    6.4.1 Chemical Oxidation Method 113
    6.4.2 Physical Method 114
6.5 Structures and Properties of CNTs 116
    6.5.1 Structures 116
    6.5.2 Electrical Properties of CNTs 119
    6.5.3 Mechanical Properties 120
    6.5.4 Thermal Properties 121
    6.5.5 Thermoelectric Power 123
6.6 Applications of CNTs 124
    6.6.1 Nanocomposites 124
    6.6.2 Hydrogen Storage 125
    6.6.3 Energy Storage Devices 125
    6.6.4 Field Emission Emitters 126
    6.6.5 Microelectronics 128
    6.6.6 Electrical Brush Contacts 128
    6.6.7 Sensors and Probes 129
    6.6.8 Artificial Implants 129
    6.6.9 Water Filters 129
    6.6.10 Coatings 130
References 130

7. **Synthesis, Properties, and Applications of Nanowires** 135
7.1 VLS Method 135
7.2 Template Method 138
    7.2.1 Electrochemical Deposition 138
    7.2.2 Electroless Deposition 140
    7.2.3 Pressure Injection Technique 140
7.3 Properties of NWs 140
    7.3.1 Thermal Properties 140
        7.3.1.1 Melting Point 140
        7.3.1.2 Coefficient of Thermal Expansion 141
        7.3.1.3 Thermal Conductivity 141
    7.3.2 Optical Properties 142
    7.3.3 Electrical Properties 144
7.4 Applications of NWs 145
    7.4.1 Electronics 145
    7.4.2 Thermoelectric Generator 145
    7.4.3 Sensors 146
    7.4.4 Magnetic Information Storage Devices 147
    7.4.5 Solar Cell 147
    7.4.6 Nanocomposites 147
References 148

8. **Graphene** 151
8.1 Introduction 151
8.2 Synthesis of Graphene 151
    8.2.1 Mechanical Exfoliation 151

        8.2.2   Chemical Exfoliation 152
        8.2.3   Reduction of Graphene Oxide.. 152
        8.2.4   Chemical Vapor Deposition 153
        8.2.5   Dry Ice Reduction Method 154
    8.3   Properties of Graphene 155
        8.3.1   Structural Properties.. 155
        8.3.2   Mechanical Properties 156
        8.3.3   Thermal Properties... 159
              8.3.3.1   Thermal Conductivity 159
              8.3.3.2   Thermal Expansion 159
              8.3.3.3   Thermal Stability... 160
        8.3.4   Electrical Properties 160
    8.4   Properties of Graphene/Polymer Nanocomposites 161
        8.4.1   Mechanical Properties of Graphene Composites................... 161
        8.4.2   Thermal Properties of Graphene Composites 163
        8.4.3   Electrical Properties of Graphene Composites...................... 165
        8.4.4   Gas Barrier Properties.. 166
    8.5   Applications of Graphene 167
    References 167

9. Magnetic Nanomaterials... 171
    9.1   Introduction 171
    9.2   Hysteresis M–H loop 171
    9.3   Effect of Particle Size on M–H Loop 173
    9.4   Surface Modification of Magnetic Nanoparticles............................. 177
    9.5   Applications of Magnetic Nanoparticles.. 178
        9.5.1   Hyperthermia.. 178
        9.5.2   Medical Diagnostics 179
        9.5.3   Magnetic Cell Separation 179
        9.5.4   Drug Delivery 179
        9.5.5   Magnetic Refrigeration 180
        9.5.6   Removal of Heavy Metals 180
        9.5.7   Oil Removal.. 180
    References 180

10. Processing, Properties, and Applications of Polymer Nanocomposites 183
    10.1   Introduction 183
    10.2   Polymer–Clay Nanocomposites.. 184
        10.2.1   Structure of Montmorillonite... 184
        10.2.2   Fabrication of Polymer/Clay Nanocomposites.................... 185
              10.2.2.1   *In Situ* Polymerization Method.................................. 185
              10.2.2.2   Solution Method 185
              10.2.2.3   Melt-Mixing Method... 186
        10.2.3   Characterization of Polymer/Clay Nanocomposites............. 188
        10.2.4   Properties of Polymer/Clay Nanocomposites 189
              10.2.4.1   Thermal Properties 189
              10.2.4.2   Permeability 191
              10.2.4.3   Mechanical Properties.. 193
              10.2.4.4   Abrasion and Wear Resistance 196

10.3   Polymer/CNT Nanocomposites 196
     10.3.1   Electrical Properties 196
     10.3.2   Mechanical Properties 199
     10.3.3   Thermal Conductivity 202
10.4   Polymer/BN Nanocomposites 203
     10.4.1   Mechanical Properties 205
     10.4.2   Thermal Properties 205
     10.4.3   Wear Resistance 207
10.5   Polymer/BaTiO$_3$ Nanocomposites 208
     10.5.1   Introduction 208
     10.5.2   Preparation of PC/BaTiO$_3$ Nanocomposites 208
     10.5.3   Electrical Properties 209
     10.5.4   Thermal Properties 211
     10.5.5   Mechanical Properties 212
References 213

11. Polymeric Nanofibers 217
11.1   Synthesis of Polymeric Nanofibers 217
11.2   Parameter Affecting Diameter of Nanofibers 218
11.3   Properties of Nanofibers 219
     11.3.1   Mechanical Properties 219
     11.3.2   Magnetic Properties 222
     11.3.3   Thermal Properties 223
     11.3.4   Electrical Properties 225
11.4   Applications 226
     11.4.1   Nanocomposite Fibers for Structural Applications 226
     11.4.2   Nanogenerators 228
     11.4.3   Nanofilters 228
     11.4.4   Biomedical Application 228
     11.4.5   Protective Fabric for Military 229
     11.4.6   Functional Sensors 229
References 230

12. Characterization of Nanomaterials 233
12.1   X-Ray Diffraction 233
12.2   Optical Spectroscopy 238
     12.2.1   Raman Spectroscopy 239
     12.2.2   UV–vis Spectroscopy 240
     12.2.3   Photoluminescence Spectroscopy 243
     12.2.4   Fourier Transform Infrared Spectroscopy 244
12.3   Surface Area Analysis (BET Method) 245
12.4   Light Scattering Method 247
12.5   Electron Microscopy 250
     12.5.1   Scanning Electron Microscopy 251
           12.5.1.1 Basic Principle of SEM 251
           12.5.1.2 Specimen Preparation for SEM 252
     12.5.2   Transmission Electron Microscopy 253
           12.5.2.1 Basic Principle of TEM 253
           12.5.2.2 Specimen Preparation for TEM 256

12.6   Scanning Probe Microscopy...258
        12.6.1   Basic Principle of STM 258
        12.6.2   Atomic Force Microscopy 260
12.7   X-ray Photoelectron Spectroscopy 262
12.8   Thermal Analyzer...263
12.9   Zeta Potential..264
References 264
Further Reading 266

13. Corrosion Behavior of Nanomaterials 267
13.1   Introduction..267
13.2   Corrosion Resistance of Nanocrystalline Metals 268
13.3   Corrosion Resistance of Nanocomposite Coating 269
References 274

14. Applications of Nanomaterials..275
14.1   Nanofluids...275
        14.1.1   Introduction...275
        14.1.2   Stabilization of Nanofluids...275
        14.1.3   Synthesis of Nanofluids...276
        14.1.4   Applications of Nanofluids 277
                   14.1.4.1 Automotive Applications 277
                   14.1.4.2 Biomedical Applications..278
                   14.1.4.3 Coolants 279
                   14.1.4.4 Dynamic Seal 281
14.2   Hydrogen Storage 281
14.3   Solar Energy..284
        14.3.1   Introduction...284
        14.3.2   Photoelectrochemical Cells 285
        14.3.3   Thermoelectric Devices 287
14.4   Antibacterial Coating 289
14.5   Giant Magnetoresistance 291
14.6   Single-Electron Transistor (SET) 292
14.7   Construction Industry...292
14.8   Self-Cleaning Coatings..293
        14.8.1   Hydrophobic coating..293
        14.8.2   Hydrophilic coatings..293
14.9   Nanotextiles...294
        14.9.1   Textile Fabrics..294
        14.9.2   Intelligent Textiles 295
14.10   Biomedical Field 296
        14.10.1   Medical Prostheses 296
        14.10.2   Drug Delivery 296
        14.10.3   Cancer Treatment..296
        14.10.4   Wound Dressing 296
        14.10.5   Cosmetics 296
14.11   Nanopore Filters..297
14.12   Water Treatment..297
        14.12.1 TiO_2 Nanomaterials...298

        14.12.2 CNT Powder 298
        14.12.3 Zerovalent Metals 298
        14.12.4 Dendrimers 299
        14.12.5 Graphene 299
        14.12.6 Polymeric Nanofibers 299
    14.13 Nanodiamond 300
    14.14 Automotive Sector 300
        14.14.1 Introduction 300
        14.14.2 Solar Energy 301
        14.14.3 Fuel Cell 301
        14.14.4 Vehicle Radiator 302
        14.14.5 Diesel Particulate Filter 302
        14.14.6 Other Applications 302
    14.15 Catalysts 303
    References 305

15. Risks, Toxicity, and Challenges of Nanomaterials 309
    15.1 Introduction to Nanoparticle Risks and Toxicity 309
        15.1.1   Dermal Exposure 310
        15.1.2   Ingestion Exposure 311
        15.1.3   Inhalation Exposure 312
    15.2 Risks And Toxicity from Metallic Nanoparticle 312
    15.3 Risks And Toxicity from Oxide Nanoparticle 313
    15.4 Challenges 314
    References 315

Index 317

Preface

The idea of writing a book on nanomaterials and nanotechnology came after I gave a series of lectures to undergraduates/graduates students and faculty of various engineering colleges in several faculty development programs at the College of Engineering, Pune (COEP), Savitribai Phule Pune University (SPPU), Pune, Maharashtra (India) during 2008–2017. This book would serve a basic introduction to comprehensive review on recent advances in nanomaterials and nanotechnology. In the engineering and scientific field, undergraduate students and first-year graduate students will find this book useful. I hope the new readers will not face any problem in understanding the content of this book because I have tried to compile systematically all the chapters on classification, calculation of surface-area-to-volume ratio and percentage surface atoms, properties versus particle size, synthesis of nanoparticle/carbon nanotubes/nanowires and polymer nanocomposites, and their applications. I also hope the specialists in the different fields may not feel uncomfortable after reading this book. In addition, the book covers several most commonly used characterization tools to characterize the nanomaterials. There are, in total, 15 chapters. Chapter 1 deals with the introduction to nanomaterials and nanotechnology, and classification of nanomaterials based on dimensionality. Chapter 2 deals with the calculation of large surface-area-to-volume ratio and increasing number of surface atoms compared to bulk atoms with decreasing particle size. A good understanding of the surface-area-to-volume ratio of the nanoparticles is essential for the understanding of the subject and exploitation of thermal, mechanical, electrical, and optical properties of nanomaterials. Chapter 3 focuses on the effect of particle (or grain) size on thermal, mechanical, electrical, magnetic, optical, and chemical sensitivity of nanomaterials. Particular attention was paid to the change of properties of nanomaterials compared to their bulk. Chapter 4 is devoted to the inspiration from the ancient history and nature, which has given plentiful examples of nanotechnology such as stained glass, Lycurgus cup, magnetotactic bacteria, Abalone shell, spider web, chloroplast, base of geckos' feet, Spider silk, etc. Chapter 5 discusses the top-down and bottom-up approaches for the synthesis of various nanomaterials. The top-down approaches include ball milling/mechanical alloying, nanolithography, soft lithography (i.e., nanoimprint), and severe plastic deformation (SPD) to produce nanostructures. Bottom-up approaches include inert gas condensation, chemical vapor deposition, colloidal method, green chemistry, sol-gel method, atomic layer deposition (ALD), etc. Chapter 6 deals with the synthesis, purification, properties, and applications of carbon nanotubes (CNTs). The synthesis and properties of semiconducting and metallic nanowires are the subject of Chapter 7. In this chapter, the growth of one-dimensional nanostructures using vapor–liquid–solid (VLS), template, and electrochemical methods are discussed. The thermal, mechanical, and electrical properties of the nanowires and their applications are also discussed. Chapter 8 focuses on the synthesis, properties and applications of graphene. Synthesis, functionalization, and applications of magnetic nanomaterials in biomedical, drug delivery, gene therapy, hyperthermia, magnetic resonance imaging (MRI) contrast agent, and magnetic recording tape are discussed in Chapter 9. Processing, properties, and applications of polymer matrix nanocomposites reinforced with clay, graphene, carbon nanotubes, barium titanate, and boron nitride particles are discussed in Chapter 10. This chapter also deals with percolation threshold phenomena in CNT polymer and the factors which affect the percolation threshold phenomenon. Chapter 11 focuses on the

synthesis and properties of polymeric nanofibers. Chapter 12 deals with the characterization of nanomaterials using the most commonly used characterization tools such as x-ray
diffraction (XRD), scanning electron microscopy (SEM), transmission electron microscopy
(TEM), scanning probe microscopy (SPM), Raman spectroscopy, UV–visible spectroscopy,
laser particle size analyzer, specific surface area analyzer (BET), and x-ray photon spectroscopy (XPS). Chapter 14 highlights the effect of alloying elements or fillers on the corrosion behavior/oxidation resistance of nanomaterials. Chapter 14 gives some examples
of the applications of nanomaterials in nanofluids/ferrofluids, hydrogen storage, fuel cell,
antibacterial fabrics, sensors, thermal interface materials (TIMs), biomedical, thermoelectrics, nanocomposite coating, and construction industries. Chapter 15 highlights the risks,
toxicity, and challenges of nanomaterials for the human being. Utmost care has been taken
in providing a necessary conceptual understanding of various topics related to bulk to
nanomaterials. Individuals who want further depth should refer the research papers or
books cited in the list of references.

I hope the readers will find this book inspiring, and will be motivated to go deeper
into this fascinating field of nanomaterials and nanotechnology. I welcome suggestions
and constructive criticisms from readers towards improvements of future editions of this
book. This is my first solo book and I am grateful to all the editors and authors of text-
books, handbooks, and research papers of different journals because without their pub-
lished literature, this book would have not been possible in present shape.

I want to thank my family, especially my wife Manju, for her steady support while I was
writing this book. Nevertheless, my lovely daughter, Muskan, and son, Prasun, played a crucial role to inspire me to complete it. Furthermore, I thank all the authors whose works have
been referred for writing this book, without their literature, things would have been much
more difficult for me. I would like to thank Dr. J. K. Chakravartty, Steel Chair Professor,
Department of Metallurgy and Materials Science, COEP, Pune, for reading the drafts and
making important suggestions in improving the quality of this book. I also thank Professor
B. B. Ahuja, director of COEP, Pune, for his moral support and encouragement. Many of
my MTech students, Mr Shubham, Mr Sachin, and Mr Rohit have contributed by preparing flow charts, diagrams, sketches, and providing photographs as well as carefully going
through the list of references of the book. I would also like to thank team members of CRC
Press for their help, suggestions, and careful proofreading of this book. Last but not the
least, I would like to thank the almighty for showering the blessings on me.

I hope you enjoy reading the book!

Dr. Rajendra Kumar Goyal
Associate Professor
Department of Metallurgy and Materials Science
College of Engineering Pune, Shivajinagar
Pune, Maharashtra (India)

Author

Rajendra Kumar Goyal is an associate professor of metallurgy and materials science at the College of Engineering, Pune (COEP), Pune, India. He also holds a position of treasurer in the student cooperative society of COEP and professional society "The Indian Institute of Metals (IIM) Pune Chapter." He earned PhD degree from the department of Metallurgical Engineering and Materials Science, Indian Institute of Technology Bombay, Mumbai, India, in 2007. He has been awarded best PhD thesis prize in materials science for the year 2007 by the Materials Research Society of India (MRSI). He obtained BE in metallurgical engineering from Malviya Regional Engineering College (now, National Institute of Technology) Jaipur, University of Rajasthan, India, in 1996. He received "Alumni Distinguished Faculty Fellow (ADFF) Award" for the duration 2010–2012 and 2015–2017 from Alumni Association of College of Engineering, Pune. He has a lifetime membership of professional society "The Indian Institute of Metals (IIM)," "Powder Metallurgy Association of India (PMAI)," and "Society of Polymer Science." His research group works on polymer composites/nanocomposites, nanomaterials, characterization of materials, electronic materials, structure–properties relationship, etc. He has coauthored 1 book and authored more than 130 research papers in peer-reviewed journals and conferences. He has organized several faculty development programs (FDP) in the area of nanomaterial and characterization. He has delivered a dozen of expert lectures in various conferences/FDP. He has advised more than 26 MTech and 1 PhD students at the College of Engineering, Pune, Savitribai Phule Pune University, Pune, India. His students received several awards for the work related to polymeric nanocomposites. He has served for more than 12 years in research laboratory and industries, and 9 years academic and research experience. He is a member of editorial board and reviewer of many international journals. He has successfully completed several research projects funded by the national funding agencies such as ISRO, UGC, and AICTE of India.

1

Introduction to Nanomaterials and Nanotechnology

1.1 Introduction

The term *nano* is originated from the Greek word for *dwarf* or an abnormally short person. It is used as a prefix for any unit such as a second or a meter, and it means a billionth (10^{-9}) of that unit. Therefore, a nanometer (nm) is a billionth of a meter. In a broad term, nanomaterials are the materials with sizes of the individual grain or particle in the range of 1–100 nm at least in one dimension. One nanometer is approximately the length equivalent to 10 hydrogen or 5 silicon atoms aligned in a line. A nanomaterial is often characterized by a dimension linked either to the dimension of the salient nanofeatures making up the material or to their organization. When some interesting property of a material emerges from the organization or pattern of random or well-ordered nanopatterns, the resultant material is referred to as nanostructure or nanostructured material. To get an idea about how much larger nanomaterials have compared to their bulk counterparts, think about a small particle with a diameter of a little >10 μm $= 10^{-5}$ m, which itself is too small to be seen by naked eyes and compare it with a particle with a diameter of 1 nm $= 10^{-9}$ m. The ratio of the diameters of these two objects is 10^4. In other words, nanoparticle with a size of 1 nm is smaller by 10^4 times than an object of 10 μm in size. In addition, this single micrometer-sized particle is equivalent to 10^{12} nanoparticles by mass. This simple comparison makes it clear that nanoparticles are really very small and also expose a large fraction of atoms on their surfaces, that is, $>1\%$ depending upon the size, whereas bulk has insignificant amount of surface atoms (i.e., $<0.01\%$). Nanomaterials exhibit multifunctional properties which are distinctively different from that of bulk materials. For example, the crystals in the nanometer scale have a low-melting point, reduced lattice constants, different crystal structure, disappearance or shift of Curie temperatures (of ferroelectrics and magnetic materials), changed electrical conductivity of metals or oxides, increased oxidation and wear resistance, and higher sensitivity of sensors compared to their bulk counterparts. The semiconductors and metals become insulators and semiconductors, respectively, when the characteristic size of nanoparticles is few nanometers. Interestingly, gold, palladium, and platinum nanoparticles with size below $\sim$5 nm exhibit excellent catalytic properties at low temperatures. The decrease in gold particle size shows blueshift (i.e., a decrease in peak absorption wavelength). Similarly, when the size of semiconductors is reduced below their Bohr radius, a dramatic change in optical properties is found, for example, change in color. This property has been exploited to probe whether a certain DNA corresponds to a particular individual or not. Quantum dots are crystals consisting of a few hundred atoms, where the electrons are confined to widely separated energy levels. The shape and size of quantum dots determine their electronic, magnetic, and optical properties. Magnetic nanoparticles are now being used to detect the particular biological species that cause disease.

There are many methods to classify synthesis techniques of nanomaterials. Some of those methods are top-down and bottom-up approaches. Top-down includes mechanical alloying, nanolithography, electron-beam lithography, x-ray lithography techniques, soft lithography, and severe plastic deformation. The bottom-up approach includes synthesis of nanoparticles by physical vapor deposition, chemical vapor deposition, colloidal route, green chemistry route, and sol-gel method, where there is buildup of atoms or molecules layer by layer. In general, nanoparticles or nanoplatelets or nanotubes are energetically not favorable and therefore their properties have been exploited by embedding them in polymer or metal or ceramic matrices to make them thermodynamically stable nanocomposite.

In general, nanotechnology is the science and technology which involves design, fabrication, and applications of nanostructures and nanomaterials with typical size smaller than 100 nm. Nanotechnology is a broad interdisciplinary field, which requires expertise in physics, chemistry, materials science, biology, mechanical and electrical engineering, medicine, and their collective knowledge. It is the boundary between atoms/molecules and the macroworld. Nature provides numerous excellent examples of nanotechnology such as biomineralization (in abalone shell or human bones), *lotus leaves,* photosynthesis (in plants), strength and toughness of spider silk, adhesion mechanism in Gecko foots, peacock feather, and excellent antireflection property in moth eye, etc.

1.2 Nanotechnology in Past

The twenty-first century has witnessed a tremendous upsurge in the field of Nanomaterials and Nanotechnology. However, this field is not new because its seeds have been sown in the past centuries. A well-known gold ruby glass, which consists of a glass matrix with dispersion of gold nanoparticles was first produced by the Assyrians in the seventh-century BC and reinvented by Kunkel in Leipzig in the seventeenth century [1]. The stained-glass windows and Lycurgus cup are the examples of Medieval era/Roman era which consist of a few tens parts per million (ppm) of gold and silver nanoparticles in the glass matrix and exhibit unique optical properties. Chinese are known to use gold nanoparticles as an inorganic dye to introduce red color into their ceramic porcelains for more than thousand years. In 1857, a stable colloidal gold has been prepared by Faraday which was destroyed during World War II. The colloidal gold was, and is still in use to diagnose several diseases and for treatment of arthritis. The beautiful Mayan blue paint of the ancient Mayan world has long been admired for its marvelous color qualities as well as its inherent resistance to deterioration and wear over long periods of time. The particular blue color has been attributed to the dispersion of the unique combination of oxide nanoparticles, indigo molecules, and clay particles.

Although some unique examples of nanomaterials or nanotechnology can be traced back for centuries, the current excitement of nanotechnology is toward shrinking of the devices such as personal computer, laptop, smart phone, embedded processors etc. In 1959, physics Nobel Laureate Richard Feynman gave a talk entitled "There's Plenty of Room at the Bottom" at California Institute of Technology on the occasion of the annual meeting of the American Physical Society. Although, Feynman could not predict about nanotechnology, his lecture became a central point in the field of *nanotechnology.* The term *nanotechnology* was first used in 1974 by Norio Taniguchi to refer to the precise and accurate tolerances required for machining and finishing materials using top-down approach. The top-down approach describes the processes starting from large pieces of material to produce the intended structure by mechanical or chemical methods. The bottom-up approach processes are the chemical processes starting from atoms or molecules as the building blocks to produce nanoparticles, nanotubes, nanorods, thin films, or layered structures. The continued decrease in device dimensions has followed the well-known Moore's law predicted in 1965 (Figure 1.1), which states that the dimension of a device halves approximately every 18 months and today's transistors have well fallen in the nanometer range [2]. In fact, in 1965, Gordon Moore noticed that the number of components in integrated circuits had doubled every year between 1958 and 1965. Thereafter, the doubling period increased from 1 year to ca. 18–24 months. For example, there were 4500 transistors in the 8080 microprocessor (with memory size 64 KB) introduced in 1974. Compared to this, the Intel I7 QuadCore, introduced in 2008, contains 731 million transistors and the 10-Core Xeon available in 2011 has 2.6 billion transistors [3].

Nevertheless, Moore's law has powered the information technology revolution since the 1960s, but now it is nearing to saturation. The exponential growth as predicted by

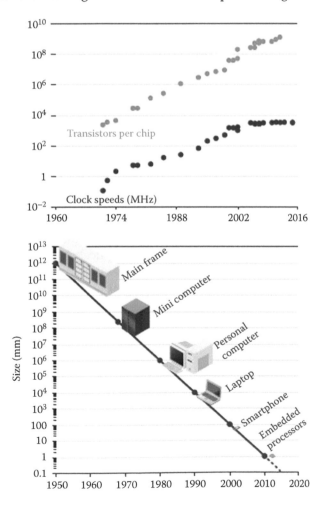

FIGURE 1.1
Miniaturization of electronic devices according to Moore's law since 1970 and its saturation. (Reprinted by permission from Springer Nature. *Nature*, Waldrop M. More than Moore. 530: 144–147, copyright 2016.)

Moore's law resulted in transformation of the first crude home computers of the 1970s into the sophisticated machines of the 1980s and 1990s, and from there gave rise to high-speed Internet, smartphones and the wired-up cars, refrigerators, and thermostats that are becoming prevalent today [2]. The twentieth century has been known for extensive investigations of heterogeneous catalysis, ultrafine powders, and thin films. However, these investigations raised several questions about the effect of the small size of particles or grains on the properties of nanostructured materials. By the start of the twenty-first century, the nanostructured materials such as nanopowders/nanocoatings of metals, alloys, intermetallics, oxides, carbides, nitrides, borides, and carbon nanostructures, nanocomposites, nanoporous materials, and biological nanomaterials have been studied extensively for several applications [4,5]. The progress in the field of nanotechnology has brought in breakthroughs in medicine and healthcare, nanoelectronics, energy, biotechnology, and information technology.

1.3 Nanotechnology in the Twenty-First Century

In the last few decades, the technologies for the fabrication of integrated circuits have made the smallest structure (i.e., <100 nm) elements of microelectronic chip devices by mass production. The gate length of solid-state transistors has reached in the range of few tens of nanometers [4]. Nowadays, microprocessors have circuit features with size of ~14 nm across, smaller than most of the viruses. In the early 2000s, the features shrunk below ~90 nm; however, this miniaturization is facing problem of heat dissipation which has restricted the speed of device (Figure 1.1) [2]. Thus, due to multifunctional miniaturization, cell phones are becoming tinier, laptops are getting lighter, and a thin strand of optical fiber is replacing thick bundles of copper telephone wire. For example, a 100 times decrease in linear size scale translates into 10^6 by volume, that is, 1 million times more circuitry could be stuffed into the same volume. Scientists have exploited the bottom-up nanotechnology of the nature and replicated a large number of examples including self-cleaning glasses, nanofabrics, nanomotors, nanobrushes, nanopaints, and hundreds of more developments by assembling and manipulating atoms, molecules, and solids. It has been well studied that as the diameter of metallic nanowires shrinks, the resistivity and failure current density increase due to increased scattering of the conduction electrons by the surface of the nanowires, internal grain boundaries (GBs), and surface-to-volume ratio. Moreover, due to the surface plasmon resonance (SPR) effect, the extinction peaks of gold nanowire arrays possessed a blueshift with decreasing diameter of the nanowires [6].

1.4 Classification of Nanomaterials

Nanomaterials are usually classified on the basis of their dimensionality, morphology, composition, uniformity, and agglomeration. Based on nanoparticle's dimensionality, nanomaterials can be classified into 0-D, 1-D, 2-D, and 3-D.

1.4.1 0-D Nanomaterials

They have all three dimensions negligibly small. They are also known as artificial atoms (or quantum dots) because their energy levels are discrete. Metallic nanoparticles include gold and silver nanoparticles, whereas semiconductor nanoparticles include quantum dots of CdS, CdSe, and CdTe. Nanoparticles may have shapes such as spherical, cubic, or polygonal with size in the range of 1–50 nm. The most common example is fullerene, which was identified by Kroto and coworkers. Fullerene (C_{60}) is the smallest and the most stable structure owing to high degree of its symmetry. C_{60} consists of 60 carbon atoms. The C_{60} molecule looks like a soccer ball (Figure 1.2a) and it is often called a soccer ball molecule. This molecule is composed of 12 pentagons and 20 hexagons whose vertexes contain carbon atoms. Each carbon atom of pentagon or hexagon is shared by three other carbon atoms; hence, total carbon atoms in C_{60} are $= 12$ (pentagonal) $\times 5$ carbon atoms $\times \frac{1}{3} + 20$ (hexagonal) $\times 6$ carbon atoms $\times \frac{1}{3} = 60$. Solid C_{60} has face-centered cubic lattice at room temperature and its density is ~1.69 g/cm³. Molecules are freely rotating due to weak intermolecular interaction. Fullerene is an allotropic modification of carbon. Besides C_{60}, other fullerene molecules are C_{70}, C_{76}, C_{78}, and C_{84}. C_{70} molecule consists of 70 carbon atoms. Fullerene molecules are quite stable; however, like in graphene, each carbon atom has only three neighbors. Therefore, there are free valences, which may be used for functionalization. It is possible to attach metal atoms or other molecules at the surface. Owing to 0-D structure, fullerene has a minimum surface energy [1,7].

1.4.2 1-D Nanomaterials

They have two dimensions of particulates in the nanometer scale and the third dimension is significant compared to other two dimensions, that is, they have length of several micrometers (or >100 nm) and diameter of only a few nanometers. Examples include nanotubes, nanofibers, nanorods, or whiskers of metals or oxides, and carbon nanotubes (CNTs) and carbon nanofibers (CNFs). These entities have a very high-surface area and aspect ratio, which are useful in nanocomposites. Imogolite is a naturally occurring 1-D silicate

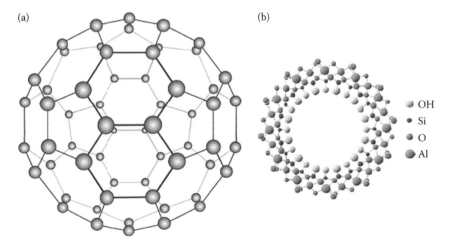

FIGURE 1.2
The atomic arrangement in (a) a fullerene C_{60}. (From Rogers B., Pennathur S., Adams J. 2011. *Nanotechnology: Understanding Small Systems*. 2nd ed. CRC Press. Reproduced with permission from Taylor & Francis Group.) (b) An imogolite fiber. (Vollath D: *Nanomaterials: An Introduction to Synthesis, Properties and Applications*. 2nd ed. 2013. Copyright Wiley-VCH Verlag GmbH & Co. KGaA. Reproduced with permission.)

having chemical formula of $Al_2SiO_3(OH)_4$. Its tubes have an inside diameter of $\sim$1 nm and outside diameters of $\sim$2 nm. The internal and outer diameters can be adjusted by varying the silicon/aluminum ratio. The tubes may be up to a few micrometers in length, and both natural and synthesized imogolite tubes form bundles with diameters ranging from 5 to 30 nm due to their high-surface area-to-volume ratio. Figure 1.2b shows the atomic arrangement in a cross section of an imogolite tube. The structure of imogolite consists of aluminum, silicon, oxygen, and $(OH)^-$ ions arranged in rings. The $(OH)^-$ ions present on the surface allow its functionalization by the organic molecules. They have surface area of $\sim$1000 $\pm$ 100 m^2/g and the Mohs hardness of ca. 2–3. Although they have high-aspect ratio, their lower strength and hardness limit their applications for the nanocomposites [1].

1.4.3 2-D Nanomaterials

They have one dimension in the nanometer scale and other two dimensions are significantly large compared to third dimension (i.e., thickness). Examples are graphene, nanolayers, nanoclays, nanosheets, nanofilms, platelet-like structure, nanoflakes, or nanoplatelets and silicate nanoplatelets. The area of the nanofilms or coating is of the order of several square centimeters, but the thickness is in the range of 1–100 nm. Montmorillonite (MMT) clay is a hydrous alumina silicate mineral whose lamellae are constructed from an octahedral alumina sheet sandwiched between two tetrahedral silicate sheets (2:1 layer silicate). In a natural state, the clay platelets are $\sim$1 nm in thickness and other dimensions (i.e., length and width) can be of the order of 150 nm to 2 μm.

1.4.4 3-D Nanomaterials

All dimensions of these materials are outside the nanometer range. They are generally referred to as equiaxed nanoparticles or nanocrystals. The best example is the nanostructured bulk, which consists of individual blocks with the size in the range of 1–100 nm. 3-D nanomaterials are also known as bulk nanomaterials, which are not confined to the nanoscale in any dimension. These materials have all three dimensions >100 nm. These materials possess a nanocrystalline structure or features at the nanoscale, that is, bulk nanomaterials consist of a multiple arrangement of nanosized crystals typically in different orientations. In other words, 3-D nanomaterials can contain dispersions of nanoparticles, nanowires, nanotubes, or multinanolayers, etc. in the matrix.

Figure 1.3 shows the sizes of several natural nanostructured materials, which are comparable to those of viruses, DNA, and proteins, while microstructures are comparable to cells, organelles, and larger physiological structures. For example, a red blood cell is $\sim$7 μm wide, a hair is 60 μm, pollen is 100 μm, while lung alveoli are $\sim$400 μm. The nanoparticles contain tens to hundreds of atoms, with dimensions at the scale of nanometers, which are comparable to the size of the viruses. For example, a human immunodeficiency virus (HIV) particle is 100 nm in diameter and called as nano-organism. Similar to viruses, some nanoparticles can penetrate lung or dermal skin barriers and enter the circulatory and lymphatic systems of humans and animals, reaching most bodily tissues and organs, and potentially disrupting cellular processes and causing disease [8].

Based on the morphology, nanomaterials are generally classified into high-aspect-ratio and low-aspect-ratio nanoparticles. Nanotubes and nanowires with various shapes such as helices, zigzags, and belts are the examples of high-aspect-ratio nanoparticles, whereas helical-, spherical-, cubic-, pillar-, and oval-shaped nanoparticles are the examples of low-aspect-ratio nanomaterials. Most of these nanoparticles occur in the form of powder, suspensions,

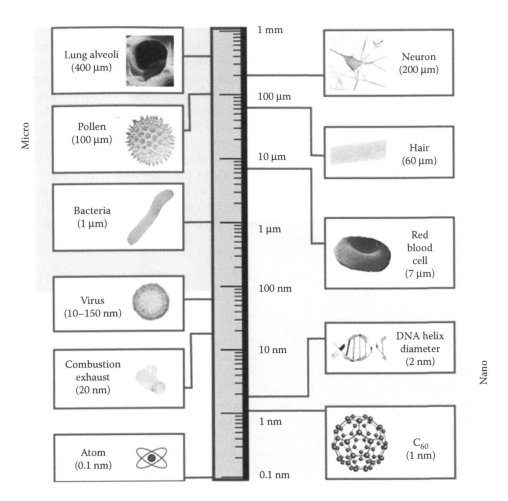

FIGURE 1.3
Logarithmical length scale showing size of nanomaterials compared to biological species. (From Buzea C., Pacheco I. I., Robbie K. 2007. *Biointerphases*, 2(4): 17–71. Open access.)

or colloids. In addition to this, nanomaterials can be made of either a single constituent material or a composite of several materials such as metals, alloys, polymers, or ceramics. Nanoparticles produced by natural processes are often agglomerations of various compositions. Pure single-composition nanoparticles (engineered nanomaterials) can be synthesized by various processes, such as mechanical processes, gas-phase processes, vapor deposition synthesis, coprecipitation, and so on. Due to their surface energy, the nanoparticles come together and tend to form clusters or agglomerates, which can be avoided with the proper chemical treatment that changes the surface energy and distributes them uniformly.

1.5 Size Effect on GBs

Nanocrystalline materials are the solids consisting of approximately equiaxed nanocrystallites (grains) separated by grain boundaries (GBs) with the thickness of 0.5–1 nm. Crystal lattices of grains are misoriented relative to each other. Nanocrystalline materials are single- or multiphase polycrystals with nanoscale (1–250 nm) grain size. The materials with the grain size of 250–1000 nm are called "ultrafine grain size" materials. Nanocrystalline materials are structurally characterized by a large volume fraction of GBs or interfaces, which may significantly alter their physical, mechanical, and chemical properties in comparison with conventional coarse-grained (grain size: 10–300 μm) polycrystalline materials. Figure 1.4a shows 2-D model of a nanocrystalline material, where the grain boundary atoms are shown by white circles and crystal atoms by black circles. As the grain size is decreased, an increasing fraction of atoms can be attributed to the GBs. Figure 1.4b shows a variation in volume fraction of intercrystal regions (or GBs) and triple junctions as a function of grain size. The GB atoms have a variety of interatomic spacing. As the nanocrystalline material contains a high density of interfaces, a significant fraction of atoms lie in the interfaces. For spherical- or cubic-shaped grains, the volume fraction of the GBs can be calculated from the following equation:

$$F_{gb} = \frac{3\delta}{d} \tag{1.1}$$

where d is the average size of the crystallites and δ is the average thickness of the interfaces which is known to be on the order of three or four atomic layers (or 0.5–1.0 nm thick). Thus, the volume fraction of atoms in the GBs can be as much as 50% for 5 nm grains, ~30% for 10 nm grains, and 3% for 100 nm grains. A nanocrystalline metal with a crystallite size of 10 nm contains typically ~6×10^{19} interfaces/cm³ with random orientation, and consequently, a substantial fraction of the atoms lie in the interfaces. In contrast, for the coarse-grained materials with a grain size of >1 μm, the volume fraction of the atoms in the GBs is negligibly small. Whether the structure of the GBs in nanocrystalline materials is different or similar to that in coarser grained conventional material is not clear [9–11].

Due to significant increase of volume fraction of GBs, nanocrystalline materials may exhibit increased strength/hardness, enhanced diffusivity, improved toughness, reduced

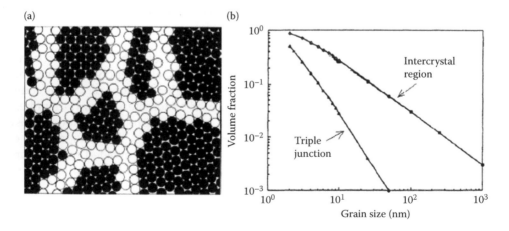

FIGURE 1.4
(a) Two-dimensional model of a nanocrystalline material showing grains and the grain boundaries and (b) the effect of grain size on calculated volume fractions of intercrystal regions (or grain boundaries) and triple junctions, for a grain boundary thickness of 1 nm. (Reprinted from *Prog. Mater. Sci.*, 51, Meyers M. A., Mishra A., Benson D. J., Mechanical properties of nanocrystalline materials, 427–556, Copyright 2006, with permission from Elsevier.)

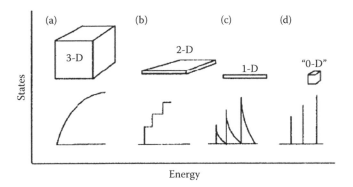

FIGURE 1.5
Density of states $N(E)$ for charge carriers as a function of the dimensionality of the semiconductor: (a) 3-D semiconductor, (b) 2-D quantum well, (c) 1-D quantum wire, and (d) 0-D quantum dot. (Reprinted with permission from Alivisatos A. P. 1996. Perspectives on the physical chemistry of semiconductor nanocrystals. *J. Phys. Chem.*, 100: 13226–13239. Copyright 1996, American Chemical Society.)

elastic modulus and ductility, higher-specific heat, enhanced coefficient of thermal expansion (CTE), and superior soft magnetic properties in comparison with conventional polycrystalline materials. GBs influence the mechanical properties of nanocrystalline materials significantly which are essentially different from those of conventional coarse-grained polycrystalline materials for the same chemical compositions, that is, nanocrystalline materials exhibit superhardness, superstrength, and high-wear resistance due to the role of GBs as effective obstacles for lattice dislocation slip. However, for the finely grained materials, that is, ca. <30 nm, the conventional lattice dislocation slip is hampered in the nanocrystalline materials but the GBs provide the effective action of deformation mechanisms. As a consequence, the crystallite refinement leads to competition between the lattice dislocation slip and deformation mechanisms associated with the active role of GBs. Nanostructured materials present an attractive potential for technological applications with their novel properties [9–11].

In addition to mechanical, magnetic and thermal properties, the optical properties (i.e., energy spectrum) of the nanomaterials are also affected by the reduction in particle size. As shown in Figure 1.5, the density of states $N(E)$ in a 3-D semiconductor is a continuous function; however, a decrease in the dimensionality of the materials results in a change of the energy spectrum from continuous to discrete due to its splitting. In 2-D quantum well structure, the charge carriers are restricted in the direction normal to layers, and may move freely in the plane of the layer. In 1-D quantum wires, the charge carriers are restricted in two directions and move freely only along the wire axis. Interestingly, in a 0-D quantum dot structure, the charge carriers are restricted in all three directions and are characterized by a completely discrete energy spectrum [5,12].

References

1. Vollath D. 2013. *Nanomaterials: An Introduction to Synthesis, Properties and Applications*. 2nd ed. Wiley-VCH Verlag GmbH & Co. KGaA, Germany.
2. Waldrop M. 2016. More than Moore. *Nature* 530: 144–147.

3. Ngô C., Van de Voorde M. (Eds.). 2014. *Nanotechnology in a Nutshell*. Atlantis Press, Netherlands.
4. Gleiter H. 1992. Materials with ultrafine microstructure: Retrospectives and perspectives. *Nanostruct. Mater.* 1: 1–19.
5. Gusev A. I., Rempel A. A. 2008. *Nanocrystalline Materials*. Cambridge International Science Publishing, United Kingdom.
6. Yao H., Duan J., Mo D., Yusuf Günel H., Chen Y., Liu J., Schäpers T. 2011. Optical and electrical properties of gold nanowires synthesized by electrochemical deposition. *J. Appl. Phys.* 110: 094301.
7. Rogers B., Pennathur S., Adams J. 2011. *Nanotechnology: Understanding Small Systems*. 2nd ed. CRC Press, Boca Raton, FL.
8. Buzea C., Pacheco I. I., Robbie K. 2007. Nanomaterials and nanoparticles: Sources and toxicity. *Biointerphases* 2(4): 17–71.
9. Rupp J., Birringer R. 1987. Enhanced specific-heat-capacity (C_p) measurements (150–300 K) of nanometer-sized crystalline materials. *Phys. Rev. B.* 36(15): 7888–7890.
10. Suryanarayana C., Koch C. C. 2000. Nanocrystalline materials—Current research and future directions. *Hyperfine Interact.* 130: 5–44.
11. Meyers M. A., Mishra A., Benson D. J. 2006. Mechanical properties of nanocrystalline materials. *Prog. Mater. Sci.* 51: 427–556.
12. Alivisatos A. P. 1996. Perspectives on the physical chemistry of semiconductor nanocrystals. *J. Phys. Chem.* 100: 13226–13239.

2

Length Scale and Calculations

The properties of the materials, particularly that of nanomaterials, are influenced by not only the chemical composition but also by the size, shape, and surface characteristics of the ultrafine particles to nanoparticles. For instance, infinitesimally small particles have an extraordinary specific surface area and large fraction of surface atoms. Therefore, it is expected that, these nanoparticles or nanostructured materials subsequently have different optical, electrical, mechanical, thermal, and magnetic properties compared to their bulk counterparts. In this chapter, both cubical and spherical shaped particles have been considered to understand the effect of size reduction from micrometer size to nanometer size on their specific surface area and percentage of surface atoms.

2.1 Size Effect on Surface Area of Cubical Particles

The surface area of a material object depends on its size and geometric shape. In this section, the effect of subdivision of a parent material on the overall surface area is discussed. Figure 2.1 shows the division of parent cube with edge length of 1 μm into smaller cubes with edge lengths of 0.1 μm (or 100 nm) and 0.01 μm (or 10 nm). The number of smaller cubes can be calculated from the mass balance equation with the assumption of no loss of materials during the change in size. We can see how much surface area is increased if an original cube with edge length of 1 μm is divided into smaller cubes of equal size with edge length of 0.1 μm (Figure 2.1). If we continue to divide each of these cubes into smaller equal-sized cubes with edge length of 0.01 μm, we can see the inverse relationship between particle size and surface area (Figure 2.2). This inverse relationship between particle size and surface area is an important key factor in the field of nanotechnology.

From mass balance equation, if there is no loss of mass during conversion from large-size to small-size cubes,

$$\text{Mass of large cube} = \text{Total mass of smaller cubes}$$

$$\text{Volume of large-sized cube}(V) \times \text{density}$$
$$= \text{Total volume of smaller-sized cubes} \times \text{density}$$

$$V = N_c \cdot v$$

or

$$N_c = \frac{V}{v} = \left[\frac{L}{l}\right]^3 \qquad (2.1)$$

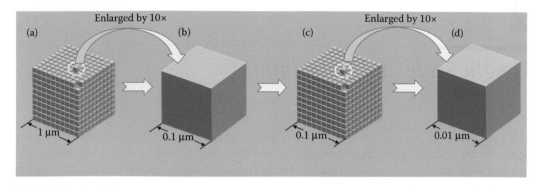

FIGURE 2.1

The division of parent cube with edge length of 1 μm into (a) smaller cubes with edge lengths of 0.1 μm and (c) 0.01 μm. The images in (b and d) show magnified view of 1/10th size of original cube shown in (a) and (c), respectively.

where v and l are the volume and edge length of each small-sized cube, N_c is the number of smaller cubes, and L is the edge length of large-sized cube. Thus, the reduction of the size of the cube from 1 to 0.1 μm and 0.01 μm will result in 10^3 and 10^6 cubes, respectively.

Similarly, the increase in the surface area due to the reduction of the cube size can be determined from the ratio of the cumulative surface area (a) of the cubes after the size reduction to the surface area of a cube before size reduction. For example,

$$\text{Ratio of increase in surface area} = \frac{N_c \cdot a}{A} = N_c \left[\frac{l}{L} \right]^2 \quad (2.2)$$

where a is the surface area of a single smaller size cube. Thus, the reduction of size of the cube from 1 to 0.1 μm and 0.01 μm will result in surface area of 10 and 100 times, respectively, compared to the surface area of original cube. Figure 2.2 shows exponential increase in specific surface area of alumina powder as the powder size is reduced from

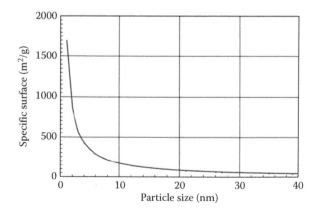

FIGURE 2.2

Variation in specific surface area of spherical alumina particles as a function of the particle size. Vollath D: *Nanomaterials: An Introduction to Synthesis, Properties, and Applications.* 2nd ed. 2013. Copyright Wiley-VCH Verlag GmbH & Co. KGaA. Reproduced with permission.

>40 to <2 nm. It is clearly seen that there is a dramatic increase in surface area when the particle size is below 10 nm compared to those of larger size.

As discussed above, compared to microparticles, nanoparticles have a very large surface area and high-particle number per unit mass. For example, a particle having mass of 0.3 µg with a diameter of 60 µm has a surface area of 0.01 mm^2. The same mass of nanomaterial with a diameter of 60 nm consists of 10^9 nanoparticles and a surface area of 11.3 mm^2, that is, the surface-area-to-volume ratio for a nanoparticle with a diameter of 60 nm is 1000 times larger than a particle with a diameter of 60 µm.

Simple geometric progression from successive division of a parent cube can explain the power of nano on collective surface area. For example, if the cubes with edge length of 1 m are divided into eight equivalent cubes (or equally halves) with each step, the surface area doubles after each division. For n equal to the number of cubes, surface area proceeds from 6 (for $n = 1$ with edge length of 1 m) to 12 ($n = 8$), 24 ($n = 64$), 48 ($n = 512$), and 96 m^2 ($n = 4096$). The length of a side of $n = 4096$ cube is only 6.25 cm. One can see Example 1 to understand the significant increase of total surface area when the size of a particle is reduced from 1 m to 1 nm.

EXAMPLE 1

How many cubes with each side of 1 nm can be carved out from a cube with each side of 1 m? Find the collective surface area of the nanometer-sized cubes.

Solution

First, we need to calculate the number of nanometer-sized cubes that are possible from a 1-m-sized cube.
   Volume of nanocube $= (1 \times 10^{-9}\,\text{m})^3 = (1 \times 10^{-27}\,\text{m}^3)$.
   Volume of larger-sized cube $= 1\,\text{m}^3$.

$$\text{Total number of nanocubes carved out from a single larger-sized cube} = \frac{1\,\text{m}^3}{(1 \times 10^{-27}\,\text{m}^3)}$$
$$= 1 \times 10^{27}.$$

Now, calculate the area of one nanocube and then sum for the total surface area.
Surface area of a 1 m cube is $= 6\,\text{m}^2$.
Surface area of a single nanocube $= (1 \times 10^{-9}\,\text{m})^2 \times 6 = 6 \times 10^{-18}\,\text{m}^2$.

$$\text{Total collective surface area of nanocubes} = \left(1 \times 10^{27}\,\text{nanocubes}\right) \times \left(6 \times 10^{-18}\,\text{m}^2\right)$$
$$= 6 \times 10^9 = 6000\,\text{km}^2.$$

$$\text{The increase in surface area after reducing the size of cube from 1 m to 1 nm} = \left(\frac{6 \times 10^9\,\text{m}^2}{6\,\text{m}^2}\right)$$
$$= 10^9\,\text{times}.$$

The total surface area of the nanocubes with side length of 1 nm is a billion times that of the cube with side length of 1 m. This is the power of nanosize.

Similarly, the division of one cube with a side length of 1 cm into nanocubes with each side length of 1 nm gives the incredible number of 10^{21} nanocubes each with a tiny surface area of $6 \times 10^{-18} \, m^2$. However, the total area of these nanocubes amounts to 6000 m^2, which corresponds to the surface area of a playground with size of 60×100 m. This demonstrates the power of surfaces at the nanoscale.

2.2 Size Effect on Surface Atoms of Cubical Particles

Let the number of atoms along edge $= n$.

The total number of atoms (N) in the cube $= n^3$.

For large N, the edge and corner corrections can be considered negligible. However, in case of nanoparticles which contain few hundred to thousand atoms, the atoms on the surface need to make the correction for double counting of atoms at the edges and corners.

The number of atoms on one surface $= n^2$.

The number of atoms on all the six surfaces including double counting at the edges $= 6n^2$.

The total number of atoms at the 12 edges which have been counted two times $= 12n$.

The total number of atoms on the corners $= 8$.

The net total number of surface atoms on the surfaces of cube $= 6n^2 - 12n$ (double-counted atoms on the edges to be removed) $+ 8$ (corner atoms to be reinstalled) $= 6n^2 - 12n + 8$.

The ratio of the number of atoms on the surface to total atoms in the particle (F)

$$F = \frac{6n^2 - 12n + 8}{n^3} = \frac{6}{N^{\frac{1}{3}}} - \frac{12}{N^{\frac{2}{3}}} + \frac{8}{N} \approx \frac{6}{N^{\frac{1}{3}}} \quad (2.3)$$

$$(\text{since}, n = N^{1/3})$$

Surface atoms do not have atoms on the top. Therefore, particles with a large fraction of atoms at the surface have a low mean coordination number compared to bulk atoms. For example, surface atoms have the highest coordination number of 9, while it is 12 for the bulk atoms. To understand the effect of size reduction on surface atoms, we can take Example 2.

EXAMPLE 2

As shown in Figure 2.3, a cube consisting of $20 \times 20 \times 20$ atoms is divided into ½ and ¼ times of its original size. What is the percentage of surface atoms on larger- and smaller-sized cubes?

Solution

a. Total number of atoms in a cube of $20 \times 20 \times 20$ atoms $= n^3 = 8000$.
   The number of atoms on the cube surface $= 6n^2 - 12n + 8 = 2168$, where n is the number of atoms along edge.
   Hence, percentage of the surface atoms $= \sim 27\%$.
b. The number of atoms along edge of a cube of ½ times of original cube $= 10$.
   Total number of atoms on the cube surface $= (6n^2 - 12n + 8) \times$ No. of cubes (i.e., 8)
   $$= 3904.$$

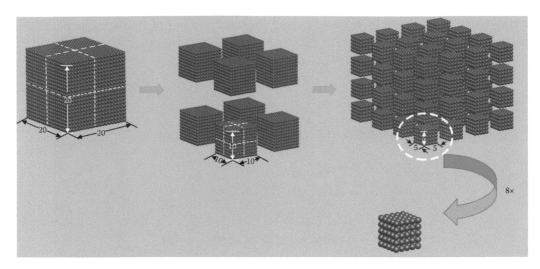

FIGURE 2.3
Increased numbers of cubes and surface atoms as a consequence of dividing a cube with $20 \times 20 \times 20$ atoms.

However, the overall number of the atoms $= 8000$.
Hence, percentage of the surface atoms $= {\sim}48.8\%$.
c. The number of atoms along the edge of a cube of ¼ times of original cube $= 5$.
Total number of atoms on the cube surface $= (6n^2 - 12n + 8) \times$ No. of cubes (i.e., 64)
$$= 6272.$$
However, the overall number of the atoms $= 8000$
Hence, percentage of the surface atoms $= {\sim}78.4\%$.

This indicates that the percentage of surface atoms increases significantly as the size of the particle decreases. In a nanoparticle with a size of few nanometers, nearly every atom is a surface atom that can interact with the outer species.

2.3 Size Effect on Surface Atoms of Spherical Particles

Let the total number of atoms in a spherical particle $= N$.
The volume of a particle $= V_c$.
The number of atoms on the surface of a particle $= N_s$.
The radius of a particle $= R_c$.
Hence,

$$\text{Volume of the particle } (V_c) = N \cdot V_a \tag{2.4}$$

where V_a is the volume of an atom.
For a spherical-shaped particle, the volume of a particle $=$ total volume of the atoms having atomic radii of r_a, that is,

$$V_c = \frac{4}{3}\pi R_c^3 = N\left(\frac{4}{3}\pi r_a^3\right) \tag{2.5}$$

Thus, radii of the particle $(R_c) = N^{1/3}r_a$ and the total number of atoms in the particles in terms of radii are

$$N = \left[\frac{R_c}{r_a}\right]^3 \tag{2.6}$$

The surface area of the particle (S_c) can be calculated by using

$$\begin{aligned}
S_c &= 4\pi R_c^2 \\
&= 4\pi\left(N^{2/3}r_a^2\right) \\
&= N^{2/3}S_a
\end{aligned} \tag{2.7}$$

$(S_a = \text{surface area of an atom} = 4\pi r_a^2)$

$$\text{Approximate number of surface atoms } (N_s) = \frac{\text{Surface area of a particle}}{\text{Cross-sectional area of an atom}}$$

$$N_s = \frac{4\pi N^{2/3}r_a^2}{\pi r_a^2} = 4N^{2/3} \tag{2.8}$$

This is the limiting expression for the surface atoms as the particle size approaches large dimensions. The fraction of atoms at the surface is called dispersion (F). In other words, the ratio of the surface atoms to total number of atoms (i.e., F) can be calculated by using

$$F = \frac{N_s}{N} = \frac{4N^{2/3}}{N} = 4N^{-1/3} \tag{2.9}$$

It can be seen clearly from Equation 2.9 that the value of dispersion (F) decreases as the total number of atoms (N) in a particle increases. For example, $F = 0.4$ for $N = 10^3$, and $F = 0.04$ for $N = 10^6$.

As shown in Figure 2.4, for spherical-shaped nanoparticles with a size of 3 nm, 50% of the atoms or ions are on the surface, which can influence bulk properties by surface effects [2]. The large surface/volume ratio for particles is responsible for a variety of electronic and vibrational surface excitations. For example, a particle with 1 μm diameter has ~1.5 × 10^{-3}% surface atoms, whereas a nanoparticle with a diameter of 10 nm has ~15% surface atoms [3].

Similarly, a cubical SiO$_2$ crystal with an edge length of about 27 cm has total surface area of ~0.44 m². After reducing its size to about 1 mm, the number of smaller cubes would be ~2 × 10⁷ with a total surface area of ~120 m². On further reducing its size to 5 nm, it would lead to ~1.6 × 10²³ nanocubes with a total surface area of ~2 km².

Volume of the layer (or shell) with a thickness "δ" at the surface of a spherical particle of a diameter "d" can be determined from the following equations:

$$V_{\text{shell}} = \frac{\pi}{6}d^3 - \frac{\pi}{6}(d-2\delta)^3 = \frac{\pi}{6}\left[d^3 - (d-2\delta)^3\right] \tag{2.10}$$

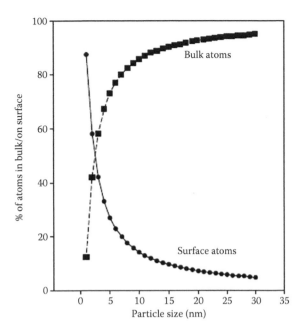

FIGURE 2.4
Calculated percentages of atoms in bulk and on particle surface (for Fe) as a function of size. (Reprinted with permission from Klabunde K. J. et al. 1996. Nanocrystals as stoichiometric reagents with unique surface chemistry. *J. Phys. Chem.*, 100: 12142–12153. Copyright 1996, American Chemical Society.)

$$V_{\text{sphere}} = \frac{\pi}{6} d^3 \tag{2.11}$$

The ratio of the volume of surface layer (or shell) to total volume of a particle can be determined by

$$R = \frac{V_{\text{shell}}}{V_{\text{sphere}}} = \frac{d^3 - (d - 2\delta)^3}{d^3} = 1 - \left(\frac{d - 2\delta}{d}\right)^3 \tag{2.12}$$

The ratio "R" approaches to 1 if $d = 2\delta$.

By using Equation 2.12, one can find out that the surface volume percentage for a 5 nm particle is 49% and 78%, if the thickness of surface layer is about 0.5 and 1 nm, respectively. In case of nanocrystalline bulk materials, one has to replace free surface by grain boundaries [1].

EXAMPLE 3

How many CdSe unit cells are in a spherical particle with size (radius) of 12 Å, if the side of unit cell is 6.05 Å?

Solution

$$\text{Number of approximate unit cells of CdSe} = \left(\frac{\text{Volume of the spherical particle}}{\text{Volume of the CdSe unit cell}}\right)$$

$$= \left(\frac{(4/3)\pi r^3}{l^3}\right) = \left(\frac{(4/3) \times \pi \times 12^3}{6.05^3}\right) \simeq 4$$

EXAMPLE 4

How many CdSe unit cells are in a cubical particle with size of 115 Å, if the edge length of the unit cell is 6.05 Å and it has the zinc-blende structure?

Solution

$$\text{Number of approximate unit cells of CdSe} = \left(\frac{\text{Volume of the cubical particle}}{\text{Volume of the CdSe unit cell}} \right)$$

$$= \left(\frac{L^3}{l^3} \right) = \left(\frac{115^3}{6.05^3} \right) \simeq 6868$$

EXAMPLE 5

How many CdSe unit cells are in cubic crystal with an edge length of 100 μm? If the edge length of the unit cell is 6.05 Å.

Solution

$$\text{Number of approximate unit cells of CdSe} = \left(\frac{\text{Volume of the cubical crystal}}{\text{Volume of the CdSe unit cell}} \right)$$

$$= \left(\frac{L}{l} \right)^3 = \left(\frac{10^6}{6.05} \right)^3 = 4.515 \times 10^{15}$$

EXAMPLE 6

How many atoms are on a surface of silicon nanocube which has an edge length of 10 nm? The silicon has a diamond unit cell with lattice parameter of 5.43 Å as shown in Figure 2.5.

Solution

Given, Si has a diamond structure with a lattice parameter = 5.43 Å.

$$\text{Number of unit cells in a Si nanocube with side of 10 nm} = \left(\frac{100 \times 100 \times 100}{5.43 \times 5.43 \times 5.43} \right) = 6246$$

$$\text{Number of unit cells along each side} = \frac{\text{Length of cube}}{\text{Length of unit cell}} \simeq 18$$

Surface area of one face of the unit cell = side × side = 5.43 × 5.43 = 29.49 Å².
Surface area of the nanocube = 6 × 100 × 100 = 6 × 10⁴ Å².

$$\begin{array}{l} \text{Number of unit cells exposed on the cube} \\ \text{surface } (a) \end{array} = \frac{\text{Surface area of the nanocube}}{\text{Surface area of one face of the unit cell}}$$

$$= \left(\frac{6 \times 100 \times 100}{5.43 \times 5.43} \right)$$

$$= 2035$$

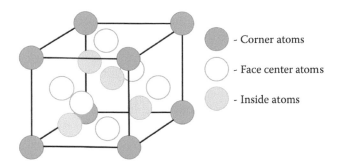

FIGURE 2.5
Diamond unit cell of silicon.

Effective number of atoms on each surface of the unit cell $(b) = \frac{1}{4} \times 4 + 1 = 2$

Number of atoms left on corners $(c) = \frac{1}{4}$(additional contribution of corner atoms)$\times 8$
$$= 2$$

Total number of approximate atoms on the cube surface = (Number of unit cells exposed on the cube surface × effective number of atoms on surface of the unit cell) + Number of atoms on corners which were not counted

$$= (a \times b) + c$$
$$= (2035 \times 2) + 2$$
$$= 4072$$

EXAMPLE 7

One gram of highly porous, low-density metal-organic framework (MOF) material can have a surface area of 4500 m². How does this value compare with the surface area of a 1 g cube of gold?

Solution

According to the question, density of gold = 19.3 g/cm³, surface area of MOF = 4500 m², and mass of gold = 1 g.

Assumption: The surface area of the sides of the sheet is negligible.
Volume of 1 g gold = mass/density = 0.0518 cm³ = 5.18 × 10⁻⁸ m³.
The side length of a cube having volume of 0.0518 cm³ = (0.0518)¹ᐟ³ = 0.382 cm.
The surface area of this cube = 6 × 0.382 × 0.382 = 0.87 cm² = 0.87 × 10⁻⁴ m².
The surface area of MOF = 4500 m².

Compared to MOF, the surface area of a gold cube of 1 g $= \dfrac{\text{Surface area of MOF}}{\text{Surface area of 1 g of gold}}$

$$= \frac{4500}{0.87 \times 10^{-4}}$$
$$= 51,720,000 \text{ times less}$$

Hence, the surface area of a cube of 1 g gold is about 51 million times less than that of MOF.

Unsolved problems

1. What are the volume and surface area of the bacteria *Escherichia coli* of about 3 μm long and 1 μm diameter?

2. What are the volume and surface area of a eukaryotic cell of about 24 μm diameter?

3. Calculate the number of smaller-sized spheres and increase in surface area if a sphere with a diameter of 1 μm is converted into equivalent spheres with a diameter of 1 nm with the same mass. (Assumption: There is no loss of materials.)

4. Calculate the number of smaller-sized cubes and increase in surface area if a cube with side of 1 μm is reduced into equivalent cubes with sides of 1 nm with the same mass. (Assumption: There is no loss of materials.)

5. If a metallic ball has a diameter of 6 cm, how small is a particle having diameter of 6 nm compared to the metallic ball?

6. How many carbon atoms with diameter of about 1.4 Å would be arranged side by side in a line of 1 nm length?

7. How many atoms are present in a face-centered cubic cobalt particle with diameter ~1.6 nm if the lattice constant of cobalt is 0.355 nm?

8. The specific surface area of an alumina powder determined by BET method is 60.0 m^2/g. What is the average particle size of alumina if its density is 3.65 g/cm^3?

9. Calculate the number of atoms in a critically sized (diameter = 1.944 nm) nucleus for the homogeneous nucleation of pure iron. The lattice constant of iron is 0.28664 nm.

10. How many atoms are in a 4 nm diameter gold nanoparticle, if the unit cell edge length for gold is 0.408 nm?

Answers: (1) 2.355 μm^3, 10.99 μm^2, (2) 7234.6 μm^3, 1808.64 μm^2, (3) 10^9, 1000 times, (4) 10^9, 1000 times, (5) 10^7, (6) 7, (7) ~192, (8) 27 nm, (9) 327 atoms, and (10) ~1973 atoms.

References

1. Vollath D. 2013. *Nanomaterials: An Introduction to Synthesis, Properties, and Applications*. 2nd ed. Wiley-VCH Verlag GmbH & Co. KGaA, Germany.
2. Klabunde K. J., Stark J., Koper O., Mohs C., Park D. G., Decker S., Jiang Y., Lagadic I., Zhang D. 1996. Nanocrystals as stoichiometric reagents with unique surface chemistry. *J. Phys. Chem.* 100: 12142–12153.
3. Theodore L. 2006. *Nanotechnology: Basic Calculations for Engineers and Scientists*. John Wiley & Sons, New Jersey.

Further Reading

1. Kickelbick G. (Ed.). 2007. *Hybrid Materials: Synthesis, Characterization, and Applications*. Wiley-VCH Verlag GmbH & Co. KGaA, Betz-Druck GmbH, Darmstadt.
2. Rogers B., Pennathur S., Adams J. 2011. *Nanotechnology: Understanding Small Systems*. 2nd ed. CRC Press, Boca Raton, FL.

3

Effect of Particle Sizes on Properties of Nanomaterials

As discussed in Chapter 1, the nanomaterials are materials which have at least one dimension of the order of 1–100 nm. Due to high-surface area to volume ratio, significant fraction of surface atoms, reduced grain size, and their significant volume fraction of grain boundaries and triple junctions, nanomaterials exhibit many unusual thermal, mechanical, electrical, chemical, and electrochemical properties compared to conventional polycrystalline or amorphous materials. In this chapter, the effect of particle or grain size on the thermal, mechanical, electrical, magnetic, optical, and catalytic properties of nanomaterials is discussed.

3.1 Thermal Properties

3.1.1 Melting Point

Melting point is the temperature when atoms, ions, or molecules of a crystalline material change their periodic ordered state to the disordered state. A number of studies reveal that the melting point of metals such as In, Sn, Pb, Bi, Cd, Al, Ag, and Au decreases with decreasing their size particularly below 30 nm. The melting initiates from the surface of the materials and is characterized by the increased mobility of the atoms or molecules in the top surface layers. The diffusion coefficient of these atoms approaches liquid-like values at temperatures much lower than the melting point of the bulk material [1–6]. This is because of the high-surface area to volume ratio of the nanoparticles which in turn have high-surface energies; hence, the activation energy required for the melting of the surface atoms is lower than the bulk. An example of a decrease in melting point of aluminum (Al) as a function of Al clusters is shown in Figure 3.1. There is a decrease in the melting point as the cluster size decreases. A reduction of 140°C has been reported for the Al clusters with radii of ∼2 nm [5].

Several researchers have taken care to avoid oxidation of Al and Fe by isolating these nanoparticles using an inert material. The melting behavior of nanocrystalline powders synthesized by mechanical attrition in three different atmospheres: argon, hydrogen, and oxygen is shown in Figure 3.2. The smallest grain size obtained in an oxygen atmosphere was 13 nm, while it was 22–25 nm for the Al nanoparticles synthesized in argon and hydrogen atmospheres. Figure 3.2a and b shows that both melting point and enthalpy of fusion of nanoparticles are linearly proportional to reciprocal of grain size, respectively. The melting point decreases considerably, reaching a minimum value of 836 K (563°C) for a grain size of 13 nm. The impurity of an iron due to attrition after 80 h of milling was <0.1 wt%, hence the effect of low level of Fe impurity on melting point of Al nanoparticles can be ignored. Enthalpy of fusion is also found to drop significantly with decreasing

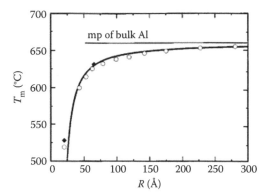

FIGURE 3.1
Melting point as a function of Al cluster size, where T_m is the melting point and R is the radii of Al cluster. (Reproduced from Lai S. L., Carlsson R. A., Allen L. H. 1998. *Appl. Phys. Lett.* 72(9): 1098–1100. With permission.)

grain size of the samples. The stored enthalpy of cold-work is also a possible source of decrease of melting point. However, the bulk melting point of Al was not recovered even after remelting. Thus, it is unlikely that the melting point depression was caused by stored enthalpy of cold-work, so this effect should be removed upon melting of the samples. A decrease in enthalpy of fusion indicates that the nanoparticles of the metals have a higher free energy relative to that of the liquid metals, that is, the nanoparticles become more like liquid. However, Ge nanocrystals embedded in silica glass melt at a temperature of almost 200°C, which is above the melting point of bulk Ge, and similarly lead nanoparticles embedded in an Al matrix exhibit superheating. This is probably due to increase in volume during heating of the nanoparticles which increases pressure on the particles by the matrix. Hence, the melting point of nanoparticles embedded in a bulk matrix increases with decreasing size of particles [6].

Moreover, the melting point of gold (Au) nanoparticles decreases dramatically as the particle size decreases particularly below 10 nm. This is the reason that the sub-10 nm particles sinter at temperatures much less than required for larger particles [7]. Figure 3.3 shows normalized melting point (T_m/T_{mB}) versus particle size for the Au nanoparticles,

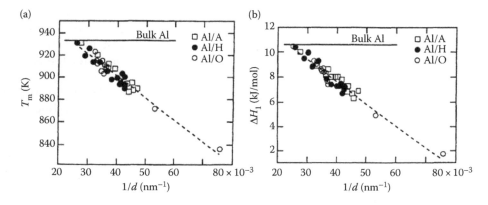

FIGURE 3.2
Melting point and enthalpy of fusion of Al nanoparticles versus reciprocal of grain size. (Reprinted from *Nanostruct. Mater.*, 2, Eckert J. et al., Melting behavior of nanocrystalline aluminum powders, 407–413, Copyright 1993, with permission from Elsevier.)

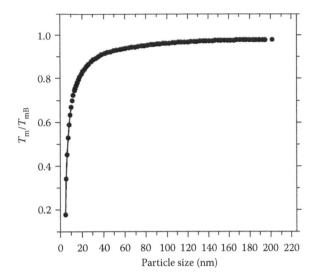

FIGURE 3.3
Normalized melting point (T_m/T_{mB}) versus particle size for the Au nanoparticles, where T_m is the melting point and T_{mb} is the melting point of bulk Au. (From Ristau R., et al. 2009. *Gold Bulletin* 42(2): 143–152. Open access.)

where T_m is the melting point and T_{mB} is the melting point of bulk Au [7]. For example, the melting point of 3 nm gold nanoparticles is more than 300°C lower than the melting point of bulk Au. The bulk Au has melting point of about 1336°C [8]. The approximate melting point of the nanoparticles may be obtained by using equation; $T_m = T_{mB} [1 - (\alpha/d)]$, where α is a constant which depends on the density, heat of melting and surface energy of the material. The d is the diameter of the nanoparticles. The melting point and heat of fusion of SnAg alloy nanoparticles decreases to about 195°C (vs. ~220°C for bulk) and to about 10 J/g, respectively for the nanoparticle with radius of about 5 nm [1]. Similar to nanoparticles, Zn nanowires also show a shift of an endothermic peak to lower temperature indicating decrease in melting point with a decrease in nanowire diameter [4].

Table 3.1 summaries depression in melting points of various metals and alloys as a function of their sizes. From the Table 3.1, the following important observation are drawn.

TABLE 3.1

Summary of Size-Dependent Melting Behavior of Metal/Alloy Nanoparticles and Nanowires

Metal/ Alloy	Atomic Radius (nm)	$T_{mB}{}^a$ (°C)	$T_m{}^b$ (°C)	ΔT^c (°C)	Bulk Particle Diameter (nm)	Onset Particle Diameter (nm)	End Particle Diameter (nm)	Ref.
Sn-Ag	Sn = 0.151 Ag = 0.144	220	195	25	35	20	5	[1]
Bi	0.150	271	150	121	100	10	2	[3]
Zn	0.133	420	409	11	225	50	25	[4]
Al	0.143	660	500	160	60	20	5	[5]
Au	0.144	937	187	750	200	20	5	[7]

[a] T_{mB} is bulk melting point,
[b] T_m is the melting point for NPs or nanowires (NWs),
[c] $\Delta T = (T_{mB} - T_m)$.

1. In case of Al clusters, ΔT of 160°C was reported for the particle size varying between 5 and 20 nm. In contrast to this, for the same range of particle size (5–20 nm) a small depression of 25°C was found for Sn–Ag alloy. For Bi nanoparticles, a depression of 121°C was reported for the particle size varying between 2 and 10 nm.

2. Zn nanowire arrays (diameter: 50–25 nm) show a very small (~11°C) decrease in the melting point.

3. It is interesting to see a decrease of 750°C in the melting point of the Au nanoparticles when its size is 5–20 nm diameter. This is a point of concern for Au used for interconnection applications in the electronics industry.

4. Author of this book could not correlate relationship between the atomic radii of different elements and the depression in melting point of metals. For example, Al and Au atoms have almost similar atomic diameters, FCC structures, and coordination number, but for the same size (~20 nm) of nanoparticles Al shows a depression in melting point by 160°C, while Au shows 750°C.

Based on the data obtained by various authors, it can be concluded that the melting point of the bulk crystals and small particles with a size >10 nm is almost the same. However, a strong decrease of the melting point was found for the nanoparticle when the size is below 10 nm. The largest decrease of the melting point of clusters of Sn, Ga, and Hg was 152, 106, and 95 K, respectively, whereas no melting of clusters of In, Pb, and Cd was detected. Several hundreds of degrees decrease in the melting point was found for colloid CdS nanoparticles with a radius of 1–4 nm. In case of Au nanoparticle, the linear relationship between the melting point of nanoparticle and inverse of particle radius breaks down at a particle size of ~1.6 nm or less. However, one can realize a significant deviation from the inverse linear relationship at larger sizes of 6.5 nm for lead. A depression or decrease in melting point with decrease in grain size of metals can be explained by a number of ways:

- According to classical thermodynamics model, the surface of a solid cluster begins to melt at temperatures below the bulk melting point forming a liquid-like shell surrounding a solid core. Once the liquid layer thickness exceeds a critical value, the whole cluster melts homogenously [5]. According to some researchers, a surface layer of roughly few nanometer thick melts first followed by melting of core in second step. It means the smaller particles with a radius smaller than this surface layer may melt whole particle at once and the particle thicker than this surface layer will melt in two steps. This may be attributed to the different degree of order of atoms in the interior and exterior of the particles, that is, it decreases from center to surface [9].

- Increase in specific surface area with a decrease in grain size of the nanoparticles leads to increased surface energy which makes the nanoparticles "unstable" or "metastable."

- The volume fraction of grain boundary increases dramatically with a decrease in grain size. Grain boundaries are regions of disordered atomic arrangement because the atomic orientation across two different grains changes at the grain boundary. This disorder-induced increase in free energy can be considered to be nucleated at the grain boundaries which may lead to vibrational instability on heating, hence resulting in melting.

- Coordination number is the number of nearest atoms of an atom in the core or surface of a crystal. The maximum coordination number of an atom at the surface is 9, whereas it is 12 for an atom at the core of a crystal. As the surface area increases with a decrease in grain size, more and more atoms are brought to the surface. This results in decreased average coordination number for the crystal and hence, this leads to an increase in dangling bonds of the atoms. As a result, the cohesive energy, that is, the energy required to break all the bonds associated with an atom, decreases. With decreased cohesive energy, the nanoparticles become thermally unstable as compared to their bulk counterparts. This leads to a decrease in the melting point.

The depression in melting point of metals and alloys can be exploited for some useful applications. For example, Sn and its alloys are used to interconnect materials in on-chip and off-chip applications. Owing to high-melting point (220–240°C) of bulk Sn, a high-reflow temperature is needed in the electronics manufacturing process which may have adverse effects like warpage and thermal stresses in the components being fabricated. A depression in melting point and thus, the processing temperatures of nano-Sn and its alloys can alleviate these problems. Similarly, sintering of silver paste can be done at a much lower temperature than that of bulk silver.

3.1.2 Heat Capacity

The specific heat of a material is closely related to its vibrational and configurational entropy, which is significantly affected by the nearest-neighbor configurations. In the temperature range of 10 K $\leq$ T $\leq$ Debye temperature, the heat capacity of nanopowders is about 1.2–2 times higher than that of the bulk materials. The increased heat capacity of the nanopowders is contributed by the large surface area. The heat capacity (C_p) of nanocrystalline-compacted sample of nano-Pd (grain size: 60 nm) and nano-Cu (grain size: 8 nm) was studied in the temperature range from 150 to 300 K. The relative density of the nano-Pd specimens was equal to 80% and that of the nano-Cu specimens was 90% of the density of pore-free polycrystalline coarse-grained palladium (Pd) and copper (Cu), respectively. The C_p of nano-Pd and nano-Cu specimens was found to be 29%–53% and 9%–11% higher than the C_p of bulk Pd and bulk Cu, respectively (Figure 3.4). Similarly, the low-temperature C_p of the bulk nanocrystalline-compacted Cu with a grain size of 6.0 or 8.5 nm in the temperature range from 0.06 to 10.0 K is 5–10 times higher than that of coarse-grained Cu. Moreover, the heat capacity of nano-Ni at temperature $\leq$22 K was found about 1.5–2 times higher than the heat capacity of coarse-grained Ni. The C_p value of the metallic glass (i.e., $Pd_{72}Si_{18}Fe_{10}$) is higher than that of polycrystalline Pd by about 8%, which is originated from the different atomic structure and the deviation in the chemical composition from the Pd. The higher heat capacity of the nanocrystalline materials has been attributed to large fraction of grain boundaries which contains free volumes and presence of some impurities like hydrogen. The lower relative density of nano-Pd in comparison to nano-Cu suggests a more open atomic structure of the grain-boundary component in the nano-Pd, and hence weaker interatomic coupling, which enhances C_p and indicates that the enhancement of C_p is primarily due to the grain-boundary component [10,11].

3.1.3 Curie Temperature

Figure 3.5a shows the variation of the critical Curie temperature (T_c) as a function of grain-size of $BaTiO_3$ (embedded in a V_2O_5–Bi_2O_3 glass matrix) at 100 kHz. It indicates a shift of T_c to lower values as particle size decreases from 200 to 20 nm. The average Curie temperature for a heat-treated sample at 24 h is very close to Curie temperature of pure $BaTiO_3$

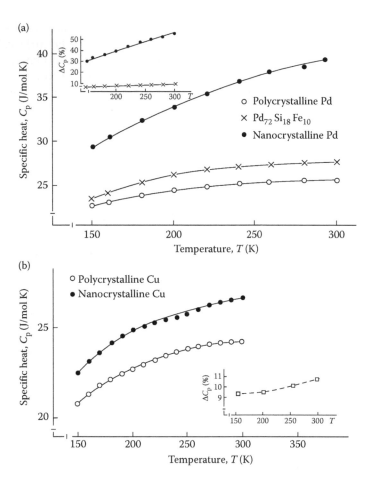

FIGURE 3.4
Effect of grain size on the temperature dependence of heat capacity C_p of (a) palladium and (b) copper. (Reprinted with permission from Rupp J., Birringer R. 1987. Enhanced specific-heat-capacity (C_p) measurements (150–300 K) of nanometer-sized crystalline materials. *Phys. Rev. B* 15: 7888–7890. Copyright 1987 by the American Physical Society.)

with particle size of 200 nm [12]. Similarly, the Raman spectrum indicated that the Curie temperature is shifted from 140°C for a BaTiO$_3$ single crystal to about room temperature for BaTiO$_3$ nanoparticles with size of 20–30 nm [13]. In contrast, Uchino et al. reported the Curie temperature shift to the lower temperature in the particle size range of 80–100 nm. The difference in the critical size for BaTiO$_3$ may be attributed to the different measurement techniques used for measuring the particle size and the crystal symmetry. Goyal et al. [14] confirmed the change in Curie temperature of BaTiO$_3$ powders milled for 0–20 h using planetary ball milled at 300 rpm in dry condition. Its crystallite size measured by XRD peak broadening was reduced to 13 nm and the particle size measured by BET method was found to be 27 and 23 nm after 10 and 20 h, respectively. Modulated differential scanning calorimetry shows that the Curie temperature (Figure 3.5b) of as received BaTiO$_3$ powder is ∼125°C corresponding to the tetragonal to cubic phase transition.

As shown in Figure 3.5b, this peak is clearly seen in both the heating cycles. However, for 10-h milled powder as shown in Figure 3.5c, a peak was observed at about 95°C during the first heating cycle, but this peak was disappeared during the second heating cycle

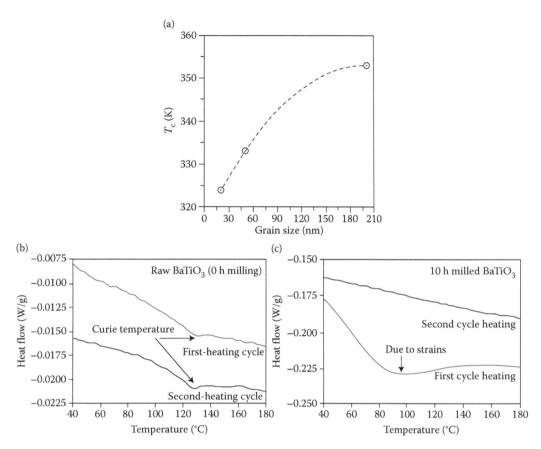

FIGURE 3.5
Effect of grain size on the Curie temperature (T_c) of (a) $10BaTiO_3$-$70V_2O_5$-$20Bi_2O_3$ glass–ceramic nanocomposite (With kind permission from Springer Science+Business Media: *J. Mater. Sci.: Mater.*, Grain-size effects on the structural, electrical properties and ferroelectric behavior of barium titanate-based glass–ceramic nanocomposite, 24, 2013, 784–792, Al-Assiri M. S., El-Desoky M. M.) and heat flow versus temperature showing Curie temperature of $BaTiO_3$ before, (b) ball milling, and (c) after 10 h ball milling. (From Pendse S., Goyal R. K. 2016. *J. Mater. Sci. Surf. Eng.* 4(3): 383–385.)

indicating that the peak observed in the first cycle is probably due to stresses induced during the milling process which are released after absorbing thermal energy during first heating cycle. Similarly, disappearance of the Curie temperature scanned during 0–180°C was found for the 20-h milled powder (not shown here). Disappearance of the Curie temperature may be attributed to transition of tetragonal phase to cubic phase for the reduced particle size after ball milling [14].

3.1.4 Coefficient of Thermal Expansion

As discussed in previous chapter, nanocrystalline materials have a large amount of interfacial volume; hence, coefficient of thermal expansion (CTE) is expected to be higher than coarse-grained material. For example, the linear CTE and volume CTE of the nanocrystalline selenium increase by ~21% and 31%, respectively, with a decrease in crystallite size from 46 to 13 nm, as shown in Figure 3.6. The nanocrystalline Cu with a mean crystallite

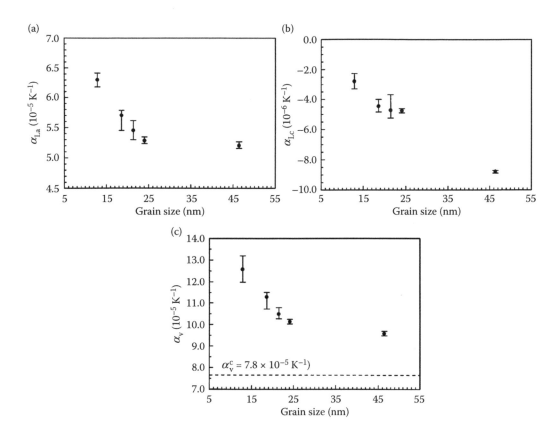

FIGURE 3.6
Grain-size dependence of the linear CTE along the a-axis (a), c-axis (b), and the volume CTE of selenium (c). (Reprinted with permission from Zhao Y. H., Lu. K. 1997. Grain-size dependence of thermal properties of nanocrystalline elemental selenium studied by x-ray diffraction. *Phys. Rev. B* 56(22): 14330–14337. Copyright 1997 by the American Physical Society.)

size of 8 nm has a CTE of $31 \times 10^{-6}\,K^{-1}$, which is twice as large as the value of $16 \times 10^{-6}\,K^{-1}$ for coarse-grained Cu. The high value of CTE for nanocrystalline materials has been attributed to the large fraction of grain boundaries, which have a considerably higher thermal expansion coefficient than that of the micrometer-sized crystallites. For example, the CTE of grain boundaries (i.e., $40.80 \times 10^{-6}\,K^{-1}$) is about 2.5–5 times higher than the value of the coarse-grained Cu [10,15]. Similarly, an increase in CTEs was found in nanocrystalline Pd, Ni–P, $Fe_{78}B_{13}Si_9$, and TiO_2 samples (Table 3.2) [16].

TABLE 3.2

CTEs (in $\times\,10^{-6}\,K^{-1}$) of Nanocrystalline, Amorphous, and Coarse-Grained Materials

Material	Temperature Range (K)	Condition		
		Nanocrystalline	Amorphous	Coarse-Grained
Ni-P	300–400	21.6	14.2	13.7
$Fe_{78}B_{13}Si_9$	300–500	14.1	7.4	6.9
Cu	110–293	31	–	16

Source: With kind permission from Springer Science+Business Media: *Hyperfine Interact*, Nanocrystalline materials—Current research and future directions, 130, 2000, 5–44, Suryanarayana C., Koch C. C.

3.2 Electrical Properties

The electrical conductivity of the materials is the inverse of their electrical resistivity. Specific electrical resistivity of nanocrystalline Cu with grain size of 7 nm measured at temperature below 275 K is 7–20 times higher than that of conventional coarse-grained Cu. As shown in Figure 3.7, at temperature >100 K, the specific electrical resistivity of coarse-grained Cu and nanocrystalline Cu increases linearly with increasing temperature, but the temperature coefficient of resistivity for nanocrystalline Cu is equal to $17 \times 10^{-9}\,\Omega \cdot \text{cm K}^{-1}$, which is higher than $6.6 \times 10^{-9}\,\Omega \cdot \text{cm K}^{-1}$ of conventional Cu. The coefficient of electron scattering at the grain boundaries in nanocrystalline Cu is higher by a factor of two in comparison to coarse-grained Cu. This difference is due to the higher volume fraction of grain boundaries, and different width and structure of the grain boundaries in nanocrystalline and coarse-grained Cu. The high electrical resistivity of nanocrystalline Cu is caused primarily by electron scattering at the grain boundaries and short mean free path (λ) of electrons, that is, $\lambda \approx 4.7$ nm for nanocrystalline Cu in comparison to $\lambda \approx 44$ nm for coarse-grained Cu. Moreover, a decrease in the crystallite size increases the degree of localization, decreases the concentration of charge carriers and, hence, increases specific electrical resistivity. The specific electrical resistivity of submicrocrystalline Cu, Ni, and Fe at 250 K is 15%, 35%, and 55% higher than that of the same coarse-grained metals [10,17].

The oxide-ion conductors are important functional materials used in fuel cells, oxygen pumps, oxygen sensors, water electrolysis, and oxygen-separating ceramic membrane. Among several ion conductors, La-deficient La_2GeO_5, that is, $La_{1.61}GeO_{5-\delta}$, is highly interesting because of its high oxide ion conductivity over a wide range of oxygen partial pressure and unique structure. The transport number of the oxide ion is nearly unity in the O_2 partial pressure ranging 1–10^{-21} atm and its conductivity is comparable to that of fast oxide-ion conductors. Figure 3.8a and b shows the effect of the film thickness on the oxide ionic conductivity of $La_{1.61}GeO_{5-\delta}$ as a function of temperature and partial pressure of oxygen (PO_2) [18].

The films were deposited on dense polycrystalline Al_2O_3 substrates by a pulsed laser deposition (PLD) method. The films were annealed to get crystalline films. It was found

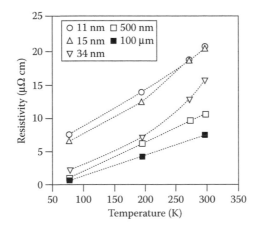

FIGURE 3.7
Resistivities of nanocrystalline Ni compared to normal crystalline Ni. (Reprinted from *Nanostruct. Mater.*, 6, Erb U., Electrodeposited nanocrystals: Synthesis, properties and industrial applications, 533–538, Copyright 1995, with permission from Elsevier.)

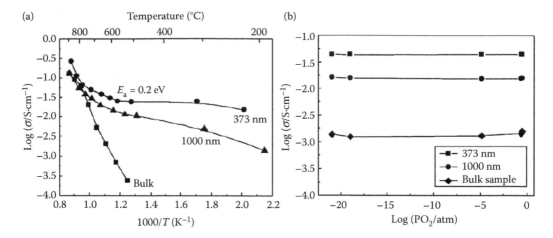

FIGURE 3.8

(a) Arrhenius plots of $La_{1.61}GeO_{5-\delta}$ thin films and that of bulk $La_{1.61}GeO_{5-\delta}$ sample and (b) PO_2 dependence of the electrical conductivity in $La_{1.61}GeO_{5-\delta}$ thin film with various thicknesses at 873 K. (Reprinted from *Solid State Ionics*, 177, Ishihara T., Yan J., Matsumoto H., Extraordinary fast oxide ion conductivity in $La_{1.61}GeO_{5-\delta}$ thin film consisting of nano-size grain 1733–1736, Copyright 2006, with permission from Elsevier.)

that the annealed $La_{1.61}GeO_{5-\delta}$ film exhibited extraordinarily high oxide ionic conductivity. The oxide ion conductivity of the thin films increased with decreasing thickness as compared to that in bulk $La_{1.61}GeO_{5-\delta}$. The improvement in conductivity of the film at low temperature was significant, that is, the electrical conductivity of the film with a thickness of 373 nm is as high as 0.05 S.cm^{-1} at 573 K. As shown in Figure 3.8b, for a given film, the insignificant change in ion conductivity indicates that the increased conductivity with temperature is not due to oxygen vacancies, but due to improved mobility of oxide ion by the local stress caused by the mismatch in lattice parameter between the film and the substrate [18,19]. Figure 3.9 shows the electrical conductivity of polycrystalline cerium

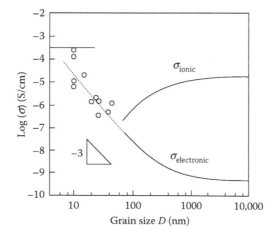

FIGURE 3.9

Electrical conductivity of polycrystalline cerium oxide as a function of grain size. The result of the space charge model (solid lines) for σ_{el} is extrapolated (dotted line) based on the flat band limit. (With kind permission from Springer Science+Business Media: *J. Electroceram*, Grain size dependence of electrical conductivity in polycrystalline Cerium Oxide, 7, 2001, 169–177, Tschoepe A., Birringer R.)

oxide as a function of grain size. Under the given conditions, the electrical conductivity of cerium oxide with large grain size is dominated by the ionic contribution. As the grain size decreases, the ionic partial conductivity decreases due to a defect in accumulation/ depletion in space charge layers whereas the electronic conductivity increases, resulting in a transition from predominantly ionic to electronic conductivity at a grain size of about 60 nm. The absolute value of electronic conductivity for a bulk cerium oxide is about 1×10^{-7} S/cm [20].

Electronic conductivity above this value could be achieved in the volume phase of cerium oxide only by doping with penta- or hexavalent donor ions. One can see the best quantitative agreement between experimental and theoretical results obtained for a space charge potential of 0.55 V. The space charge potential is caused by the standard chemical potential of oxygen vacancies, which is reduced by ~1.8 eV in the core of grain boundaries as compared to the bulk [20]. The electrical conductivity of NiO is strongly related to the formation of microstructural defects inside the NiO crystallites, such as nickel vacancies and interstitial oxygen. It was found that the increasing substrate temperature leads to a more perfect crystalline structure, which leads to decrease in the carrier (i.e., electron holes) concentration and thus decreases the electrical conductivity [21]. As the grain size of SnO_2 decreases, its electrical conductivity increases and the increase in the conductivity with grain size has been attributed to the larger fraction of the grain boundary volume which enables the formation of charged states of oxygen. The low-frequency relaxation was observed due to the migration of charged particles of SnO_2 such as O^{2-}, O^-, and O_2^- or bulk species, such as oxygen defects V_O^{2+} across the grain boundaries [22].

3.3 Lattice Constant

Figure 3.10a shows the effect of decreasing particle size on the lattice parameter of gold and palladium. It can be clearly seen that there is a contraction of about 3% in lattice parameter of both gold and palladium when the nanoparticle size is about 0.6 nm. Similarly, Ag nanoparticles with size of 3.1 nm and Pt nanoparticles with a size of 3.8 nm showed the

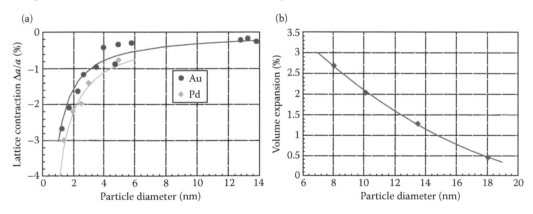

FIGURE 3.10
Effect of particle size on (a) lattice constant of gold and palladium nanoparticles and (b) volume expansion of γ-Fe$_2$O$_3$. (Vollath D: *Nanomaterials: An Introduction to Synthesis, Properties, and Applications.* 2nd ed. 2013. Copyright. Wiley-VCH Verlag GmbH & Co. KGaA. Reproduced with permission.)

lattice constant contraction of about 0.7% and 0.5% in comparison to their bulk counterparts. A decrease in lattice constant of about 1.5% was found for Al nanoparticles when its size is reduced from 20 to 6 nm. However, a decrease in the size of Cu clusters to 0.7 nm decreased the lattice constant by 2% in comparison with bulk Cu. The cause of a decrease in lattice parameter or contraction may be attributed to either hydrostatic pressure [9] or surface relaxation [10]. For a spherical nanoparticle, the surface energy causes a hydrostatic pressure of the order of few thousands bar, which is probably responsible for the lattice parameter change due to the deformation. The surface relaxation is the most probable reason for the decrease of the lattice constant of the small nanoparticles in comparison with bulk. We have seen in previous chapter that as the particle size decreases in the nanoscale, there is significant increase in surface atoms compared to bulk atoms. Surface atoms are not surrounded by the top planes; as a result of this, the distances between the atomic planes in the vicinity of the surface of the particle decrease because surface relaxation takes place. However, the lattice constant of Si nanoparticles increases by 1.1% when its size decreases from 10 to 3 nm. Similarly, an increase of the lattice constant of CeO_2 and γ-Fe_2O_3 oxides was found with a decrease of the particle size from 25 to 5 nm and 18 to 8 nm, respectively. The expansion of CeO_2 was attributed to the adsorption of water. As shown in Figure 3.10b, the cause of expansion in γ-Fe_2O_3 was attributed to a change in the lattice structure at the surface with decreasing particle size. The cations at the surface of an oxide are terminated by oxygen ions or other anions, such as $(OH)^-$. Therefore, the surface of an oxide is covered with ions, all bearing negative electrical charges. These negative charges repel each other and hence, the lattice expands [9,10].

In the case of $BaTiO_3$, the lattice constants "c" and "a" of barium titanate ($BaTiO_3$) powders can be obtained by fitting the overlapping (002) and (200) diffraction peaks with two distinct peaks by XRD. Li and Shih reported that the value of true c/a decreases with decreasing grain size of tetragonal phase $BaTiO_3$ particles. As the particle size decreases, the "c" decreases and "a" increases; hence, the "c/a" ratio finally approaches close to a value of 1.0 corresponding to cubic phase at particle size of about 50 nm, as shown in Figure 3.11. Hence, the tetragonality of the 100-nm or lesser sized nanoparticles is considered to be lower than that of a single crystal at room temperature [13,23].

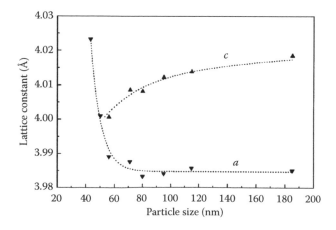

FIGURE 3.11
Effect of particle diameter on lattice constant of $BaTiO_3$. (Li X., Shih W. H.: Size effects in barium titanate particles and clusters. *J. Am. Ceram. Soc.* 1997. 80. 2844–2852. Copyright Wiley-VCH Verlag GmbH & Co. KGaA. Reproduced with permission.)

3.4 Phase Transformation

In the case of zirconia, the monoclinic-tetragonal transformation is affected by the size of zirconia and yttria (Y_2O_3) particles. At room temperature, zirconia with particle size in the range of micrometer has monoclinic phase and transforms to the tetragonal phase and the cubic phase at 1475 K and 2650 K, respectively. However, the addition of small amounts of Y_2O_3 or MgO shifts these transformations to lower temperatures as zirconia dissolves Y_2O_3, and the Y_2O_3 stabilizes its tetragonal phase. Figure 3.12a shows that the temperature of transformation decreases as the particle size decreases. For a given particle size, an increasing Y_2O_3 content decreases the transformation temperature significantly. For example, at a particle size of about 120 nm, the transformation temperature for pure zirconia is about 1100 K, which reduces to about 720 K for 1 wt% Y_2O_3-doped zirconia. As shown in Figure 3.12b, the transformation temperature almost linearly decreases with decreasing particle size. The decrease in transformation temperature with decreasing particle size is found for both free particles as well as sintered samples [9]. The partially stabilized zirconia is used to toughen the ceramics.

3.5 Mechanical Properties

3.5.1 Elastic Modulus

The elastic modulus of a material is proportional to the bond strength between atoms or molecules. The higher the bond strength, the higher will be the elastic modulus. The elastic properties of crystalline materials are usually considered to be microstructure independent. However, a large increase in vacancy and defect concentration are expected to decrease the elastic modulus. Due to very high-defect concentration, the elastic modulus of nanomaterials was found to be reduced by ~30%–50% in comparison to bulk materials. The significant drop in elastic modulus of nanocrystalline materials is due to the presence

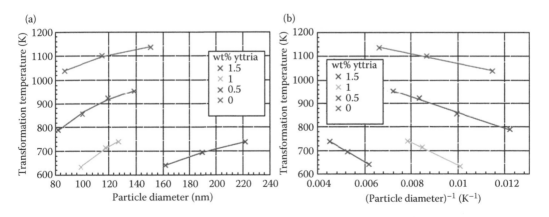

FIGURE 3.12
Temperature of monoclinic-tetragonal transformation of yttria-doped zirconia as a function of (a) particle size and (b) inverse particle size. (Vollath D: *Nanomaterials: An Introduction to Synthesis, Properties, and Applications.* 2nd ed. 2013. Copyright Wiley-VCH Verlag GmbH & Co. KGaA. Reproduced with permission.)

of large volume fraction of grain boundary with a thickness of 1 nm or more. The elastic modulus of grain boundary is only 12% of the bulk counterpart. However, for the nano-crystalline materials with a grain size smaller than 5–10 nm, the fraction of atoms, located in the grain boundaries, is very large and these materials may have atomic structure close to amorphous sample. The elastic modulus of amorphous materials is approximately half the modulus of similar coarse-grained crystalline materials. Therefore, the elastic modulus of nanomaterials with a grain size of 5–10 nm or smaller is considerably lower [10] in comparison to bulk. In contrast to nanograined materials, the elastic modulus of carbon nanotubes (CNTs) increases with decrease in tube diameter. The increase of apparent elastic modulus for smaller diameter CNTs is attributed to the surface tension effects.

3.5.2 Hardness and Strength

Conventionally, the plastic deformation of crystalline materials is due to the movement of the dislocations. The movement of dislocations is obstructed by the grain boundaries in polycrystalline materials. A decrease in grain size of given materials leads to an increase in volume fraction of grain boundaries (i.e., as high as 10^{13}–10^{14} cm^{-3} for 10 nm nanocrystal-lite) which in turn increases the hardness and strength, as represented by the *Hall–Petch* equation (3.1):

$$\sigma = \sigma_0 + k \times d^{-1/2} \tag{3.1}$$

where σ is the flow stress of the material with grain size d. The σ_0 and k are constant parameters, which are specific for the materials. The σ_0 is the friction stress and the constant k reflects difficulty in slip transfer from one grain to another. The Equation 3.1 indicates an increase in strength with decreasing grain size of the particles. Hence, materials with grain size of <100 nm exhibit substantially higher strengths than the conventional material with grain sizes >10 μm. The yield stress of palladium with a grain size of 14 nm is about 259 MPa, which is approximately five times that of coarse-grained (grain size ~50 μm) palladium (i.e., 52 MPa). Moreover, nanocrystalline Cu shows a yield strength and ultimate tensile strength of around 800 MPa and 1100 MPa, respectively, which are much higher than those of an annealed high-purity bulk Cu [17].

It can be seen that the microhardness determined by load versus displacement curves of nano-Ni is almost four times that of micro-Ni or bulk Ni. It is also found that 10 nm grains have higher hardness than 20 nm grains indicating grain-boundary strengthening in the nanocrystalline range. Similarly at 300 K, the microhardness of nanocrystalline materials is 2–7 times higher than that of coarse-grained materials [10]. A decrease in the grain size of coarse-grained Cu from 25 to 5 μm is accompanied by an increase in microhardness. Microhardness of nanocrystalline Cu with grain size of 16 nm is about 2.5 times higher than that of Cu with a grain size of 5 μm. Similarly, in the case of intermetallic compound, titanium aluminum (TiAl) with grain sizes above ~20 nm, an increase in hardness was found with a decreasing grain size. In polycrystalline materials, on decreasing the grain size from tens of microns to the critical grain size (d_c), the grain boundaries density increases which are thought of as a strengthening factor because they act as barriers to the dislocation motion. In general, the critical value of d_c varies between 10 and 30 nm, depending on material and structural parameters. The stress concentration at the grain boundaries must be sufficient to nucleate slip in the next grain and a large applied stress will be needed to yield the boundaries, that is, the material is strengthened. The stress needed to yield a grain boundary is dependent on the grain boundary structure and energetic state. For example, for a high-angle grain boundary, the force needed to yield it and to start a successive motion of dislocations in neighboring grains will be larger than that in the case of a small-angle boundary [9]. However, the microhardness of submicro-crystalline Cu and Pd with a mean grain size of 200–300 nm and 150 nm, respectively, and measured at room temperature after annealing at the temperature range of 500–600 K decreases rapidly by almost a factor of 3 due to the grain growth and partial annealing of dislocations [10].

In contrast to above case, when the grain size of TiAl is further reduced below 20 nm, a decrease in the hardness was found [9]. It has been theoretically as well as experimentally shown that for the nanomaterials with grain sizes in the range between 10 and 50 nm, Frank-Reed sources for dislocation generation are almost impossible in most of the cases. Hence, deformation of these nanomaterials via dislocations is not possible. Therefore, further decrease in the grain size of nanocrystalline Cu from 16 to 8 nm decreases the microhardness by about 25%. A similar decrease of microhardness was also found with a decrease in the grain size of nanocrystalline Pd from 13 to 7 nm. Similarly, a decrease in the microhardness of nanocrystalline alloys of Ni–P, TiAlNb, TiAl, and NbAl$_3$ was found with a decrease in the grain size from 60–100 nm to 6–10 nm. This smallest range of grain size follows an inverse Hall–Petch relationship.

A nanocrystalline electrodeposited Ni with an average grain size of 26 nm shows a tensile strength >2.2 GPa and Vickers hardness of about 600 MPa. However, below 26 nm, a downward deviation from the values predicted from the Hall–Petch equation can be seen clearly in Figure 3.13. A negative slope is seen for Ni with a grain size less than ~11 nm. However, below grain size of 6 nm, the measured value of the hardness is decreased by rapid room temperature creep and follows roughly inverse Hall–Petch equation [17]. This is due to the fact that below the critical grain size (d_c), grain boundaries sliding and/or diffusion flow may become important deformation mechanism even at room temperature, greatly increasing the deformation rate. These deformation mechanisms provide either saturation or a decrease of the hardness or strength with decreasing grain size (d) in the

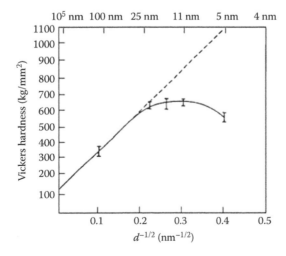

FIGURE 3.13
Hall–Petch plot for electrodeposited Ni. (Reprinted from *Nanostruct. Mater.*, 6, Erb U., Electrodeposited nanocrystals: Synthesis, properties and industrial applications, 533–538, Copyright 1995, with permission from Elsevier.)

range of $d < d_c$. In brief, there is critical size where the competition between the lattice dislocation slip and grain boundary (GB) diffusional creep (Coble creep) balances leading to saturation and becomes responsible for the inverse Hall–Petch relationship below the d_c [17].

3.5.3 Toughness

Impact toughness is one of the major mechanical properties of structural materials. It is generally observed that the impact toughness of coarse-grained materials decreases with decreasing testing temperature. High-impact toughness is desired to prevent catastrophic failure during an impact loading, particularly at low-service temperatures. Metals and alloys (steel, Ti, and Al) usually have lower impact toughness with decreasing temperature, which may limit their applications at low-service temperatures. For example, the sinking of the Royal Mail Ship Titanic was attributed to the freezing sea temperature (−2°C), at which steel hull plate of the ship became brittle. Therefore, both high strength and high ductility are required to have high-impact toughness. The impact toughness of nanostructured Ti processed by severe plastic deformation (SPD) (ECAP + cold rolling) followed by low-temperature annealing (at 300°C for 1 h) is increased significantly at low temperatures of −70°C and −196°C, which contradicts the observations in coarse-grained materials (Figure 3.14). The increased impact toughness may be attributed to the increased strength and ductility of nanostructured Ti as well as smaller fracture dimples at lower temperatures (Insets a and b of Figure 3.14) [24].

From 20°C to 100°C, the impact toughness of nanostructured Ti is slightly higher than coarse-grained Ti and follows a trend similar to that of coarse-grained Ti, that is, it decreases with decreasing temperature. Interestingly, below room temperature, the impact toughness of nanostructures Ti increases with decreasing temperature, that is, it increases from ∼3.1 (for coarse-grained Ti) to 5.6 kgm/cm² (for nanostructured Ti) at −196°C. SEM images of nanostructured-Ti reveals that the dimples formed at −196°C are much smaller (Figure 3.14b) and have much higher density than those formed at 20°C (Figure 3.14a). The

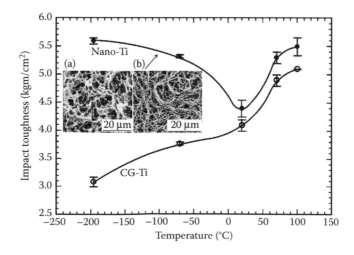

FIGURE 3.14
Impact toughness of nanostructured and coarse-grained Ti versus testing temperature. The inset shows SEM images of fracture surfaces of nanostructured Ti after impact test at (a) 20°C and (b) −196°C. (Reproduced from Stolyarov V. V. and Valiev R. Z. 2006. *Appl. Phys. Lett.* 88: 041905. With permission.)

high density of small dimples requires much higher energy to produce than large dimples, thus rendering the nanostructured Ti a highly tough material at −196°C than at 20°C. Moreover, nanostructured Ti exhibits an increase in both strength and ductility at lower temperatures. The higher strength, higher ductility, and larger amount of dimple ligaments are main factors contributing to the enhanced impact toughness of nanostructured Ti with decreasing temperature [24]. The equal channel angular pressing (ECAP) method refines the grains structure of Cu, which leads to a strength (i.e., >350 MPa) and an elongation to failure larger than (i.e. ~15%) (Figure 3.15), or at least comparable with, those of Cu heavily cold rolled at room temperature. Compared to ECAP-Cu, the strength in the range of 300–400 MPa and only a few percent elongation of room temperature rolled Cu is due to the elongated grains with low-angle subgrain structures (or dislocation of cell blocks). The additional plastic deformation of the ECAP-Cu by cold rolling at the liquid nitrogen temperature (LNT) by approximately 1180% enhances the strength to more than 450 MPa and an elongation to about 20%. Further cold rolling of ECAP-Cu at LNT to 1340% followed by annealing at 100°C for 2 h increases the elongation by about 30% with a slight decrease in strength compared to the ECAP + 1180% LNT rolled Cu. The decrease in strength is probably due to the decrease in dislocation density after low-temperature annealing [25].

Figure 3.16 shows stress–strain curves for coarse-grained and nanocrystalline Cu processed by two different methods. The *in situ* consolidated nanocrystalline Cu (nc-Cu) sample consists of equiaxed nanograins of about 23 nm which are oriented randomly. The coarse-grained polycrystalline Cu sample has an average grain size larger than 80 μm and a nanocrystalline-Cu sample prepared by an inert gas condensation (IGC) and compaction has a mean grain size of 26 nm. The artifact-free bulk nanocrystalline Cu sample with a narrow grain size distribution and mean grain size of 23 nm exhibited tensile yield strength about 11 times higher than that of conventional coarse-grained

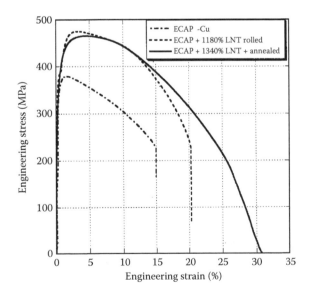

FIGURE 3.15
Tensile engineering stress–strain curves for three Cu samples: As received ECAP Cu, after additional LNT rolling to 1180%, and after LNT rolling to 1340% plus annealing at 100°C for 2 h, where LNT represents liquid nitrogen temperature. (Reproduced from Wan Y. M., Maa E., Chen M. W. 2002. *Appl. Phys. Lett.* 80(13): 2395–2397. With permission.)

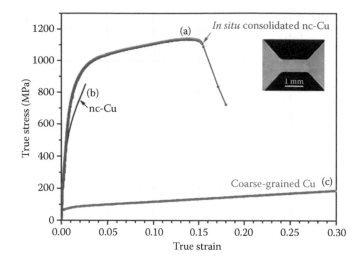

FIGURE 3.16
A typical tensile stress–strain curve for the bulk *in situ* consolidated nanocrystalline Cu sample (a), nanocrystalline Cu sample prepared by an inert gas condensation (b), and coarse-grained polycrystalline Cu sample (c). (Reprinted with permission from Youssef K. M., et al. 2005. Ultrahigh strength and high ductility of bulk nanocrystalline Cu. *Appl. Phys. Lett.* 87: 091904. Copyright 2005, American Institute of Physics.)

Cu, while retaining a 14% uniform tensile elongation. A comparison of the mechanical behavior during tensile testing reveals that nc-Cu exhibits extremely high yield and ultimate tensile strengths. The 0.2% offset yield strength and the ultimate tensile strength were 791 ± 12 MPa and 1120 ± 29 MPa, respectively. This yield strength is at least one order of magnitude higher than that of coarse-grained pure Cu samples and the ultimate tensile strength is about five times higher. In addition to these extraordinarily high strengths, the nc-Cu sample shows a significant tensile ductility, with 14% uniform elongation and 15.5% elongation to failure. This ductility is much larger than that of nc-Cu samples prepared by an IGC method, which show ductility as low as 2%. *In situ* dynamic straining in TEM shows a dislocation-controlled deformation mechanism that allows for the high strain hardening. It is found that the dislocations are trapped in the individual nanograins [26].

The nanoparticles are also useful in improving the toughness of the ceramics. The fracture toughness of ceramics can be significantly increased by dispersing a nanosized second phase reinforcements such as whiskers, platelets, or long fibers in the ceramic matrix. These nanocomposites or hybrids show enhanced fracture toughness and strength up to very high temperatures. In contrast, the properties of coarse-grained composites normally degrade. Further, nano reinforcements increase the creep resistance by suppressing grain boundary sliding. Diamond is the hardest material with a Vickers hardness up to about 100 GPa, highest bulk modulus (443 GPa), and the highest shear modulus (535 GPa) among all materials. However, diamonds are brittle with a low-fracture toughness of 3–5 MPa m$^{1/2}$ for single crystals, which limits their applications in harsh environments of dynamic impacts and high-stress concentrations. The fracture toughness (measured at loading force of 98 N) of diamond–SiC nanocomposites has been increased significantly by incorporating 70–80 wt% diamond crystals in SiC matrix using high-energy ball milling followed by rapid reactive sintering at high pressure and temperature. The measured fracture toughness K_{IC} of the synthesized nanocomposites has been increased up to 50% (from

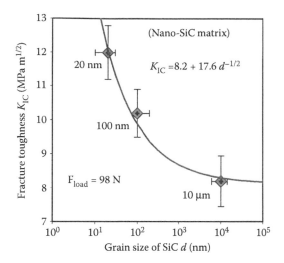

FIGURE 3.17
Fracture toughness of the diamond–SiC nanocomposites as a function of grain size of SiC matrix. The diamonds indicate the measured fracture toughness values for the corresponding SiC grain sizes. (Reproduced from Zhao Y. et al. 2004. Enhancement of fracture toughness in nanostructured diamond–SiC composites. *Appl. Phys. Lett.* 84(8): 1356–1358. With permission.)

8.2 to 12.0 MPa m$^{1/2}$) as the crystal size of the SiC matrix decreases from 10 μm to 20 nm (Figure 3.17). The Vickers hardness lies in the range of 60–80 GPa [27].

In these nanocomposites, the nanomatrix (of SiC) results in a situation where the size of a microcrack is bigger than the size of the matrix nanocrystals. To grow further, a crack would have to grow around the crystals, which would decrease its ability to propagate. This mobility reduction thus leads to an increase in the fracture toughness of the diamond–SiC nanocomposites [27]. Similarly, SiC nanoparticles improve the fracture toughness and hardness of MgO/SiC nanocomposites. The mullite-based 3M fiber retains 85% tensile strength at 1200°C. This is due to a two-phase nanocomposite of 55 wt% mullite and 45 wt% Al_2O_3, with 20% 200 nm mullite needles randomly dispersed in a matrix of 100 nm Al_2O_3 particles. Such super ceramics may find use in high efficiency gas turbines, and in aerospace and automotive components. The WC–Co nanocomposites have also shown microhardness as high as 2200 VHN with increased toughness. Thus, the abrasive wear resistance of nanostructured WC–Co is higher than the abrasive wear resistance of conventional WC–Co for a given hardness. Thus, WC–Co nanocomposites cutting tools are expected to have more than double lifetime than conventional coarse-grained composites [16].

3.5.4 Fatigue and Creep

The tensile strength of the free-standing thin films with grain sizes in the range of 9–67 nm prepared by electrodeposition method is shown in Figure 3.18a as a function of grain size. It can be seen that the rate of increase in tensile strength of films is slower when the grain size increases from 15 to 67 nm, whereas the rate of increase in tensile strength is comparatively higher for the films with grain size 9–15 nm. The increase in tensile strength can be attributed to pileup of dislocations at the grain boundaries. The dislocation motion within grains is the predominant mechanism of plastic deformation. However, the strength is increased with decreasing grain size due to impeding dislocation motion at the grain boundaries, but this resulted in lower ductility. For the films with grain size 9–15 nm, the

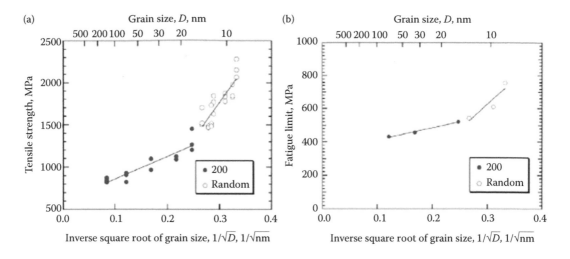

FIGURE 3.18
(a) Tensile strength and (b) fatigue strength of the free-standing thin films of Ni. (Reprinted from *Procedia Eng.*, 10, Tanakaa K., Sakakibara M., Kimachi H., Grain-size effect on fatigue properties of nanocrystalline nickel thin films made by electrodeposition, 542–547, Copyright 2011, with permission from Elsevier.)

ductility was found to increase with decreasing grain size compared to the films with grain size 15–67 nm. This is due to the fact that there is no dislocation accumulation and the grain boundary sliding is responsible for enhanced ductility. As shown in Figure 3.18b, the fatigue strength also increases with decreasing grain size. The fatigue strength of the nanofilms is much larger than the bulk nickel, that is, the fatigue strength corresponding to 10^7 cycles of nanocrystalline films reaches to 757 MPa, which is about two times larger than that of a nanocrystalline nickel of 20–40 nm grain size. In addition to the grain size, some other microstructural features are expected to be responsible for the improvement of the fatigue strength [28].

Figure 3.19 shows the creep behavior of the Ni–P samples with an average grain size of 28 and 257 nm which were investigated in the temperature range from 543 to 573 K. Under

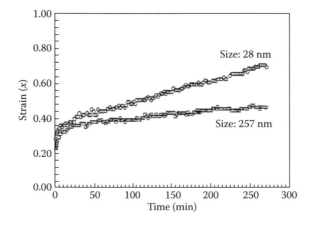

FIGURE 3.19
Creep behavior of micro- and nanocrystalline Ni–P alloy at 573 K and 146 MPa. (Reprinted from *Mater. Sci. Eng. R.*, 16, Lu K., Nanocrystalline metals crystallized from amorphous solids: Nanocrystallization, structure, and properties, 161–221, Copyright 1996, with permission from Elsevier.)

the similar test condition, the creep rate of the nanocrystalline (28 nm) sample was found much larger than that of the coarse-grained sample (i.e., 257 nm). The apparent activation energies of creep for the nanocrystalline and the coarse-grained samples are measured to be 0.71 and 1.1 eV, respectively. It clearly shows that grain boundary diffusion mechanism dominates in the creep process of the nanocrystalline sample, while lattice diffusion is controlling the creep in the coarse-grained sample [29].

3.6 Magnetic Properties

Materials can be classified based on their response to external magnetic field into diamagnetic, paramagnetic, and ferromagnetic. Du et al. have studied the magnetic properties of ultrafine nickel particles with size in the range of 2–99 nm. The specific saturation magnetization (σ) and the coercivity (H_c) of fine nickel particles measured at 7 kOe are shown in Figure 3.20a and b, respectively. It is seen that the specific saturation magnetization first decreases slowly with decreasing particle size and below 15 nm, the magnetization drastically reduces which implies that most particles turn to superparamagnetism at room temperature. The Ni particle with size of 85 nm exhibits specific saturation magnetization of about 43 emu/g which is less than that of bulk Ni (i.e., 54 emu/g) at room temperature. H_c increases to a highest value of 250 Oe (for single domain sized Ni nanoparticles) and then decreases on reducing the particle size from 85 nm to <15 nm. Its value approaches to 0 when the particle size is close to critical particle size of superparamagnetic particles (i.e., ~15 nm) (Chapter 9). The value of effective anisotropy constant for Ni nanoparticle is 5.8×10^5 erg/cc which is much larger than the value of bulk Ni (i.e., about 3.4×10^4 erg/cc). The reason for the higher value of effective anisotropy constant for Ni nanoparticles has been attributed to *surface anisotropy, shape anisotropy* and the change of *crystal anisotropy*. The Curie temperature for 9 nm Ni nanoparticles is about 300°C which is less than that of bulk Ni (T_c ~358°C). This is due to the contraction of bond length with a decreasing cluster size of Ni. This decreases the interval between the Ni atoms, thus results in smaller exchange integral. This in turn causes a reduction in the Curie temperature [30].

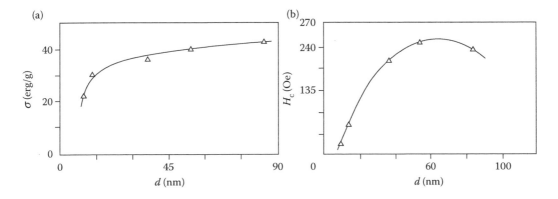

FIGURE 3.20
(a) Specific saturation magnetization and (b) coercivity as a function of average diameter of nickel particles. (Reprinted with permission from Du Y. W. et al. 1991. Magnetic properties of ultrafine nickel particles. *J. Appl. Phys.* 70: 5903–5905. Copyright 1991, American Institute of Physics.)

Nanostructured magnetic materials are already used to some extent as magnetic recording materials in both the information storage media and in the "write" and "read" heads. Thin-film recording media (e.g., disks) typically have thickness of 10–100 nm and at very high bit densities, longitudinal media must be very thin (10–50 nm), while perpendicular recording media can be much thicker (50–250 nm). The magnetic tapes used in high-density applications consist of an acicular iron core passivated by an oxide surface. The iron particles are about 200 nm long and about 20 nm diameter. For both disks and tapes, the magnetic single domains must be of smaller size to achieve the higher densities with satisfactory signal-to-noise ratios but not so small as to become paramagnetic [16].

Table 3.3 shows some fundamental properties of nanostructured and coarse-grained metals/alloys. The value of Curie temperature of nanostructured nickel processed by severe plastic torsion straining (SPTS) is 595 K (vs. 631 K for bulk) which is less than that of coarse-grained nickel by ~36 K. The saturation magnetization of the nanostructured nickel after the SPD is about 38.1 Am^2/kg (vs. 56.2 for bulk), which is less than that of the coarse-grained nickel by 31%. The decrease in Curie temperature and the saturation magnetization have been attributed due to reduced grain size and significant distortion of the crystal lattice. Both values recover to a value close to that of bulk when the nanostructured nickel is heated from 300 to 600 K. The Debye temperature of the nanostructured nickel obtained by SPD is 293 K, which is significantly less than the value of 375 K for bulk nickel. Similarly, the Debye temperature of nanostructured Cu is 233 K, which is less than the bulk Cu by 23%. The Debye temperature of nanostructured iron at its near-boundary regions is 240 K, where coarse-grain iron has 467 K. Compared to bulk, the lower Debye temperatures for nanostructured metals have been attributed to an increased amplitude of thermal vibrations of atoms in grain boundaries and an increased concentration of point defects in near-grain boundary regions. Moreover, nonequilibrium grain boundaries and long-range elastic stress fields are also responsible for a decrease in Debye temperatures for nanostructured materials processed by SPD. It is well studied that the value of Debye temperature in near-boundary regions of nanostructured nickel is about 127 K, which is almost 200 K less than the corresponding value for the bulk nickel. As shown in Table 3.3, the Debye temperature of the grain boundary phase in the nanostructured iron is about

TABLE 3.3

Some Fundamental Properties of Metals in Nanostructured (NS) and Coarse-Grained (CG) States

Properties	Materials	Value	
		NS	CG
Curie temperature, K	Ni	595	631
Saturation magnetization, Am^2/kg	Ni	38.1	56.2
Debye temperature, K	Fe	240[a]	467
	Ni	293	375
Diffusion coefficient, m^2/s	Cu in Ni	1×10^{-14}	1×10^{-20}
Ultimate solubility at 293 K, %	C in α-Fe	1.2	0.06
Young's modulus, GPa	Cu	115	128

Source: Reprinted from *Prog. Mater. Sci.*, 45, Valiev R. Z., Islamgaliev R. K., Alexandrov I. V. Bulk nanostructured materials from severe plastic deformation, 103–189, Copyright 2000, with permission from Elsevier.

[a]   For near boundary regions.

240 K, which is lower than that of coarse-grained iron by 227 K. The lower Debye temperature in near-boundary regions of nanostructured SPD materials shows the increased dynamic activity of their atoms. This resulted in a significant increase in diffusion rates. For example, the diffusion coefficient of Cu in nanostructured Ni is about six orders of magnitude more than in a coarse-grain nickel, due to reduced diffusion path length in nanostructured nickel [31].

The Young modulus of nanostructured Cu is about 115 GPa (vs. 128 GPa for bulk Cu) which is less than the value of coarse-grained Cu, which is approximately lowered by 10% than the bulk value. It is due to the fact that the value of elastic moduli in the near-boundary regions is 15%–17% of the value of the elastic moduli in coarse-grained metal. Transmission electron microscopy confirmed that the value of moduli is recovered to a value close to that of bulk by transformation of nonequilibrium grain boundaries (present in nanostructured materials) to the equilibrium grain boundaries (of bulk) [31].

3.7 Optical Properties

The size effects of the optical properties are important for the nanoparticles with size smaller than 10–15 nm, which is significantly smaller than the wavelength (λ) of light. In semiconductors, the energy of interatomic interactions is high. The electronic excitation of semiconductor crystal may lead to the formation of a weakly bonded electron-hole pair, that is, the Mott–Vanie exciton. The region of delocalisation of these excitons is considerably greater than the lattice constant of the semiconductor. When the size of semiconductor crystal is comparable with the size of the exciton (or the Bohr radius), the optical properties of the particle are affected significantly. In semiconductors, the Bohr radius of the exciton varies from 0.7 for CuCl to 10 nm for GaAs. The energy of the electronic excitation of a molecule is usually higher than the energy gap in the macroscopic semiconductor. In other words, a decrease in the size of the semiconductor nanoparticles is generally accompanied by displacement of the absorption band to the low-wavelength range, that is, a blueshift of the absorption band of semiconductor nanoparticles occurs with a decrease in their size. The semiconductor CdS shows a blueshift of the absorption band for nanoparticles with size smaller than 10–12 nm. For another semiconductor CdTe nanoparticles, the energy band increases when its size reduces from a bulk crystal to 4 (1.5 eV vs. 2.0 eV) and 2 nm (1.5 eV vs. 2.8 eV). The energy of the maximum of the absorption band is inversely proportional to the square of the radius of the nanoparticles. In addition, the distance between the peaks tends to increase with decreasing particle size due to the spreading of the energy levels. The extent of blueshift depends upon the particle size distribution, size, shape, and aspect ratio of the nanoparticles. To a certain extent, the presence of impurities may also affect it. Figure 3.21 shows how the energy bandgap (or energy level spacing) changes with decreasing particle size. The quantum size confinement effect becomes significant particularly when the particle size becomes comparable to or smaller than the Bohr exciton radius (α_B). The exciton Bohr radius for GaAs is around 11.8 nm with an exciton binding energy of 4.6 meV, while the Bohr radius of CdS is around 2.4 nm and particles with radius smaller or comparable to 2.4 nm show strong quantum confinement effects (i.e., a blueshift). Similarly, the absorption spectra of CdSe nanoparticles show a dramatic blueshift with decreasing particle size. The emission spectra usually show a similar blueshift with decreasing size [32].

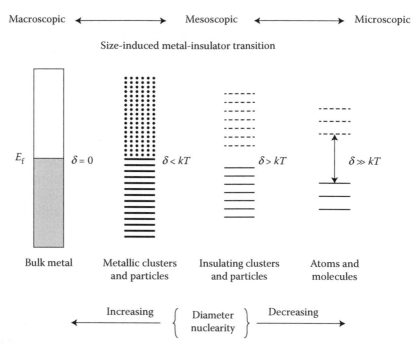

FIGURE 3.21

Evolution of the bandgap and the density of states as the number of atoms in a system increase from right to left. (Roduner E. 2006. Size matters: Why nanomaterials are different. *Chem. Soc. Rev.* 35: 583–592. Reproduced by permission of Royal Society of Chemistry.)

There are some interesting differences between semiconductors and metals in terms of quantum confinement. A major difference is in their electronic band structure. The bandgap which plays a key role in semiconductors is much less pronounced in metals, since electrons are free to move even at fairly low temperature in their half-filled conduction band. As the particle size increases, the center of the band develops first and the edge last. In metals with Fermi level lying in the band center, the relevant energy level spacing is very small even in relatively small particles, whose properties resemble those of the bulk. Hence, the quantum size confinement effect is expected to become important at much smaller sizes for metal than for semiconductor nanoparticles. For very small metal nanoparticles, a significant bandgap can be observed and the metal nanoparticles may become semiconductors or even insulators at the temperatures where bandgap energy dominates the thermal energy. For example, gold is known for its metallic shine, noble nature, and yellow color that does not tarnish. It has a face-centered cubic structure and melts at 1336 K. However, a small particle of the same gold with size of about 10 nm absorbs green light and thus appears red. Moreover, gold ceases to be noble, and 2–3-nm-sized nanoparticles are excellent catalysts which exhibit considerable magnetism. The metallic particles with size smaller than 2 nm become insulators at low temperature. For example, Hg clusters show a nonmetallic bandgap which decreases with increase in cluster size. However, this bandgap is closed for the clusters containing >300 atoms.

An important threshold is reached when the gap between the highest occupied and the lowest unoccupied state (called the Kubo gap δ) equals thermal energy, where the electrons get thermally excited across the Kubo gap. In other words, a low-temperature insulator becomes a semiconductor at higher temperatures. Blueshift is the most probable reason of using colloidal gold nanoparticles in stained glass windows of cathedrals and palaces in the middle ages as dyes. For example, gold particle with size of 10 nm absorbs green light and thus appears red due to surface plasmons (i.e., a collective excitation of electrons near the surface represents standing waves on a surface) [32].

This size-dependent optical behavior can be explained by the increase in energy level spacing for semiconductor nanoparticles, whereas surface plasmon resonance (SPR) for the metals. The SPR is caused by the coherent motion of all the free electrons within the conduction band leading to in-phase oscillation, when the metallic particles under study interact with an electromagnetic field. When the size of the nanocrystal is smaller than the wavelength of the incident radiation, an SPR is generated. The electric field of an incoming light induces a polarization of free electrons relative to the cationic lattice. The net charge difference occurs at the nanoparticle surface which acts as a restoring force. Thus, a dipolar oscillation of the free electrons is created with a certain frequency. The SPR is a dipolar excitation of the entire particle between the negatively charged electrons and the positively charged lattice. The energy of the SPR depends on both the free electron density and the dielectric medium surrounding the nanoparticle. The width of the resonance varies with the characteristic time before electron scattering. For larger nanoparticles, the resonance sharpens as the scattering length increases and for noble metals the resonance frequency is in the visible light range [33].

The decrease in size below the electron mean free path gives rise to intense absorption in the UV–visible range. Optical excitation of the SPR gives rise to the SPR. According to Mie, the larger the particles are, the more important are the higher order modes as light cannot polarize the nanoparticles homogenously. These higher order modes show peak at lower energies (or higher wavelengths). Thus, the plasmon band shifts toward higher wavelength (also called a redshift) with increasing particle size [34,35]. As shown in Figure 3.22, the suspension of gold nanoparticles in a solvent will show a red solution if this solution strongly absorbs lights of green wavelength (~520 nm) and blue wavelength (~450 nm). In general, the color of the solution is complementary to the color of light absorbed by the solution. The characteristic SPR bands for typical gold and silver nanoparticles are around 520 and 400 nm, respectively. The SPR bandwidth and peak position both are sensitive to the aspect ratio of the nanorods, which exhibit two distinct extinction bands corresponding to electron oscillations across (known as transverse mode, TM) and along the long axis (known as longitudinal mode, LM). In such anisotropic nanoparticles, the blue band is due to the transverse plasmon resonance while the redder band is due to longitudinal plasmon resonance. The redder band shifts with the increasing nanorod length or aspect ratio [33,36].

3.8 Wear Resistance

For many engineering applications, wear resistance is one of the most important tribological properties because wear accounts for more than 50% loss of all materials in service. The grain size refinement has shown significant increase in hardness, yield strength, tensile strength, and wear resistance of the nanocrystalline materials. For example, nanostructured WC–Co composites with WC grain size of about 70 nm showed nearly double the abrasive wear resistance compared to the conventional cermets. Nanocrystalline nickel with 10–20 nm grain size made by electrodeposition showed 100–170 times higher wear resistance and 45%–50% lower friction coefficient than bulk polycrystalline nickel measured by pin-on-disk tribometer. The

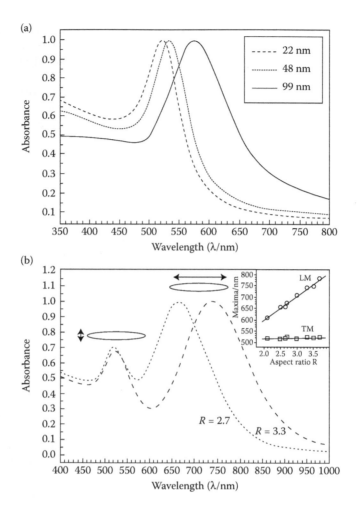

FIGURE 3.22
(a) Optical absorption spectra of 22, 48, and 99 nm spherical gold nanoparticles. The broad absorption particles correspond to the surface plasmon resonance and (b) optical absorption spectra of gold nanorods with mean aspect ratios (R) of 2.7 and 3.3. (Reproduced from Link S., El-Sayed M. A. 2000. *Int. Rev. Phys. Chem.* 19: 409–453. With permission from Taylor & Francis Group.)

Vickers microhardness of the nanocrystalline nickel coating increases considerably with decreasing grain size from 90 μm to 13 nm. This in turn increases the average abrasive wear resistance of the nanocrystalline coatings (62 and 13 nm) measured by Taber wear tester. In addition, the steady state wear rates of the nanocrystalline coating were reached after about 6000 wear cycles, whereas it was reached only after 9000 cycles for the bulk counterpart [37].

3.9 Chemical Sensitivity

According to Fick's law, the diffusion coefficient of the species could be determined using Equation 3.2. The equation can be rewritten in the form of Equation 3.3 which shows that the square of mean diffusion path $(\bar{x})^2$ of the species or atoms is proportional to the time (t) and diffusivity of the atom in a material.

$$C = C_0 \text{erfC} \frac{x}{2\sqrt{Dt}} \tag{3.2}$$

$$(\bar{x})^2 = D \cdot t \tag{3.3}$$

$$D = D_0 \cdot e^{-Q/k_BT}$$

where D is the diffusion coefficient which depends exponentially on the temperature (T). The C_0 and C are the concentrations of species on the surface and depth of "x" distance, respectively. The diffusion will get faster with increasing temperature. The quantity Q is the activation energy and k_B is the Boltzmann constant. For a given temperature, doubling of the particle diameter (i.e., diffusion path) leads to a fourfold time needed for diffusion. Conventional materials generally have grain sizes of about 10 μm or more and the homogenization time of these materials at elevated temperatures is in the range of many hours. In contrast, the homogenization time for nanomaterials with grain sizes of about 10 nm is reduced by a factor of 10^6, that is, nanomaterials need homogenization time of milliseconds indicating the instantaneous homogenization. This phenomenon is often called "instantaneous alloying." This instantaneous diffusion through nanoparticles can be exploited for some useful applications including in gas sensor [9] and non-equilibrium alloys. As discussed in chapter 2, nanomaterials contain a very large fraction of atoms on the surface and at the grain boundaries. The large fraction of grain boundaries of nanomaterial provides a high density of reduced diffusion paths compared to single crystal or conventional coarse-grained polycrystalline materials. Thus, nanomaterials are expected to exhibit an enhanced diffusivity with the same chemical composition. For example, the diffusivity for 8 nm-grain sized copper samples measured at room temperature is 2.6×10^{-20} m^2/s while it is 4.8×10^{-24} m^2/s for grain boundary diffusion and 4×10^{-40} m^2/s for lattice diffusion. This indicates that the diffusivities in nanocrystalline copper are about 14–20 orders of magnitude higher than lattice diffusion and about 2–4 orders of magnitude larger than grain boundary diffusion. The increased diffusivity (or reactivity) leads to increased solid solubility limits, for example, the solid solubility of Hg in nanocrystalline Cu is reported to be 17 at.% which is much higher than that of equilibrium value (i.e., 1 at.%) [16].

Some of the gas sensor works on variation in the electric conductivity due to changes in the stoichiometry of transition metal oxides (i.e., oxygen/metal ratio). The stoichiometry of the small sensing particles (or nanoparticles) is immediately changed due to the oxygen potential in the surrounding atmosphere. In contrast to conventional gas sensors, the time response is now controlled by the gas diffusion through the small diffusion paths in between the nanoparticles. A typical design of a gas sensor is set up on a conductive substrate on a carrier plate. The surface of this conductive layer is covered completely with the transition metal oxides of TiO_2, SnO_2, Fe_2O_3, etc. On the top of this layer, a gas-permeable conductive layer is applied as a counter electrode. Any variation in the oxygen potential in the surrounding atmosphere changes the stoichiometry of these oxides, and, thus the electrical conductivity. This process is reversible. It has been investigated very well that the response of the nanosensor is faster and the signal is much better than the conventional sensor due to the diffusion scaling law. The time constant depends primarily on the diffusion of the gas molecules in the open-pore network and through the conducting

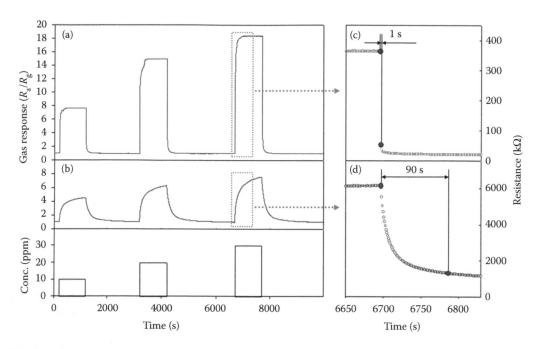

FIGURE 3.23

Ethanol sensing characteristics at 400°C (a) gas response of hierarchical SnO_2 spheres (b) gas response of dense SnO_2 spheres (c) change in resistance of the hierarchical spheres after exposure to 30 ppm ethanol (d) change in resistance of the dense spheres after exposure to 30 ppm ethanol. (Reprinted from *Sensor. Actuat. B.*, 136, Kim H. R. et al., Highly sensitive and ultra-fast responding gas sensors using self-assembled hierarchical SnO_2 spheres, 138–143, Copyright 2009, with permission from Elsevier.)

cover layer [9]. Diffusion through the nanoparticles is fast due to reduced diffusion path. Sensors are based on the principle that changes in electrical conductivity are caused by the changes in stoichiometry of oxides. Due to high percentage of surface atoms on nanoparticles, gas nanosensors containing hierarchical tin oxide (SnO_2) spheres can exhibit high sensitivity and ultrafast response to ethanol at 400°C. Figure 3.23a and b shows the change in the gas response (R_a/R_g resistance in air/resistance in gas) to 10–30 ppm ethanol as a function of sensing time. The sensitivity (R_a/R_g) values of hierarchical spheres are about two times higher than those of their dense (coarser) counterparts. It was observed that 90% response time of the hierarchical spheres to ethanol was extremely short (i.e., ~1 s) while that of the dense spheres was about 90 s [38,39].

It is believed that a change in the surrounding oxygen potential causes change in the stoichiometry of the oxide nanoparticles. In other words, the oxide/metal ratio is changed with time in presence of outer atmosphere. Semiconductor gas sensors are widely used for the detection of toxic or foul smelling gases, chemical process control, monitoring of air pollution in the environment, etc. [39]. The response of a sensor is heavily dependent on the size of the SnO_2 particles used as the sensing material. Figure 3.24 shows clearly an increase in the sensitivity of detection for carbon monoxide (CO) with decreasing grain size of SnO_2 particles. The sensitivity increases from 3 units for SnO_2 particles with grain size ~13.2 nm to ~15 units for SnO_2 nanoparticles with grain size less than ~4.8 nm. The significant increase in sensitivity of sensors with significantly smaller-sized SnO_2 nanoparticles can be attributed to a reduced diffusion time for outer species and huge surface area of nanoparticles, thereby accelerating exchange with the surrounding atmosphere. The high

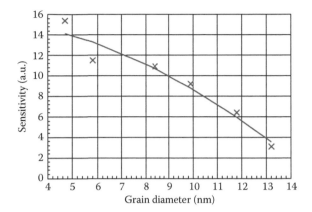

FIGURE 3.24
Effect of particle size on sensitivity of CO determination of a gas sensor. (Vollath D: *Nanomaterials: An Introduction to Synthesis, Properties, and Applications.* 2nd ed. 2013. Copyright Wiley-VCH Verlag GmbH & Co. KGaA. Germany. Reproduced with permission.)

porosity of the sensing thick-film layer facilitates rapid diffusion of the gas species to electrode. Depending on the molecule's size, the time response for different elements depends on the thickness of the surface coating [9].

3.10 Dielectric Constant

The dielectric constant (or permittivity) of an insulating material is defined as the ratio of the charge stored by an insulating material placed between two parallel metallic plates to the charge that can be stored when the insulating material is replaced by air or vacuum. It indicates the ability of an insulator to store the electrical energy. The influence of particle size effect on the dielectric constant of the $BaTiO_3$ nanoparticles has been studied by several authors. Figure 3.25a shows the grain size (from 1.1 to 13 μm) dependence of the permittivity at 1 kHz. The permittivity of the $BaTiO_3$ ceramics increased with decreasing the grain size from 13 μm in the conventional method (or from 5.9 μm in two-step sintered samples). The highest permittivity value of the $BaTiO_3$ reported so far is 7700 for a grain size of 1.1 μm. The increase of the permittivity with decreasing the grain size was attributed to decrease in size of 90° domain configurations. Figure 3.25b shows the effect of domain size on the permittivity of the $BaTiO_3$ ceramics. The increase of permittivity with decreasing grain size is due to the domain size effect. The permittivity of the $BaTiO_3$ ceramics increased with decreasing the domain size and it is the highest of 7700 at ~100 nm. The high-domain density enhanced the orientational polarizability due to the domain-wall vibration and the ionic polarizability due to the lattice vibration [40]. Figure 3.26 also shows the variation of average particle sizes (17–500 nm) of $BaTiO_3$ on the dielectric constant. The dielectric constant of the particles was measured for a $BaTiO_3$ suspension using modified powder dielectric measurement method. It can be seen that the dielectric constant of $BaTiO_3$ suspension increases with decreasing particle sizes down to 70 nm. The dielectric constant of $BaTiO_3$ particles with a size of around 68 nm exhibited a maximum value of over 15,000. However, thereafter the value of dielectric constant of $BaTiO_3$ suspensions decreases with decreasing particle sizes below 68 nm [41].

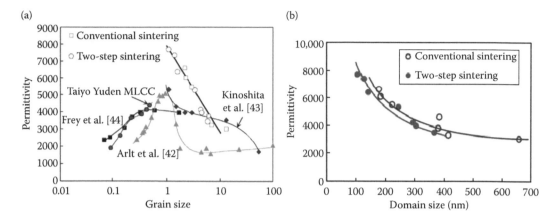

FIGURE 3.25
(a) Grain size and (b) domain size dependence of permittivity of BaTiO₃ ceramics. (The source of the material Hoshina T. et al., Domain size effect on dielectric properties of barium titanate ceramics. *Jpn. J. Appl. Phys.*, 2008, Institute of Physics is acknowledged.)

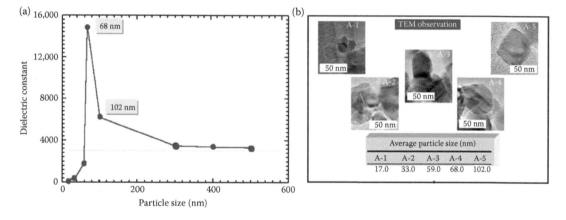

FIGURE 3.26
(a) Variation of particle sizes on dielectric constant of BaTiO₃ fine powder and (b) TEM images of BaTiO₃ nanoparticles with sizes from 13 nm to 102 nm. (From Kakimoto M.-a. et al. 2006. *Materials Science and Engineering B* 132:74–78. With permission.)

Ohno et al. reported that the intrinsic dielectric constant (estimated from the Raman spectra by using the LST relation) for the BaTiO₃ nanoparticles with grain size of 105 nm along the c-axis (ε_c) and a-axis (ε_a) was 42 and 1255, respectively. The intrinsic dielectric constant increases with decreasing particle diameter. BaTiO₃ nanoparticles with size of 30 nm show largest value, which is more than the single crystal. These results indicate that the intrinsic dielectric constant increases drastically with decreasing particle diameter, suggesting the possibility for high dielectric materials with thin layers of ferroelectric nanoparticles [13].

References

1. Li Y., Lu D., Wong C. P. 2010. *Electrical Conductive Adhesives with Nanotechnologies*. Springer XI, New York, USA, 50, 25–79.

2. Shin H. S., Yu J. 2007. Size-dependent thermal instability and melting behavior of Sn nanowires. *Appl. Phys. Lett.* 91: 173106.
3. Olson E. A., Efremov M. Y., Zhang M., Zhang Z., Allen L. H. 2005. Size-dependent melting of Bi nanoparticles. *J. Appl. Phys.* 97: 034304.
4. Wang X. W., Fei G. T., Zheng K., Jin Z., Zhang L. D. 2006. Size-dependent melting behavior of Zn nanowire arrays. *Appl. Phys. Lett.* 88: 173114.
5. Lai S. L., Carlsson R. A., Allen L. H. 1998. Melting point depression of Al clusters generated during early stages of film growth: Nanocalorimetry measurements. *Appl. Phys. Lett.* 72(9): 1098–1100.
6. Eckert J., Holzer J. C., Ahn C. C., Fu Z., Johnson W. L. 1993. Melting behavior of nanocrystalline aluminum powders. *Nanostruct. Mater.* 2: 407–413.
7. Ristau R., Tiruvalam R., Clasen P. L., Gorskowski E. P., Harmer M. P., Kiely C. J. 2009. Electron microscopy studies of thermal stability of gold nanoparticles arrays. *Gold Bulletin* 42(2): 143–152.
8. Buzea C., Pacheco I. I., Robbie K. 2007. Nanomaterials and nanoparticles: Sources and toxicity. *Biointerphases* 2(4): 17–71.
9. Vollath D. 2013. *Nanomaterials: An Introduction to Synthesis, Properties, and Applications.* 2nd ed. Wiley-VCH Verlag GmbH & Co. KGaA, Germany.
10. Gusev A. I., Rempel A. A. 2008. *Nanocrystalline Materials.* Cambridge International Science Publishing, United Kingdom.
11. Rupp J., Birringer R. 1987. Enhanced specific-heat-capacity (Cp) measurements (150–300 K) of nanometer-sized crystalline materials. *Phys. Rev. B* 15: 7888–7890.
12. Al-Assiri M. S., El-Desoky M. M. 2013. Grain-size effects on the structural, electrical properties and ferroelectric behavior of barium titanate-based glass–ceramic nano-composite. *J. Mater. Sci.: Mater. Electron.* 24: 784–792.
13. Ohno T., Suzuki D., Suzuki H., Ida H. 2004. Size-effects for barium titanate nanoparticles. *Kona* 22: 195–201.
14. Pendse S., Goyal R. K. 2016. Disappearance of Curie temperature of $BaTiO_3$ nanopowder synthesized by high energy ball mill. *J. Mater. Sci. Surf. Eng.* 4(3): 383–385.
15. Zhao Y. H., Lu K. 1997. Grain-size dependence of thermal properties of nanocrystalline elemental selenium studied by x-ray diffraction. *Phys. Rev. B* 56(22): 14330–14337.
16. Suryanarayana C., Koch C. C. 2000. Nanocrystalline materials—Current research and future directions. *Hyperfine Interact.* 130: 5–44.
17. Erb U. 1995. Electrodeposited nanocrystals: Synthesis, properties and industrial applications. *Nanostruct. Mater.* 6: 533–538.
18. Ishihara T., Yan J., Matsumoto H. 2006. Extraordinary fast oxide ion conductivity in $La_{1.61}GeO_{5-\delta}$ thin film consisting of nano-size grain. *Solid State Ionics* 177: 1733–1736.
19. Aliofkhazraei M. 2011. *Nanocoatings, Engineering Materials.* Springer-Verlag, Heidelberg, Berlin.
20. Tschoepe A., Birringer R. 2001. Grain size dependence of electrical conductivity in polycrystalline Cerium Oxide. *J. Electroceram.* 7: 169–177.
21. Fasakia I., Koutoulakia A., Kompitsasb M., Charitidis C. 2010. Structural, electrical and mechanical properties of NiO thin films grown by pulsed laser deposition. *Appl. Surf. Sci.* 257: 429–433.
22. Bose C. A., Balaya P., Thangadurai P., Ramasamy S. 2003. Grain size effect on the universality of AC conductivity in SnO_2. *J. Phys. Chem. Solids* 64: 659–663.
23. Li X., Shih W. H. 1997. Size effects in barium titanate particles and clusters. *J. Am. Ceram. Soc.* 80(11): 2844–2852.
24. Stolyarov V. V. and Valiev R. Z. 2006. Enhanced low-temperature impact toughness of nanostructured Ti. *Appl. Phys. Lett.* 88: 041905.
25. Wan Y. M., Maa E., Chen M. W. 2002. Enhanced tensile ductility and toughness in nanostructured Cu. *Appl. Phys. Lett.* 80(13): 2395–2397.
26. Youssef K. M., Scattergood R. O., Murty K. L., Horton J. A., Kocha C. C. 2005. Ultrahigh strength and high ductility of bulk nanocrystalline copper. *Appl. Phys. Lett.* 87: 091904.

27. Zhao Y., Qian J., Daemen L. L., Pantea C., Zhang J., Voronin G. A., Zerda T. W. 2004. Enhancement of fracture toughness in nanostructured diamond–SiC composites. *Appl. Phys. Lett.* 84(8): 1356–1358.

28. Tanakaa K., Sakakibara M., Kimachi H. 2011. Grain-size effect on fatigue properties of nanocrystalline nickel thin films made by electrodeposition. *Procedia Eng.* 10: 542–547.

29. Lu K. 1996. Nanocrystalline metals crystallized from amorphous solids: Nanocrystallization, structure, and properties. *Mater. Sci. Eng. R.* 16: 161–221.

30. Du Y. W., Xu M. X., Shi Y. B., Lu H. X., Xue R. H. 1991. Magnetic properties of ultrafine nickel particles. *J. Appl. Phys.* 70: 5903–5905.

31. Valiev R. Z., Islamgaliev R. K., Alexandrov I. V. 2000. Bulk nanostructured materials from severe plastic deformation. *Prog. Mater. Sci.* 45: 103–189.

32. Roduner E. 2006. Size matters: Why nanomaterials are different. *Chem. Soc. Rev.* 35: 583–592.

33. Link S., El-Sayed M. A. 2000. Shape and size dependence of radiative, non-radiative and photothermal properties of gold nanocrystals. *Int. Rev. Phys. Chem.* 19: 409–453.

34. Cao G. 2004. *Nanostructures & Nanomaterials: Synthesis, Properties & Applications.* Imperial College Press, London.

35. Prasad M. J. N. V., Ghosh P., Chokshi A. H. 2009. Synthesis, thermal stability and mechanical behavior of nano-nickel. *J. Indian Inst. Sci.* 89(1): 43–48.

36. Zhang J. 2009. *Optical Properties and Spectroscopy of Nanomaterials.* World Scientific Publishing Co. Pte. Ltd, Singapore.

37. Jeong D. H., Gonzalez F., Palumbo G., Aust K. T., Erb U. 2001. The effect of grain size on the wear properties of electrodeposited nanocrystalline nickel coatings. *Scripta Mater.* 44: 493–499.

38. Yang Z., Huang Y., Chen G., Guo Z, Cheng S., Huang S. 2009. Ethanol gas sensor based on Al-doped ZnO nanomaterial with many gas diffusing channels. *Sensor. Actuat. B.* 140: 549–556.

39. Kim H. R., Choi K. I., Lee J. H., Akbar S. A. 2009. Highly sensitive and ultra-fast responding gas sensors using self-assembled hierarchical SnO$_2$ spheres. *Sensor. Actuat. B.* 136: 138–143.

40. Hoshina T., Takizawa K., Li J., Kasama T., Kakemoto H., Tsurumi T. 2008. Domain size effect on dielectric properties of barium titanate ceramics. *Jpn. J. Appl. Phys.* 47(9): 7607–7611.

41. Kakimoto M.-a., Takahashi A., Tsurumi T.-a., Hao J., Li L., Kikuchi R., Miwa T., Oono T., Yamada S. 2006. Polymer-ceramic nanocomposites based on new concepts for embedded capacitor. *Materials Science and Engineering B* 132:74–78.

42. Arlt G., Hennings D., De G. 1985. Dielectric properties of fine-grained barium titanate ceramics. *J. Appl. Phys.* 58:1619.

43. Kinoshinata K., Yamaji A. 1976. Grain-size effects on dielectric properties in barium titanate ceramics. *J. Appl. Phys.* 47: 371.

44. Frey M. H. Xu Z., Han P., Payne D. A. 1998. The role of interfaces on an apparent grain size effect on the dielectric properties for ferroelectric barium titanate ceramics. *Ferroelectrics* 206: 337.

4

What Have We Learnt From the Nature and History?

4.1 Nanotechnology of Nature

Nanotechnology field seems to be a new field, but it is not an entirely new field because the nature has given us many multifunctional biological structures ranging from nano to macro in scale and inspired the human beings to create multifunctional devices. Nature has long been using the bottom-up nanofabrication method to form self-assembled nano-materials that are much stronger and tougher than man-made materials synthesized by top-down approach. Thus, nature can be called a "school of learning." For example, photosynthesis harnesses solar energy to support plant life. In plants, light-harvesting molecules, that is, chlorophyll, are arranged within the cells on the nanometer to micrometer scales. These structures capture light energy, and convert it into the chemical energy that drives the biochemical machinery of plant cells. The biological molecular nanomachine (flagella) which rotates at over 10,000 rpm is driven by the proton flow caused by the electrochemical potential differences across the membrane. The *lotus leaves exhibit* superhydrophobic and dirt-resistant properties due to their unique nanostructure (~120 nm), which has inspired the scientific community for the development of self-cleaning windows, windshields, exterior paints for buildings and navigation ships, utensils, roof tiles, and textiles. Nature-made nacre and bones are the best examples of the nanocomposites made by bottom-up approach, and exhibit strength and toughness higher than those of their constituents. The presence of pigment particles typically of ~500–800 nm in the cells of animals makes animal's body to appear dark or colored. The animals use this feature for signaling purpose and/or to hide themselves.

In brief, bioinspired nanostructures provide interesting multifunctional properties such as hydrophilicity, superhydrophobicity (in lotus leaf), excellent mechanical properties (in spider silk), adhesion properties (in Gecko), photosynthesis (in plants), excellent antireflection (in moth eye), and so on, which have inspired the scientific community to the innovative developments such as smart textile, self-cleaning windows, antireflective coating, solar cells, etc [1–14].

4.1.1 Lotus Leaf

Among many plant leaves, lotus leaf (or water lily) is one of the most important leaf which has been a symbol of purity in Asian religions for over 2000 years. Lotus leaves have dirt-resistant properties because of their unique structure. As shown in Figure 4.1, lotus leaf surfaces possess randomly distributed micropapillae with size in the range of 5–9 μm and covered by fine branchlike nanostructures with an average diameter of ~120 nm (Figure 4.1b and c). The cooperation of the multiscale surface structures and hydrophobic

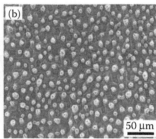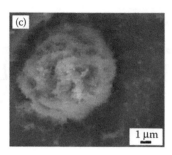

FIGURE 4.1
(a) Water droplets picking up dirt particles and rolling down across the lotus leaf surface, indicating its self-cleaning effect, (b) SEM image of the lotus leaf surface, showing the micropapillae, and (c) SEM image of a single papilla. (Reprinted from *Nano Today*, 6, Liu K., Jiang L. Bio-inspired design of multiscale structures for function integration. Copyright 2011, with permission from Elsevier.)

epicuticular wax confers a high-water contact angle and a small-sliding angle, exhibiting superhydrophobic and low-adhesion functions. Water droplets on its surface are almost spherical and can roll freely in all directions and then pick up dirt particles if they are lying on the leaf surface. Thus, this phenomenon is known as self-cleaning effect or lotus effect. The lotus effect has inspired scientists to develop many superhydrophobic self-cleaning surfaces. Similar to lotus leaf, rice leaf has hierarchically structured papillae, which are arranged in quasi-1-D order parallel to the leaf edge. This special feature makes the rice leaf both superhydrophobic and anisotropic in wettability, that is, the water droplet can roll more easily along the direction parallel to the rice leaf edge than along the perpendicular one [1]. Solga et al. [2] developed a prototype textile which stays dry for about 4 days when submerged in water (Figure 4.2). The containers or cannulas filled with resin paint can be emptied without any residues. The manufacturer of powder has also developed a nanoparticle powder for multipurpose applications for generating self-cleaning films on various substrates [2].

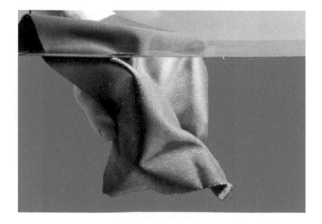

FIGURE 4.2
Prototype of a superhydrophobic textile holding an air film under water. (The source of the material Solga A. et al., The dream of staying clean: Lotus and biomimetic surfaces, *Bioinsp. Biomim.*, 2007 and Institute of Physics is acknowledged.)

(a)   (b)   (c)

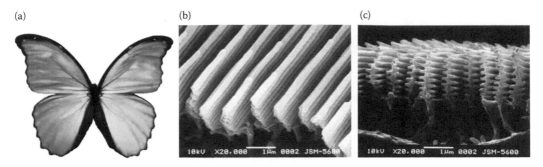

FIGURE 4.3
(a) *Morpho didius* SEM images of (b) an oblique view and (c) a cross section of a ground scale of the butterfly *Morpho didius*. (From Liu K., Jiang L. 2011. *Nano Today* 6: 155–175; Reprinted from *Proc. R. Soc. Lond. B*, 269, Kinoshita S., Yoshioka S., Kawagoe K., Mechanisms of structural color in the Morpho butterfly: Cooperation of regularity and irregularity in an iridescent scale, 1417–1421, Copyright 2002, with permission from Elsevier.)

4.1.2 Morpho Butterfly

Morpho butterfly is a bright insect living in Central and South America and is famous for its brilliant blue iridescent colors, which arise from multiscale photonic structures with size range of nanometer to millimeter. Its wings consist of two distinct layers of different scales. The underside layer is planar and featureless, while the top and externally visible surface has an intricate microstructure. This top surface exhibits one of several forms of ridging that extends longitudinally from one end of the scale to the other. The majority of ridges are connected at intervals by a series of cross-ribs. Spacing between ridges lies in the range 0.5–5.0 µm depending on type of species (Figure 4.3). Interference of light between these air cuticles is the mechanism by which the stunning iridescent blues and violets are produced [3,14]. In addition, it exhibits superhydrophobicity, self-cleaning, and selective chemical-sensing capabilities. The highly ordered photonic structures exhibit a different optical response to different individual vapors such as water, methanol, ethanol, and isomers of dichloroethylene. The highly selective vapor response of iridescent scales is superior to that of existing nanoengineered photonic sensors [1].

4.1.3 Peacock Feather

The peacock tail feather (of male) is one of the largest and most attractive display feathers in nature. The basic structure of the peacock tail feather in the eye region is shown in Figure 4.4. The tail feather has a central stem with an array of barbs on each side. In addition, these barbs are covered with a large number of barbules. Each barbule contains around 20 round indentations along its length. These individual indentations produce color. The male-peacock feather displays iridescent colors and intricate eye patterns on the tail feathers (Figure 4.4a and b). In addition to the astonishing structural color, peacock feathers also exhibit superhydrophobicity (Figure 4.4c) by the water droplets which have water contact angle >90°. The tiny and intricate 2-D photonic crystal structure is responsible for the color production in peacock feathers (Figure 4.4d and e). The variations in the lattice constant and the number of periods in the photonic crystal structure lead to a variation in color [1]. As shown in Figure 4.4, the transverse cross section reveals that a barbule consists of a medullar core of ~3 µm surrounded by a cortex layer. The cortex of all differently colored barbules contains a 2-D photonic crystal structure made

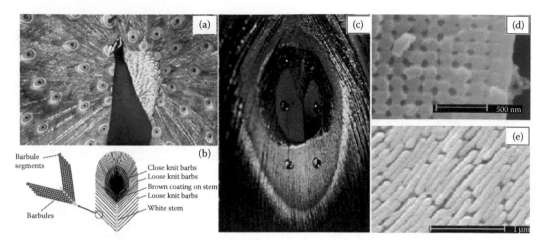

FIGURE 4.4
(a) Display of peacock tail, (b) structure of peacock tail feathers. (From Burgess S. C., King A., Hyde R. 2006. *Opt. Laser Technol.* 38: 329–334.) (c) Iridescent peacock feathers with superhydrophobicity, (d) SEM images of transverse cross section of the cortex for the green barbule, and (e) SEM image of longitudinal cross section of the green barbule after the removal of the surface keratin layer. (Reprinted from *Nano Today*, 6, Liu K., Jiang L., Bio-inspired design of multiscale structures for function integration, 155–175, Copyright 2011, with permission from Elsevier.)

up of melanin rods (length: $\sim$0.7 μm) embedded in keratin (Figure 4.4d and e). Photonic crystal structures in all differently colored barbules are quite similar. For the blue, green, and yellow barbules, the lattice structure is nearly square, while it is rectangular lattice for the brown barbule. However, there is difference of the lattice constant (i.e., rod spacing) and the number of periods (i.e., melanin rod layers) along the direction normal to the cortex surface, which is main cause of the diversified colors in peacock feather. One or two layers of melanin rods are buried in the surface keratin layer. The lattice constants for the blue, green, and yellow barbules are $\sim$140, 150, and 165 nm, respectively. In the brown barbule, the lattice structure is less regular, with the lattice constants of $\sim$150 and 185 nm along the directions parallel and perpendicular to the cortex surface, respectively. The number of periods is $\sim$9–12 for the blue and green barbules, 6 for the yellow barbule, and $\sim$4 for the brown barbule [4].

Thin-film interference takes place in the keratin layers of the barbule segments and produces bright and iridescent colors that change with the angle of view. Each barbule is about 60 μm wide and 5 μm thick. The foam core is around 3 μm thick surrounded by the keratin layers having thickness of $\sim$0.4–0.5 μm. The wavelength of visible colors is between 0.4 and 0.8 μm. Therefore, optical interference takes place when light reflects off the different keratin layers, causing reflected colors. The color depends on the thickness of the layers and the refractive index of the keratin. The thickness of the keratin layers is optimal for producing the brightest thin-film colors. It has been found that the keratin layers contain one or two layers of melanin rods joined by keratin. Color is dependent on the spacing of the melanin rods and the number of melanin rod layers [5].

4.1.4 Cuttlefish

Cuttlefish possess neurally controlled, pigmented chromatophore organs that allow rapid changes in skin patterning and coloration in response to visual cues. The chromatophore contains a multiscale assembly of nanostructured pigment granules encapsulated within a cell that give rise to its functionality. Isolated pigment granules fluorescence between the wavelengths of 650 and 720 nm. When the granular structure is altered, the emission spectrum is also changed to blueshift [6].

4.1.5 Water Strider

Water striders are insects with length of 1 cm and weight of 10 dynes that reside on the surface of ponds, rivers, and the ocean. Their weight is supported by the surface tension force generated by curvature of the free surface, and they propel themselves by driving their central pair of hydrophobic legs in a sculling motion. The strider transfers momentum to the underlying fluid through hemispherical vortices shed by its driving legs [7]. Water strider (*Gerris remigis*) can stand and move effortlessly and quickly on water by using their legs. It is due to the special design of strider's legs which are covered with numerous oriented and needle-shaped micrometer-sized setae. The setae are arranged at an inclined angle of about 20° from the surface. Each seta consists of many elaborate helical nanogrooves, which can trap tiny air bubbles (Figure 4.5). This multiscale structure of water strider's legs provides durable and robust superhydrophobicity. The maximal supporting force of a single leg is about 15 times the total body weight of a water strider. The corresponding volume of water ejected is roughly 300 times that of the leg itself indicating its superwater repellency. Therefore, water striders can dance up and down even in turbulent conditions (Figure 4.5a) [1,8].

4.1.6 Spider Silks

Over the past few decades, spider silks have attracted the attention of the scientific community for their amazing mechanical properties and endurance under stress. Spider can synthesize continuous, insoluble, and lightweight orb webs for prey capture. At ambient temperature and pressure from an aqueous solution, spiders produce different types of silks for different purposes, which can withstand rain, wind, and sunlight. The spider silk possesses multiscale structures from nano- to macroscale. Spider dragline silks exhibit remarkable strength and toughness, which are superior to almost all natural and man-made high-performance materials. In addition, spider dragline silk exhibits wetting-induced supercontraction, resulting in an increase in diameter and shrinkage in length, and also exhibits unrivalled torsional qualities that stop the spider twisting and swinging. Capture silk is the sticky spiral in the webs of orb-weaving spiders and five times as elastic as dragline silk. Spider capture silk is a multifunctional material possessing high strength, elasticity, stickiness, and directional water-collection ability from moist air [1].

Compared to steel beam, it is difficult to break a beam made of the spider silk. It is about 100 times tougher than the steel. The spider silk has a tensile strength (i.e., 1.5 GPa) comparable to that of steel with much lower density than the steel. Silk's unusual combination of high strength and stretch leads to toughness values never attained in synthetic high-performance fibers. Figure 4.6 compares the stress-strain curves of spider silk with the Kevlar. Silk consists of very repetitive blocks of mainly glycine and alanine (glycine and alanine are types of amino acids—the molecules that make up the long protein chains). These are the simplest and smallest amino acids allowing the strands to be packed together. Strands are "glued" together by hydrogen bonds to form tightly packed, highly ordered crystalline regions. The hard crystals make up only 10%–15% of the total volume. Other regions contain bulky amino acids like tyrosine or arginine that prevent the close packing. These

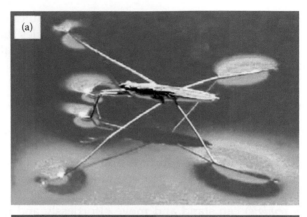

FIGURE 4.5
The superhydrophobic water strider leg and its biomimetic materials; (a) digital photograph of a water strider standing on the water surface, (b) SEM image of the leg with oriented spindly microsetae. The inset in (b) shows the nanoscale groove structure on a single seta. (Reprinted with permission from Feng X. Q. et al. 2007. Superior water repellency of water strider legs with hierarchical structures: Experiments and analysis. *Langmuir* 23: 4892–4896. Copyright 2007, American Chemical Society.)

regions form amorphous, disordered areas that allow the silk to stretch. The interplay between the hard crystalline segments and the elastic regions gives spider silk its extraordinary properties [9].

Silk is an exceptionally strong, extensible, and tough material made from simple protein building blocks. The molecular structure of dragline spider silk repeat units consists of semiamorphous and nanocrystalline β-sheet protein domains. The silk repeat units are scaled up to create macroscopic silk fibers with outstanding mechanical properties despite the presence of cavities, tears, and cracks. The geometric confinement of silk fibrils to diameters of 50 ± 30 nm is critical to facilitate a powerful mechanism by which hundreds of thousands of protein domains synergistically resist deformation and failure to provide enhanced strength, extensibility, and toughness at the macroscale. Through this mechanism, silk fibers exploit the full potential of the nanoscale building blocks, despite the presence of large defects. The silk fibers are composed of silk fibril bundles with diameters in the range of 20–150 nm. The spider silk (or an ancient biological protein material) exhibits intriguing mechanical properties, in spite of its simple protein building blocks. Out of many types of silk, the dragline silk of orb-weaving spiders is extremely strong, extensible, and tough. In addition to the relatively large ultimate strength of spider silk,

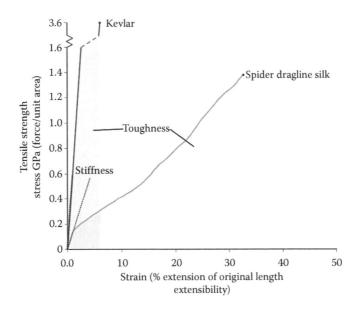

FIGURE 4.6
Tensile strength of spider silk. (From Powers A. 2013. *Berkeley Sci. J.*, 18(1): 46–49. Open Access.)

comparable to that of steel, silk features a strength-to-density ratio that is up to 10 times higher than that of steel because of the material's small density ($\sim$1.3 g/cm^3). This is due to a hierarchical structure of silk, which consists of a β-sheet nanocrystals ($\sim$2–4 nm) embedded in semiamorphous protein domains [10].

4.1.7 Beetles

The beetle's adhesive structures are present on the whole body from wing to tiptoe. Beetle exhibits unique and complex surface architectures with or without liquid secretion. The beetle has ability to use appropriate adhesion system like hair interlocking, mushroom-shaped pads, oil-assisted spatula-shaped pads, and claws under different surface conditions and situations such as flying, resting, mating, crawling locomotion, and wet-assisted protection. In contrast to the highly organized hierarchical structures of gecko toepad, the beetle's adhesion is mediated via capillarity or van der Waals forces with relatively simple structures depending on the specific function [11].

4.1.8 Glass Sponges (Euplectella)

Glass sponges, also called Venus Flower Basket, are a group of deep sea organisms. Their skeletons exhibit complex hierarchical structures from nanometer to macroscopic length scales, which comprise at least seven hierarchical levels such as silica nanospheres, lamellar spicules, bundled beams, square-grid cells, etc. At each level, the structural hierarchy of the glass sponges has a solution to overcome their brittleness. Thus, this successive hierarchical assembly results in an outstanding mechanical rigidity and stability. The remarkable hierarchical organization of *Euplectella* is regarded as a textbook of mechanical engineering resembling advanced structural engineering masterpieces including the Swiss Re Tower in London and Eiffel Tower in Paris. *Euplectella* consists of a lattice of fused spicules. The spicule has a concentric lamellar structure. These layers are separated from one another by organic interlayers, which are made up of silica nanospheres and arranged in a cylindrical fashion around a central proteinaceous filament. In addition to their outstanding mechanical properties, spicules possess remarkable fiber-optical properties [12].

4.1.9 Gecko

Geckos are the world's supreme climbers who are capable of attaching and detaching their adhesive toes in milliseconds while running recklessly on vertical and inverted surfaces. This ability of Geckos can be attributed to the nanostructures (spatulae) present on their feet hair (setae). Dry adhesion mechanism in gecko lizards has attracted much attention since it provides strong, yet reversible attachment against surfaces of varying roughness and orientation. Such unusual adhesion capability is attributed to arrays of millions of fine microscopic foot hairs (setae), splitting into hundreds of smaller, nanoscale ends (called as spatulae) (Figure 4.7). Spatulae are $\sim$0.2 μm in length and width at the tip, which form intimate contact to various surfaces by van der Waals forces with strong adhesion ($\sim$10 N/cm^2).

The presence of millions of adhesive setae on the gecko's toes enables them responsible to climb-up or climb-down vertical surfaces at speeds of over 1 m/s. All 6.5 million setae attached simultaneously on the toes of one gecko could lift 133 kg. Moreover, the geckos are able to detach their feet in 15 milliseconds without external forces. The van der Waals forces are responsible for the adhesion between gecko feet and wall. At the nanoscale, the electrostatic forces of attraction are more dominant than gravitational forces, as the gravitational forces are mass dependent and the mass of interacting nanobodies is extremely small. The gecko setae are strongly hydrophobic with a water drop contact angle of 160.9°. The gecko has inspired development of artificial dry adhesive based on polymer and carbon nanotubes. The dry adhesives based on micropatterned CNT arrays with optimized

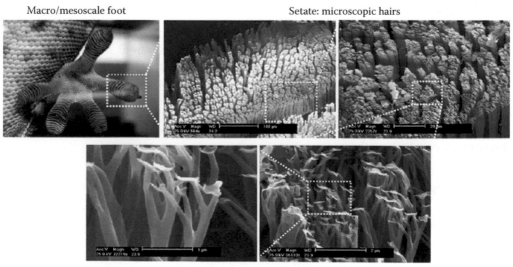

Macro/mesoscale foot   Setate: microscopic hairs

Spatulae: nanoscale tips

FIGURE 4.7
Photo/images of gecko foot hairs in increasing order of magnifications indicating millions of fine setae on the attachment pads which split into hundreds of spatulae. (Reprinted from *Nano Today*, 4, Jeong H. E., Suh K. Y., Nanohairs and nanotubes: Efficient structural elements for gecko-inspired artificial dry adhesives, 335–346, Copyright 2009, with permission from Elsevier.)

geometry exhibit four to seven times higher shear adhesion ($\sim$36 N/cm^2) strength than nonpatterned CNT arrays [11]. However, conventional pressure-sensitive adhesives (PSAs) are either strong and difficult to remove (i.e., duct tape) or weak and easy to remove (i.e., sticky notes). Conventional PSAs, made of polymers, degrade, foul, self-adhere, and attach accidentally to inappropriate surfaces. In contrast, gecko toes have angled arrays of branched, hair-like setae formed from stiff, hydrophobic keratin that act as a bed of angled springs with similar effective elastic modulus to that of PSAs. Setae are self-cleaning and maintain function for months during repeated use in dirty conditions. Setae exhibit anisotropic frictional adhesive enabling either a tough bond or spontaneous detachment [13].

The geckos or little lizards have a remarkable ability to cling to any surfaces (i.e., smooth or wet) when they are upside down. The gecko's sticking abilities stemmed from the 200-nm-wide keratin hairs that coated the soles of their feet. Capillary forces cause hairs to stick to films of wet surfaces and equally strong van der Waals forces enable them to attach to dry surfaces. Each hair exerts only 10^{-7} N of force, but they are densely packed enough to collectively have an adhesive force of 10 N/cm^2. This force is enough to suspend a 100 kg mass from a patch of 10 cm on each side. Inspired by these findings, a team of scientist from the University of Manchester's Centre for Mesoscience and Nanotechnology in England attempted to reproduce the gecko hairs using an array of plastic fibers, but the team achieved an adhesion of 3 N/cm^2, an almost 30% that of the real gecko. This adhesion strength would be sufficient to suspend a man with just adhesive gloves covering his palms [11–13].

4.1.10 Chameleon

Cephalopods have often been referred to as the chameleons of the sea. Chameleons (*Chamaeleonidae*) are a distinctive and highly specialized clade of Old World lizards, which have the ability to change colors of their different body parts. They have two superimposed layers within their skin that control their color and thermoregulation. The top layer contains a lattice of guanine nanocrystals. The spacing between the nanocrystals decides the wavelengths of light which is reflected and/or absorbed. In a relaxed state, the nanocrystals reflect blue and green light, but in an excited state the lattice increases the distances between the nanocrystals, which in turn reflects longer wavelengths (i.e., yellow, orange, green, and red). The presence of yellow pigments in its skin combines with the blue reflected light (in relaxed crystal lattice) and results in green color, which is common in many chameleons in their relaxed state. The deeper layer of skins primarily controls the amount of near-infrared light that is absorbed or reflected, and therefore may influence thermoregulation.

4.1.11 Abalone Nacre

The nacreous portion of the abalone shell is composed of calcium carbonate ($CaCO_3$) crystals interleaved with layers of polymers (i.e., viscoelastic proteins). Nacre is found in the shiny interior of many mollusk shells. It is also known as mother-of-pearl. It is the best example of a natural nanocomposite made by bottom-up approach and exhibits excellent strength and toughness, despite the brittle nature of its constituents (i.e., $CaCO_3$) [15]. The robustness of the abalone shell lies in the brick- and mortar-like shell design. The shell is a biological composite consisting of $\sim$95% aragonite (i.e., $CaCO_3$) bricks held together by a few percentage of mortar (or polymer). The brick (or mineral) provides strength, while the mortar acts like a lubricant between the bricks, allowing the bricks to slide

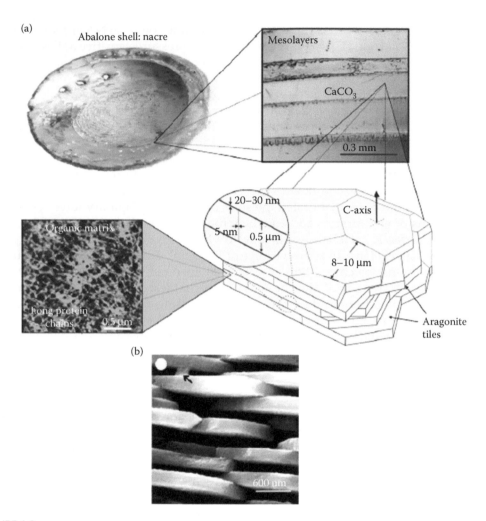

FIGURE 4.8
Overall view of hierarchical structure of abalone shell, showing (a) mesolayers, mineral tiles, and tile pullout in a fracture region. (Reprinted from *J. Mete. Behav. Biomed. Mater.*, 1; Meyers M. A. et al., Mechanical strength of abalone nacre: Role of the soft organic layer, 76–85, Copyright 2008, with permission from Elsevier). (b) SEM image of fracture section of nacreous. (Reprinted with permission from Li X. D. et al. 2004. Nanoscale structural and mechanical characterization of a natural nanocomposite material: The shell of red abalone. *Nano Lett.* 4: 613–617. Copyright 2004, American Chemical Society.)

over each other. This $CaCO_3$/polymer composite is well structured with an extraordinary beauty and high-quality mechanical properties. The aragonite platelets have a thickness of ~400 nm and an edge length of ~5 μm, and are separated by a 5–20 nm thick polymer. The cross section of aragonite platelet resembles a brick wall. The thin polymer layer is porous with pore size of ~50 nm. These pores enable the formation of mineral bridges between adjacent tiles, thus the size of bridges is ~50 nm. Due to highly ordered hierarchical structure, the nacreous layer exhibits excellent mechanical properties, that is, nacre shows a two-fold increase in strength and a 1000-fold or more increase in toughness over its constituent ($CaCO_3$). The deformability of the aragonite platelets together with the crack deflection, platelet slip, and polymer (or organic adhesive interlayer) contributes to the nacre's tremendous fracture toughness. Figure 4.8 shows the photo of inside surface of an abalone shell. Its cross section shows the mesolayers, which are spaced ~300 μm apart and separated by thick regions that consist of polymer (primarily organic material) with minerals embedded into it.

The fracture strength of nacre measured under bending was between 56 and 116 MPa depending upon the type of nacre. Its Young's modulus was of ~70 GPa for dry and 60 GPa for wet sample and the tensile strength was ~170 MPa for dry and 140 MPa for wet samples [15–17].

4.1.12 Magnetotactic Bacteria

Magnetotactic bacteria (MTB) are a diverse group of microorganisms with the ability to orient and migrate along geomagnetic field lines. This ability is based on specific intracellular nanostructures having membrane-bound crystals of the magnetic iron minerals magnetite (Fe_3O_4) or greigite (Fe_3S_4). These nanocrystals function as tiny compasses that allow the microbes to navigate using Earth's geomagnetic field and also help the bacteria in locating the growth favorable conditions within the water column or aquatic sediment. They can push themselves through the water by rotating their helical flagella and interestingly, they can swim at speeds nearly twice that of *Escherichia coli* cells [18]. Figure 4.9 shows a TEM bright-field image of a single cell of *Magnetospirillum magnetotacticum* [19].

It can be clearly seen that the magnetite chain is 1200 nm in length containing 22 crystals of 45 nm in size with separation between them of ~9.5 nm [19]. Mature magnetite crystals have typical size range of about 35–120 nm and highest lengths up to 250 nm. These sizes exhibit stable single domain (SD) and are permanently magnetic at ambient temperature. Thus, these crystals are uniformly magnetized and have maximum magnetic dipole moment. The particles with lengths above ~120 nm are nonuniformly magnetized because of the formation of domain walls, which tend to form multiple magnetic domains of opposite magnetic orientation, thereby reducing the total magnetic remanence of the crystals. Therefore, by controlling particle size, MTB have optimized the magnetic dipole moment per magnetosome.

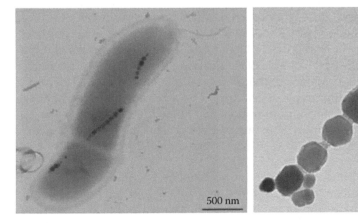

500 nm   100 nm

FIGURE 4.9
TEM image of a single cell of *Magnetospirillum magnetotacticum* at scale bar of 500 and 100 nm. (Reprinted with permission from Prozorov R. et al. 2007. Magnetic irreversibility and the Verwey transition in nanocrystalline bacterial magnetite. *Phys. Rev. B*, 76: 054406. Copyright 2007 by the American Physical Society.)

MTB are known to mediate numerous environmental biogeochemical transformations of key elements including sulfur, nitrogen, carbon, and iron. They have been used in immobilization of relatively large quantities of bioactive substances, and in cell sorting and separation [17–19]. Moreover, flagella of bacteria, which rotate at over 10,000 rpm, are an example of a biological molecular machine. The flagella motor is driven by the proton flow caused by the electrochemical potential differences across the membrane. The diameter of the bearing is about 20–30 nm, with an estimated clearance of ~1 nm.

4.1.13 Antireflective Nanostructures of Moth Eyes

Moths' eyes have an unusual property, that is, their surfaces are covered with a natural nanostructured film, which reduces reflections. This allows the moth to see well in the dark, without reflections, to give its location away to predators. The structure consists of a hexagonal pattern of bumps, each roughly 200 nm high and spaced on 300 nm centers. This kind of antireflective coating works because the bumps are smaller than the wavelength of visible light (400–800 nm). This structure has inspired the researchers for developing antireflective films. For example, moth-eye structures grown from tungsten oxide and iron oxide can be used as photoelectrodes for splitting water to produce hydrogen. The structure consists of tungsten oxide spheroids of several 100 μm size coated with a few nanometers thin iron oxide layer [11]. Antireflective wings of cicada have attracted significant interest owing to the highly ordered vertical nanonipple arrays on their surfaces, and represent a unique approach to the minimization of light reflection over a wide range of wavelengths. The subwavelength structures on the wing surface create a change in optical impedance, matching at the air-to-cuticle interface, which increases the photon collection and reduces reflectance. Such ARSs offer a continuous graded refractive index from air to the material surface, and reduce the reflection over a broad range of wavelengths and angles of incidence. In addition to their high-performance optical properties, certain cicada wings also exhibit superhydrophobicity.

4.1.14 Natural Nanocomposite

Nature is perhaps the foremost inspiration for nanoscientists and nanotechnologist. The cell membranes, and several other functional organelles within the biological cell of living beings, are of nanometer size. Proteins and enzymes together account for all the metabolic activities in the body. Figures 4.10 and 4.11 show the examples of nature-made nanostructures. Bone is the best example of a nature-made nanocomposite which consists of a fibrous biopolymer (i.e., collagen) and a mineral phase (carbonated hydroxyapatite) assembled into a complex hierarchical structure. The size of mineralized collagen fibril ranges between 50 and 200 nm. In human bone, an amorphous mineral serves as a precursor to the formation of a highly substituted nanocrystalline apatite. The amorphous calcium phosphate with size of ~100–300 nm is present in the disordered phase of trabecular bone [20].

There are two types of bone; compact and cancellous (trabecular or spongy). Cancellous bone consists of a flat lamellar structure forming platelets and rods that assemble into a porous structure. Thus, high-impact loads can be sustained by the skull, vertebrae, ribs, and the head of the femur. In contrast, compact bone consists of osteons, which are tubular structures composed of concentric lamellae that surround the blood vessel. The presence of nanoscale-mineralized collagen fibril (Figure 4.11) contributes to the excellent fracture resistance [21].

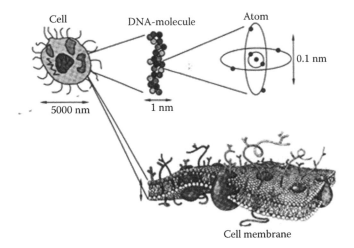

FIGURE 4.10
Biological features of DNA, cells, and membranes ranging from subnanostructure to macrostructure. (From Murty B. S. et al. 2013. *Textbook of Nanoscience and Nanotechnology*. Springer. Reproduced with permission from Universities Press (India) Pvt. Ltd.)

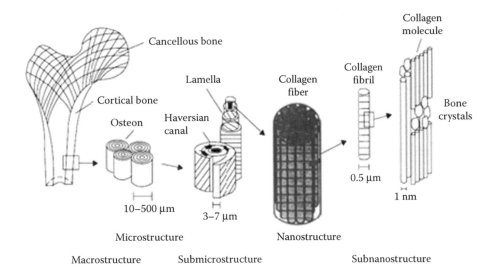

FIGURE 4.11
A schematic representation of bone consisting of subnanostructure to macrostructure. (From Gogotsi Y. 2006. *Nanomaterials Handbook*. Taylor and Francis Group. Reproduced with permission from Taylor & Francis Group.)

4.1.15 Dentin

Dentin is the naturally occurring hybrid material containing apatite and fibrils of type I collagen. It is the most abundant mineralized tissue in the human tooth. It is found beneath the enamel layer in teeth. It is a fiber reinforced composite material containing ~50% carbonated apatite and 30% organic matter (i.e., type I collagen). It has fiber-like tubules that provide reinforcement to the surrounding matrix. The lining of these tubules is composed of a highly mineralized cuff of intertubular dentin containing mostly small apatite crystals. The crystals are needle like near the pulp becoming more plate like closer to the enamel. Unique nanostructures of bones provide excellent interaction between the proteins and the bone cell in the body and also impact ability to absorb the impact load. Moreover, bones behave like a spring which can change shape on deformation within limit without cracking, to shorten and widen in compression and to lengthen and narrow in tension. If the load imposed exceeds the bone's ability to deform elastically, it can deform further and change shape permanently by plastic deformation. If both the elastic and plastic zones are exceeded, the bone fractures. Thus, bone is the material which shows good stiffness and strength along with flexibility and light weight due to its unique nano-structured design [21].

4.1.16 Natural Solar Cell

In a natural solar cell, the chlorophyll molecules (from a few nm to a few μm scales), absorb light particularly the red and blue parts of the spectrum, thus reflecting the green light. The absorbed energy is sufficient to free an electron from the excited chlorophyll. These free electrons are carried away by other molecules for photosynthesis process, which harnesses solar energy to support plant life. Gratzel et al. simulated some features of the natural photosynthesis process to develop a dye-sensitized solar cell, which consists of an array of nanometer-sized crystallites of TiO_2 and coated with light-sensitive molecules that can transfer electrons to the semiconductor particles when they absorb photons. The light-sensitive molecules, which absorb light strongly in the visible range, play a role equivalent to chlorophyll in photosynthesis [22,23].

4.2 Nanotechnology in Ancient History?

4.2.1 Damascus Sword

For more than 400 years ago, Indians were aware of the importance of wootz steels which were used to produce Damascus swords with high mechanical strength and flexibility. These swords were used as weapons in the battlefield till the nineteenth century against Muslim nations in Europe [24]. The steel of Damascus blades used by Crusaders fighting against Muslims showed a characteristic wavy banding pattern (Figure 4.12a and b) with extraordinary mechanical properties, and an exceptionally sharp cutting edge from the ultrahigh-carbon steel. It is believed that Damascus blades were forged directly from wootz steel produced in ancient India. A sophisticated thermomechanical treatment of forging and annealing was applied to refine the steel to its exceptional quality. Small additions of the alloying elements such as vanadium, chromium, manganese, cobalt, nickel, and others are known to facilitate the formation of cementite (Fe_3C) bands during thermal cycling at temperatures below the cementite formation temperature (about 800°C) [25].

Moreover, the Damascene steel contains rare earth elements and shows evidence of cementite nanowires in its microstructure. High-resolution transmission electron microscopy showed presence of multiwalled carbon nanotubes (MWNTs) with the characteristic distance of 0.34 nm and also bent MWNTs in a genuine Damascus sabre produced by the

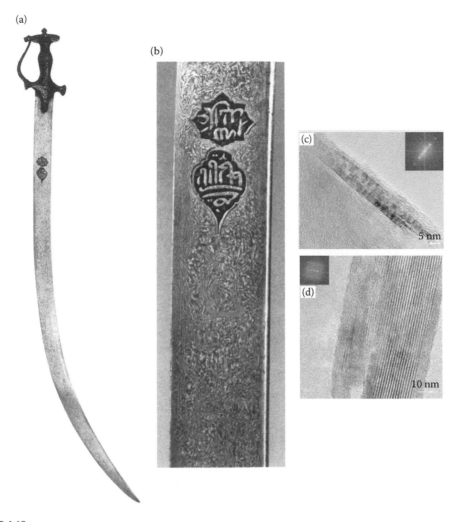

FIGURE 4.12
(a) Damascus sword, (b) wavy pattern on the sword. (From Verhoeven J. D. Genuine Damascus steel: A type of banded microstructure in hypereutectoid steels. *Steel Research*. 2002. 73:355-364. Copyright Wiley-VCH Verlag GmbH & Co. KGaA. Reproduced with permission.) (c) High-resolution TEM images of carbon nanotubes in a genuine Damascus sabre after dissolution in hydrochloric acid, and (d) remnants of cementite nanowires encapsulated by carbon nanotubes, which prevent the wires from dissolving in acid. (Reprinted by permission from Springer Nature. *Nature*, Reibold M. et al. 2006. Carbon nanotubes in an ancient Damascus sabre. 444: 286, copyright 2006.)

famous blacksmith Assad Ullah (of India) in the seventeenth century. Its microstructure consists of very small carbon nanotubes (CNTs) which were apparent only after dissolution of the sample in hydrochloric acid (Figure 4.12c). Some remnants (Figure 4.12d) show evidence of incompletely dissolved cementite nanowires indicating that these wires could have been encapsulated and protected by the CNTs. The presence of impurities (or alloying elements) probably acts as catalysts for the growth of CNTs during repeated forging, thus resulting in banding microstructure. However, the diminishing supply of some of the ores, which contain vanadium, molybdenum, chromium, manganese, cobalt, nickel, and rare earth elements from particular mines in India during the

(a)   (b)

FIGURE 4.13
The Lycurgus cup shining in (a) day light (or reflected) and (b) transmitted light. (With kind permission from Springer Science+Business Media: *Gold Bulletin*, The Lycurgus Cup—A Roman nanotechnology, 40, 2007, 270–277, Freestone I. et al.)

eighteenth century, probably could have prevented blacksmiths from practicing their ancient recipes. Thermal cycling and cyclic forging cause catalytic elements to segregate gradually into planar arrays parallel to the forging plane. These elements may give rise to the growth of CNTs, which in turn initiate the formation of cementite nanowires and coarse cementite particles. It has been investigated that the band formation in these steels resulted from the microsegregation of low levels of carbide-forming elements (i.e., V, Mo, Cr, Mn, and Nb). The presence of vanadium as low as 40 ppm by weight is quite effective in producing the bands of clustered Fe_3C particles [26].

4.2.2 Stained Glass

In this section, some of the important examples of nanotechnology throughout ancient, medieval, and modern times are discussed. The metal nanoparticles (or metal colloids) dispersed in a glass matrix of "Roman Lycurgus cup" (Figure 4.13) are the most fascinating example of ancient glass dated to the fourth century AD and have been found across the Roman Empire. This splendid cup is housed at the British Museum and depicts King Lycurgus dragged into the underworld by Ambrosia. The mythological scenes on the cup depict the death of Lycurgus, King of the Edoni in Thrace at the hands of Dionysus and his followers. The cup possesses the unique feature of changing color depending upon the light in which it is viewed. It appears green when viewed in reflected light, but looks ruby-red except for the King, which looks purple when a light is shone from inside and is transmitted through the glass. This unique and special color is due to the presence of a very small amount of tiny (~70 nm) metal crystals containing Ag and Au in an approximate molar ratio of 14:1 in the glass matrix. According to some study, the dichroism is due to the presence of nanosized (<100 nm) particles of silver (66.2%), gold

Red:
Ag (~100 nm, triangle)

Yellow:
Au (~100 nm, spheres)

Green:
Au (~50 nm, spheres)

Light blue:
Ag (~90 nm, spheres)

Blue:
Ag (~40 nm, spheres)

FIGURE 4.14
Rosace nord stained glass in the Cathédrale Notre-Dame de Chartres (France). (Horikoshi S., Serpone N. (Eds). *Microwaves in Nanoparticle Synthesis*. 2013. Copyright Wiley-VCH Verlag GmbH & Co. KGaA. Reproduced with permission.)

(31.2%), and copper (2.6%) in the glass matrix. The dichroism is attributed to the presence of minute amounts of submicroscopic crystals or colloids of gold (about 40 ppm) and silver (about 300 ppm) in the soda-lime-silica type glass. Colloidal systems can give rise to light scattering phenomena that result in dichroic effects. Light absorption and scattering by these nanoparticles determine the different colors, that is, the gold nanoparticles are mainly responsible for the reddish transmission and the silver for the greenish reflection [27,28].

Second example is of the beautiful "stained glass windows" produced in medieval times, and visible in several churches. These stained glasses are made up of a composite of glass and nanosized metal particles. The ruby-red glasses are a mixture of glass with nanosized gold powder. The purple of Cassius is a colloidal mixture of gold nanoparticles and tin dioxide in glass. The stained glass of a wonderful rose (Figure 4.14) can be seen at the world heritage Cathédrale Notre-Dame de Chartres in France. These different colors can be tuned by the size, shape, and the aspect ratio of the nanoparticles of gold and silver. Third example is of Chinese porcelain (or famille rose), which contains gold nanoparticles with size in the range of 20–60 nm. The change in color of this porcelain is due to surface plasmon effect by the metal nanoparticles. The size, shape, and aspect ratio of the metal nanoparticles influence their visible colors [28].

As the size of Au or Ag particles decreases particularly below 100 nm, there is a blue-shift and the characteristic color steadily changes to different colors depending upon its size. These effects are due to the surface plasmon resonance (SPR), that is, the frequency at which conduction electrons oscillate in response to the alternating electric field of incident electromagnetic radiation. The SPR phenomenon is exhibited in the visible spectrum by the metals such as Au, Ag, Cu, and the alkali metals which have free electrons [29]. The glasses with different colors can be obtained by varying the size of Ag nanoparticles, as shown in Table 4.1 [30].

The presence of copper and cuprous oxide (cuprite Cu_2O) nanoparticles has been reported in Celtic red enamels dated from 400 to 100 BCE. The copper and silver are still used in all luster (or glaze) decoration on ceramics since the ninth century. The green color of the glaze indeed comes from the Cu^{2+} ions. The distance between the two layers

TABLE 4.1

Particle Size of Ag Versus Color

Approx. Size of Ag Particle (in nm)	Color
<25	Blue
25–55	Green
35–45, 50–60	Yellow–green
70–80, 120–130	Brown

Source: With kind permission from Springer Science+Business Media: *Glass. Ceram.*, Nanotechnology in Glass Materials (Review), 65, 2008, 148–153, Min'ko N. I., Nartsev V. M.

was around 430 nm, which is of the same order of magnitude as visible wavelengths. The resultant double layer structure behaves as an optical network [31]. One more use of copper nanoparticles is in the famous red flambé and mixed blue–red Jun glazed porcelains from the medieval Song and Ming to Qing Chinese Dynasties as well as some Vietnamese porcelains and stoneware. The practice of using copper, silver, and gold dispersion in an optically clear matrix (i.e., of clay or glass) by potters and glassmakers is a very old and dates at least from the Bronze Age [32].

4.2.3 Maya Blue Paint

The blue color often used in pottery, murals, and ceremonial artifacts in Mayan archeological sites was different from any blue ever identified on ancient or medieval paintings from Europe or Asia. Maya blue was used in Mesoamerica and colonial Mexico probably as late as the twentieth century. In addition to its beautiful look, Maya blue is resistant to diluted mineral acids, alkalis, solvents, oxidants, reducing agents, moderate heat, and even biocorrosion. Paintings in the Bonampak archeological site have retained their blue color after centuries in the extreme conditions of the rain forest. The presence of superlattice of clay (or palygorskite crystals including impurity) which is intercalated (during mixing) with indigo molecules ($C_{16}H_{10}N_2O_2$) might be the reason of excellent corrosion resistance compared to control palygorskite. The blue color is probably due to the absorption curve peaked in the visible part of the spectrum. The metal and metal oxide nanoparticles can account at least in part for the color [33]. The murals and pottery of the ancient Mayan world show the beautiful blue paint which have long been admired for its marvelous color qualities as well as its inherent resistance to deterioration and wear over long periods of time [34].

4.2.4 Age-hardened Aluminum Alloys

In 1906, age-hardened aluminum alloys with refined microstructure have been developed by Alfred Wilm, which have fine precipitates responsible for age hardening, for example, Al–4% Cu alloys have clusters of Cu atoms, known as GP zones, and the metastable partially coherent θ' precipitate. Maximum hardness is observed with a mixture of GPII (or θ'') (coarsened GP zones) and θ' with the dimensions of the θ' plates typically about 10 nm in thickness by 100 nm diameter. In addition, the critical current density of commercial superconducting Nb_3Sn was found inversely proportional to grain size [35].

References

1. Liu K., Jiang L. 2011. Bio-inspired design of multiscale structures for function integration. *Nano Today* 6: 155–175.
2. Solga A., Cerman Z., Striffler B. F., Spaeth M., Barthlott W. 2007. The dream of staying clean: Lotus and biomimetic surfaces. *Bioinsp. Biomim.* 2: 126–134.
3. Kinoshita S., Yoshioka S., Kawagoe K. 2002. Mechanisms of structural color in the Morpho butterfly: Cooperation of regularity and irregularity in an iridescent scale. *Proc. R. Soc. Lond. B* 269: 1417–1421.
4. Zi J., Yu X., Li Y., Hu X., Xu C., Wang X., Liu X., Fu R. 2003. Coloration strategies in peacock feathers. *PNAS.* 100(22): 12576–12578.
5. Burgess S. C., King A., Hyde R. 2006. An analysis of optimal structural features in the peacock tail feather. *Opt. Laser Technol.* 38: 329–334.
6. Deravi1 L. F., Magyar A. P., Sheehy S. P., Bell G. R. R., Mathger L. M., Senft S. L., Wardill T. J., Lane W. S., Kuzirian A. M., Hanlon R. T., Hu E. L., Parker K. K. 2014. The structure–function relationships of a natural nanoscale photonic device in cuttlefish chromatophores. *J. R. Soc. Interface* 11: 20130942.
7. Hu D. L., Chan B., Bush J. W. M. 2003. The hydrodynamics of water strider locomotion. *Nature* 424(7): 663–666.
8. Feng X. Q., Gao X., Wu Z., Jiang L., Zheng Q. S. 2007. Superior water repellency of water strider legs with hierarchical structures: Experiments and analysis. *Langmuir* 23: 4892–4896.
9. Powers A. 2013. Spider Silk. *Berkeley Sci. J.* 18(1): 46–49.
10. Giesa T., Arslan M., Pugno N. M., Buehler M. J. 2011. Nanoconfinement of spider silk fibrils begets superior strength, extensibility, and toughness. *Nano Lett.* 11: 5038–5046.
11. Pang C., Kwak M. K., Lee C., Jeong H. E., Bae W. G., Suh K. Y. 2012. Nano meets beetles from wing to tiptoe: Versatile tools for smart and reversible adhesions. *Nano Today* 7: 496–513.
12. Jeong H. E., Suh K. Y. 2009. Nanohairs and nanotubes: Efficient structural elements for gecko-inspired artificial dry adhesives. *Nano Today* 4: 335–346.
13. Autumn K., Gravish N. 2008. Gecko adhesion: Evolutionary nanotechnology. *Phil. Trans. R. Soc. A* 366: 1575–1590.
14. Vukusic P., Sambles J. R., Lawrence C. R., Wootton R. J. 1999. Quantified interference and diffraction in single Morpho butterfly scales. *Proc. R. Soc. Lond. B* 266: 1403–1411.
15. Li X. D., Chang W. C., Chao Y. J., Wang R., Chang M. 2004. Nanoscale structural and mechanical characterization of a natural nanocomposite material: The shell of red abalone. *Nano Lett.* 4: 613–617.
16. Meyers M. A., Lin A. Y., Chen P., Muyco J. 2008. Mechanical strength of abalone nacre: Role of the soft organic layer. *J. Mech. Behav. Biomed. Mater.* 1: 76–85.
17. Li X. D., Chang W. C., Chao Y. J., Wang R., Chang M. 2004. Nanoscale structural and mechanical characterization of a natural nanocomposite material: The shell of red abalone. *Nano Lett.* 4: 613–617.
18. Yan L., Zhang S., Chen P., Liu H., Yin H., Li H. 2012. Magnetotactic bacteria, magnetosomes and their application. *Microbiol. Res.* 167: 507–519.
19. Prozorov R., Prozorov T., Mallapragada S. K., Narasimhan B., Williams T. J., Bazylinski D. A. 2007. Magnetic irreversibility and the Verwey transition in nanocrystalline bacterial magnetite. *Phys. Rev. B* 76: 054406.
20. Murty B. S., Shankar P., Raj B., Rath B. B., Murday J. 2013. *Textbook of Nanoscience and Nanotechnology.* Springer and Universities Press (India) Private Limited, Hyderabad, India.
21. Gogotsi Y. 2006. *Nanomaterials Handbook.* Taylor and Francis Group, USA.
22. Kalyanasundaram K., Grätzel M. 1998. Applications of functionalized transition metal complexes in photonic and optoelectronic devices. *Coord. Chem. Rev.* 77: 347–414.

23. Kickelbick G. (Ed.). 2007. *Hybrid Materials: Synthesis, Characterization, and Applications*. Wiley-VCH Verlag GmbH & Co. KGaA, Betz-Druck GmbH, Darmstadt, Germany.
24. Verhoeven J. D. 2002. Genuine Damascus steel: A type of banded microstructure in hypereutectoid steels. *Steel Research* 73(3): 355–364.
25. Verhoeven J. D., Pendray A. H., Dauksch W. E. 1998. The key role of impurities in ancient Damascus steel blades. *JOM* 50: 58–64.
26. Reibold M., Paufler P., Levin A. A., Kochmann W., Pätzke N., Meyer D. C. 2006. Carbon nanotubes in an ancient Damascus sabre. *Nature* 444: 286.
27. Freestone I., Meeks N., Sax M., Higgitt C. 2007. The Lycurgus cup—A Roman nanotechnology. *Gold Bulletin* 40(4): 270–277.
28. Horikoshi S., Serpone N. (Eds). 2013. *Microwaves in Nanoparticle Synthesis*. Wiley-VCH Verlag GmbH & Co. KGaA, Germany.
29. Liz-Marzán L. M. 2004. Nanometals: Formation and color. *Mater. Today* 7: 26–31.
30. Min'ko N. I., Nartsev V. M. 2008. Nanotechnology in glass materials (Review). *Glass. Ceram.* 65(5–6): 148–153.
31. Sciau P. 2012. Nanoparticles in ancient materials: The metallic luster decorations of medieval ceramics. In Hashim A. A. (Ed.). *The Delivery of Nanoparticles*. InTech, Croatia, pp. 525–540. ISBN: 978-953-51-0615-9.
32. Colomban P., Tourníe A., Ricciardi P. 2009. Raman spectroscopy of copper nanoparticle-containing glass matrices: Ancient red stained-glass windows. *J. Raman Spectrosc.* 40: 1949–1955.
33. Jose-Yacaman M., Rendon L., Arenas J., Serra Puche M. C. 1996. Maya blue paint: An ancient nanostructured material. *Science* 273: 223–225.
34. Ashby M. F., Ferreira P. J., Schodek D. L. 2009. *Nanomaterials, Nanotechnologies and Design*. Butterworth-Heinemann, Elsevier, USA.
35. Suryanarayana C., Koch C. C. 2000. Nanocrystalline materials—Current research and future directions. *Hyperfine Interact.* 130: 5–44.

5

Synthesis of Nanomaterials

There are two approaches, top-down approach and bottom-up approach, for the synthesis of nanomaterials and nanostructures. In the top-down approach, a suitable starting material is reduced in size using mechanical or chemical means. Bottom-up approach refers to the building of a structure atom-by-atom, molecule-by-molecule, or cluster-by-cluster. In this approach, initially the nanostructured building blocks (i.e., nanoparticles) are formed and, subsequently, assembled into the final material using chemical or biological procedures. Attrition or milling is a typical top-down approach in making nanoparticles (NPs), whereas the colloidal dispersion is a good example of bottom-up approach for the synthesis of nanomaterials. Lithography may be considered as a hybrid approach, that is, it involves the growth of thin films by bottom-up approach, whereas etching is the top-down approach. The nanolithography and nanomanipulation are commonly considered under bottom-up approach. Both approaches play very important roles in nanotechnology and have their advantages and limitations. The main problem with the top-down approach is the imperfection of the surface structure, for example, conventional lithography may cause significant crystallographic damage, defects, impurities, and structural defects on the surface of the nanowires. These surface imperfections would result in a reduced electrical conductivity. Moreover, top-down approach introduces internal stress, surface defects, and contaminations. Bottom-up approach provides nanostructures with lesser defects, more homogeneous chemical composition, and better short- and long-range ordering. The nanomaterials produced by bottom-up approach are in a state closer to a thermodynamic equilibrium state.

5.1 Top-Down Approaches

5.1.1 Mechanical Alloying

Introduction: Mechanical alloying (MA) is a solid-state powder processing technique involving repeated welding, fracturing, and rewelding of powder particles in a high-energy ball mill. Originally, it was developed to produce oxide-dispersion-strengthened (ODS) nickel- and iron-base superalloys for the applications in the aerospace industry. MA has now been used to synthesize a variety of equilibrium and nonequilibrium alloy phases. The nonequilibrium phases include supersaturated solid solutions, metastable crystalline and quasi-crystalline phases, nanostructures, and amorphous alloys [1]. MA has been used for developing dispersion-strengthened superalloys based on Ti, Ni, and Fe nanocomposites, and metallic glasses to retain the room temperature strength at high temperature for a long time [2]. In recent years, MA is also used to prepare the polymer or metal matrix nanocomposites containing a uniform dispersion of nanoparticulates in the matrix [3,4]. In case of metallic systems, MA may induce chemical reactions (i.e., mechanochemical reactions) at lower temperatures than normally required to produce pure metals and nanocomposites.

Two different terms, MA and mechanical milling (MM) or ball milling, are commonly used to represent the processing of powder particles in high-energy ball mills. MA describes the process when mixtures of powders (of different metals or alloys or compounds) are milled together. It involves material transfer to obtain a homogeneous alloy. On the other hand, MM does not involve material transfer in producing the powders of uniform composition. MM requires less time than the time required for MA to achieve the same effect and hence, MM reduces oxidation of the constituent powders due to shortened time of processing. The main purposes of MA or ball milling are to reduce the particle size, and mix and blend the fillers and matrices.

High-energy ball milling: High-energy ball milling is one of the simplest methods to synthesize NPs of metals, alloys, and composites. There are several types of mills namely planetary, vibratory, rod, tumbler, etc. used for MA. The planetary ball mill owes its name to the planet-like movement of its vials. The vials are arranged on a rotating support disk. A special drive mechanism rotates the vials around their own axes. The centrifugal force is produced by the vials rotating around their own axes. In planetary milling, usually one or more containers are used at a time to make large quantities of fine particles. Hardened steel or tungsten carbide balls are put in containers along with powder or flakes (<50 μm) of a material of interest. Photographs of a planetary ball mill along with balls, and vials are shown in Figure 5.1.

The grinding vials and balls are generally made up of agate, tungsten carbide, zirconia, silicon nitride, sintered corundum, chrome steel, Cr–Ni steel, and plastic polyamide [5].

FIGURE 5.1
Planetary ball mill; (a) Basic model Retsch PM 200, (b) vials fixed on stand, (c) empty vial, and (d) balls in vial.
(Department of Metallurgy and Materials Science, COEP, Pune, India.)

Compared to other mills, the planetary ball mill is widely used for dry or wet grinding of the materials. Once the vials are loaded with the balls and the powder, they are closed with tight lids and rotated at high speed around their own axis as well as around some central axis. When the grinding vial and supporting disk are rotated in opposite directions, the centrifugal forces are generated and act on the grinding balls and powder. The rotation of vials forces the powder to the walls and the powder is pressed against the walls. The powder mixture is milled for the suitable length of time until a steady state is reached. The size reduction is achieved by imparting mechanical energy to the sample powder. The amount of energy imparted to the sample by the balls depends on the type of mills, the materials of interest, type of milling media, milling atmosphere, milling environment (e.g., dry milling or wet milling), ball-to-powder ratio (BPR), temperature, and duration. The particle size distribution, degree of crystallinity, and stoichiometry of the final materials depend on the milling conditions.

Generally, BPR of 10:1 and less than half-filled container are advisable to get better efficiency of milling. Weight of milling balls also affects the results, that is, larger-sized balls increase the impact energy on collision and hence, produce smaller grain size but larger defects in the particles. The raw materials used for MA have particle sizes in the range of 1–200 μm. The powder particle size should be smaller than the grinding ball size. The raw powders include pure metals, master alloys, prealloyed powders, and refractory compounds. MA can be used to mill a mixtures of ductile–ductile, ductile–brittle, and brittle–brittle powders.

The transfer of mechanical energy to the powder particles results in introduction of strain into the powder through generation of dislocations and other defects which act as fast diffusion paths. In addition, there is a refinement of particles and grain sizes. During milling, a slight increase in powder temperature occurs which leads to alloying of the blended elemental powders. This may result in the formation of solid solutions, intermetallic phases, and amorphous phases with ultrafine-grained and nanostructured phases [2]. It is to be noted that the high-speed milling may lead to temperature rise in the range of 100–1000°C. The generated heat is dissipated during milling using liquid medium, called as wet grinding. If no liquid is involved, then it is called a dry grinding. In case of cryomilling, liquid nitrogen is used as medium. The wet grinding is a more suitable method than dry grinding to obtain finer-ground products because the solvent molecules are adsorbed on the newly formed surfaces of the particles and lower their surface energy. However, the wet milling has faster rate of amorphization and leads to more contaminations in the milled powder than that of dry grinding [5]. By controlling the speed of rotation, duration of milling, BPR, and other factors, it is possible to get the fine powder with size of few tens of nm. The cryomilling may produce NPs of about 2–10 nm. Moreover, nanocrystalline powders of pure metal as well as alloys like Ni–Ti, Al–Fe, and Ag–Fe can be made using ball mill in a short time of few minutes to a few hours. This mill has the advantage of ease of handling the vials with capacity from 45 to 500 mL. In contrast, the vibratory ball mill has vial capacity of about 10 mL and is used mainly for preparing amorphous alloys. In this mill, the powder and milling tools are agitated in three perpendicular directions at very high speed (i.e., as high as 1200 rpm). The ball milling process may add some impurities from the vials/balls and air or inert gas [5,6]. The ball milling of metals with fcc lattice structure is found to sinter to larger particles. A small amount of Ti in the Al–Fe system favored the formation of the amorphous phase. Cryomilling of aluminum powder leads to a matrix of aluminum with a fine grain size (<50 nm) of dispersed AlN particles. A major advantage of the MA is to produce several kilograms of nanomaterial in time up to 100 h. By high-energy ball milling, the grain size of pure bcc metals (Cr, Nb, W), hcp metals

(Zr, Hf, Co, Ru), and intermetallic compounds with CsCI structure and immiscible systems (FeAI) can be reduced to nanometer scales (~5–13 nm diameter) during longer durations of ball milling. This results in an extremely fine-grained microstructure with randomly oriented grains separated by high-angle grain boundaries [7,8]. As shown in Figure 5.2, the particle or crystallite/grain size of the milled powder decreases with increasing milling time. The optimum milling time depends on the type of mill, milling speed, size of the grinding medium, temperature of milling, ball-to-powder ratio, etc. [9,10].

There are number of different types of mills which differ in their capacity, speed of operation, and their ability to control the operation by varying the temperature of milling and the extent of minimizing the contamination of the powders. The material of the grinding vials must be same as the powder to minimize the contamination of the milled powder. Hardened steel, tool steel, hardened chromium steel, tempered steel, stainless steel (SS), WC-Co, WC-lined steel, and bearing steel are the most common types of materials used for the grinding vials or vessels. The faster the mill rotates, the higher would be the energy

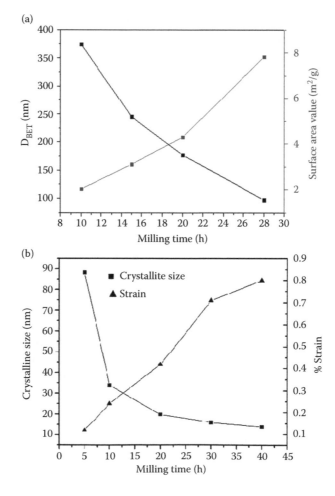

FIGURE 5.2
The effect of milling time on (a) particle or crystallite size of ZrO_2. (From Goyal R. K. et al. 2016. *J. Mater. Sci. Surf. Eng.* 4(1): 339–341.) (b) Crystallite size and lattice strain of $Al_{70}Co_{15}Ni_{15}$. (From Yadav T. P., Yadav R. M., Singh D. P. 2012. *Nanosci. Nanotechnol.* 2(3): 22–48. Open access.)

input into the powder. However, the maximum speed of rotation should be just below the critical speed to produce the maximum collision energy. High speeds of vials may accelerate the transformation process resulting in the decomposition of supersaturated solid solutions or other metastable phases formed during milling. It has been reported very well that the level of contamination increases and some undesirable phases form if the powder is milled for times longer than required. The density of the grinding medium should be high enough so that the balls create enough impact force on the powder. It is always suggested to use the grinding vials and the grinding balls made of the same material as the powder being milled to avoid cross contamination. The ratio of the weight of the balls to the powder (BPR) is an important variable in MA. This ratio has been varied from a value as low as 1:1 to as high as 220:1. In general, it has been found that a ratio of 10:1 is most appropriate when milling the powder in a small capacity mill (i.e., SPEX mill), while a ratio of 50:1 or more is used in a large capacity mill (i.e., attritor). The higher the BPR, the shorter is the time required to get final milled product. At a high BPR, because of an increase in the weight proportion of the balls, the number of collisions per unit time increases and consequently more energy is transferred to the powder particles and so alloying takes place faster. Generally, vial space of about 50% is left empty to move around the impacting balls and thus to transfer the impact energy to the powder. The powders are milled in an evacuated or inert-gas- (argon or helium) filled vials to avoid the oxidation. Sometimes, a process control agent (PCA or surfactant) is added to the powder mixture during milling to reduce the effect of cold welding. The PCA adsorbs on the surface of the powder particles and minimizes cold welding between powder particles and thereby inhibits agglomeration. The surface-active agents adsorbed on particle surfaces interfere with cold welding and lower the surface tension of the solid material. The content of PCA is about 1–5 wt% of the total powder charge. The most commonly used PCAs are stearic acid, hexane, methanol, and ethanol.

Applications of MA: The MA has been used widely for producing the oxide dispersion strengthened (ODS) materials, solder alloy, Cr–V alloys, dental filling alloys/polymer composites, thermoelectric material (i.e., $FeSi_2$), Mg-based materials for hydrogen storage, catalyst materials, inorganic pigments, fertilizers, and nanostructured materials. Pure Mg has a good hydrogen storage capacity (i.e., 7.6 wt% hydrogen as a hydride); however, it gets easily oxidized. The addition of a small amount of Ni (1–2 at%) dispersion as discrete particles in the Mg matrix, called as Mg–Ni alloys, increases its oxidation resistance with a slight decrease of hydrogen storage capacity (i.e., 6.1 wt%) compared to pure Mg metal. The ODS nickel- and iron-based superalloys have applications in the aerospace industry because these alloys are strong at room temperature and retain strength at elevated temperatures. The elevated temperature strength is mainly derived from two mechanisms. Firstly, the uniform dispersion (with a spacing of ~100 nm) of very fine (5–50 nm) oxide particles (i.e., Y_2O_3 and ThO_2) inhibits dislocation motion in the metal (Ni- or Fe-based) matrix, and increases the resistance of the alloy to creep the deformation. Secondly, the dispersed particles inhibit the recovery and recrystallization processes by pinning dislocations and hence, allow a very stable grain size. These large grains resist grain rotation during high-temperature deformation [11].

Biomedical-grade iron NPs with size of ~32 nm can be obtained using high-energy ball milling of iron powder (of 20 μm size) after 3 h of cryomilling (temperature: ~150 K) under high-purity argon atmosphere. The saturation magnetization of dilute samples of the cryomilled powder was found to be 215 emu/g, which was higher than that of room-temperature-milled sample (i.e., 120 emu/g) [12]. The magnetite (Fe_3O_4) NPs have been prepared directly from micrometer-sized (~20–40 μm) metallic iron (Fe) powder (purity ~99.99%)

in distilled water medium using a planetary ball mill for 8 h at 300 rpm. The hardened SS balls and SS vial with a BPR of 36:1 were used. The resultant magnetite NPs showed saturation magnetization of $\sim$63.7 emu/g (vs. $\sim$146 emu/g for pure Fe) [8].

The synthesis of bulk TiC by direct chemical reaction between Ti and carbon under vacuum at high temperature is an expensive process. However, TiC NPs with crystallite size of about 12–46 nm can be synthesized by MA after milling of 30–90 h followed by heat treatment at 1300°C for 60 min [13]. Ni–Cu alloy with average crystallite size of 15–40 nm has been successfully synthesized after ball milling of $\sim$30 h [14]. Ball milling of monoclinic ZrO_2 for more than 30 h stabilizes tetragonal- and cubic-ZrO_2 at room temperature and produces stabilized ZrO_2 NPs with a mean crystallite size of $\sim$10 nm [15]. MA is also used for the fabrication of metal or ceramic NPs-filled polymer nanocomposites. However, the degree of crystallinity of semicrystalline polymers was found to decrease with increasing milling time. For example, the crystallinity of pure polyethylene (PE) decreases from $\sim$36.5% to $\sim$27.0% after milling time of 40 h. In addition to this, the addition of alumina (Al_2O_3) NPs decreases the degree of crystallinity of the PE fraction in the nanocomposites. However, after 20 h milling, the degree of crystallinity of the PE fraction in the nanocomposites remains nearly constant (i.e., $\sim$24%) [4].

Pros and cons of MA: In contrast to rapid solidification processing, the mechanically alloyed nanomaterials have significantly higher solid solubility, for example, the room temperature solubility of copper in silver is increased from 0 to 100 at% by MA. Amorphization (or inverse of crystallinity) is one of the important issues of MA powder mixtures. Amorphous phases have been obtained from blended elemental powders of intermetallics or stoichiometric compounds. The percentage of crystallinity of MA semicrystalline polymers is reduced with increasing milling time. The contamination of milled powders during MA is a major problem. The small size, large surface area, and formation of new surfaces on the particles during milling contribute to the contamination of the powder. In addition, grinding balls, liquid (for wet grinding), grinding vials, time of milling, intensity of milling, etc. and the atmosphere under which the powder is milled contribute to the contamination level. In some cases, like titanium and zirconium which are reactive, the levels of contamination are high and unacceptable. The level of contaminations increases with increasing milling time and the highest levels reported are 44.8 at% oxygen in an Al-6 at% Ti powder, 26 at% nitrogen in a Ti-50 at% Al powder, and 60 at% iron in a W-5 at% Ni alloy powder. The best method to minimize the powder contamination level is the use of high-purity metals, high-purity atmosphere, balls and vials of the same material that is to be milled, self-coating of the balls with the milled material, and minimum milling times [11,12].

5.1.2 Severe Plastic Deformation

Severe plastic deformation (SPD) is the metal-forming process in which a very large plastic strain is introduced into a bulk metal in order to create the ultrafine-grained metals. SPD process is employed to produce high-strength and lightweight parts with environmental harmony. In this, a billet is heavily strained under high-imposed pressure, resulting in nanostructured materials with almost free porosity. There are two most commonly used SPD techniques: high-pressure torsion (HPT) and equal channel angular pressing (ECAP). In HPT, disc-shaped (Figure 5.3a) sample with a diameter of 10–20 mm and thickness of 0.2–0.5 mm is put between anvils and compressed under a high pressure (approximately several GPa). The lower anvil turns, and friction forces result in shear straining of the sample. Due to the high-imposed pressure, the sample under deformation does not break even at high

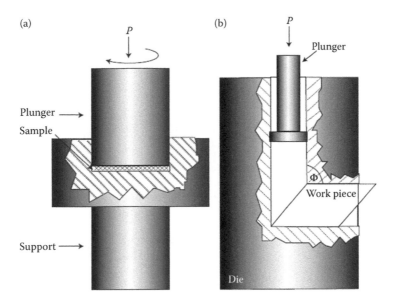

FIGURE 5.3
Schematic diagrams of severe plastic deformation techniques; (a) high-pressure torsion and (b) equal channel angular pressing. (Reprinted by permission from Springer Nature. *Nat. Mater.*, Valiev R. 2004. Nanostructuring of metals by severe plastic deformation for advanced properties. 3: 511–516, copyright 2004.)

strains and it can produce a homogeneous nanostructure with a typical grain size of about 100 nm or less. In case of ECAP, the ingot is pressed in a special die through two channels with equal cross sections intersecting at an angle φ, usually $60° < \varphi < 135°$ and often $\varphi = 90°$ (Figure 5.3b). In each pass, sample gets a strain of about 1. ECAP refines the microstructure of pure metals, alloys, and intermetallics using straining with one pass or a few [16].

The typical final grain size of a sample produced by ECA pressing is 50–300 nm or smaller depending on the material being strained and on processing parameters. Up to 70%–80% high-angle boundaries may be created in the microstructure of samples subjected to multipass ECAP or HPT with five or more revolutions at relatively low temperatures (usually <0.3 times of the melting point). It has been well studied that the homogeneous ultrafine-grained (UFG) metals produced by SPD possess a complex microstructure. This produces an ultrafine equiaxed grained structure with mainly high-angle grain boundaries, the absence of macroscopic damages and cracks in the samples. Thus, it results in significant increase in strength with loss of ductility. The ECAP method was mainly applied for nonferrous alloys (e.g., Al and Mg alloys) and some low carbon steels. The finest ferrite grain size obtained by this method is about 200 nm. The original ECAP is a discontinuous process and as such has a low production efficiency and high cost [17].

SPD-processed nanostructured TiNi alloy exhibits multifunctional extraordinary properties of very high strength, superelasticity, and shape–memory effect, which makes it different from its conventional coarse-grained counterpart. These nanostructured TiNi alloys are used for space, medical, and other applications. The SPD-processed nanostructured titanium exhibits yield tensile strength of $\geq$950 MPa at strain rate of 10^{-3}/s, endurance of >500 MPa at 2×10^7 cycles, and excellent biological compatibility. Hence, nanostructured titanium is used for medical implants in plate implants for osteosynthesis, conic screw for spine fixation, device for correction, and fixation of spinal column (Figure 5.4) [16].

Al-based commercial alloys with yield strength over 800–900 MPa are used for the motor industry and aviation. These metals and alloys with ultrafine-grained structure are

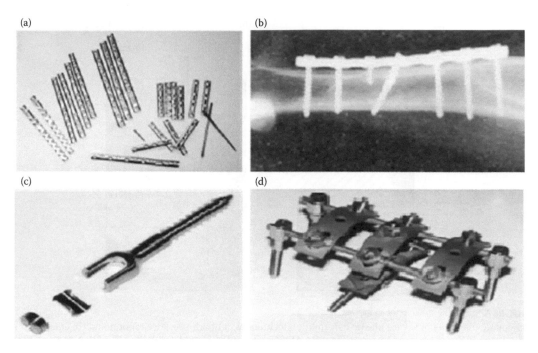

FIGURE 5.4

Medical implants made of SPD-processed nanostructured titanium; (a and b) plate implants for osteosynthesis, (c) conic screw for spine fixation, and (d) device for correction and fixation of spinal column. (Reprinted by permission from Springer Nature. *Nat. Mater.*, Valiev R. 2004. Nanostructuring of metals by severe plastic deformation for advanced properties. 3: 511–516, copyright 2004.)

also useful at cryogenic temperatures [16]. SPD can be used for structure refinement and powder consolidation with density of about 100% in bulk nanostructured samples [18]. SPD-processed UFG (grain sizes <300 nm) metals typically are 5–10 times stronger than their coarse-grained counterparts. It is to be noted that the nanocrystalline metals (including Cu) fabricated by conventional rolling exhibit a room temperature tensile elongation to failure of no more than a few percent, which is much lower than the coarse-grained counterpart (~60%). The ECAP Cu with grain size <200 nm has a strength (~500 MPa) and an elongation to failure larger (~30%) than, or at least comparable with, those of Cu heavily cold-rolled at room temperature. The typical values of strength and an elongation for cold-rolled Cu are in the range of 300–400 MPa and a few percent or less. The as-received ECAP Cu or ECAP Cu followed by cold rolling at liquid nitrogen temperature is fractured with some ductile features dominated by microvoid formation on a much finer scale [19].

In SPD approach, a large shear strain up to about 4 can be imposed by the cumulative application of plastic deformation, using multiple passes of deformation processing. However, SPD approach possesses two important limitations. First, multiple passes of deformation are required to create the large plastic strains. Second, moderate and high-strength alloys are difficult to deform at ambient temperature due to constraints imposed by the forming equipment, including durability of the tools and dies [20]. ECAP is a discontinuous process and there is a challenge in upscaling it. SPD techniques have the ability to produce UFG metallic materials through microstructure refinement in initially coarse-grained materials. The final grain size produced depends strongly on both processing regimes and the type of material. SPD under high applied pressures provides bulk porous-free nano- (grain size: <100 nm) and submicrocrystalline (grain size: 100–1000 nm)

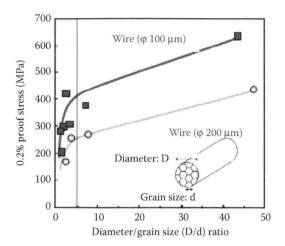

FIGURE 5.5
Effect of increasing the diameter to grain size (D/d) ratio on strength of a wire specimen. (Reprinted from *CIRP Ann. Manuf. Technol.*, 57, Azushima A. et al., Severe plastic deformation (SPD) processes for metals, 716–735, Copyright 2008, with permission from Elsevier.)

metals and alloys. For pure metals, the mean grain size obtained by HPT is typically about 100–200 nm, while it is about 200–300 nm after processing by ECAP. For alloys and intermetallics, the grain size is usually in the range of 50–100 nm or smaller. Higher-processing temperature is generally avoided to prevent the growth of grain size [21]. SPD produces bulk nanostructured materials with dense and refined microstructure, which is difficult to obtain by conventional deformation technique. The proof stress of the UFG irons is five times greater than commercially pure iron. Thus, the conventional structural metals with ultrafine grains can have high-specific strength, better corrosion resistant, and excellent fatigue properties of metals. Figure 5.5 shows the effect of grain size reduction (or increase of ratio of diameter of Cu wire to its grain size, D/d) on the proof stress of a wire sample processed by SPD. The 0.2% proof stress increases with decreasing grain size for a given wire diameter (i.e., D/d increases). In the initial stage of deformation, that is, when D/d is up to 5 the proof stress increases abruptly with decreasing grain size (or increasing D/d). The D/d ratio of >100 can provide better safety and the reliability of metals for microparts [22].

Figure 5.6 illustrates a general tendency of the change in strength and ductility during SPD for a typical material. The strength of the materials drastically increases with increasing applied strain and then gradually saturates because of the pinning of the large number of dislocations at the grain boundaries. In contrast, the ductility drops significantly with a relatively small strain, and then slightly decreases as the strain increases [22].

5.1.3 Lithography

This section involves a discussion on physical techniques particularly lithographic techniques including photolithography, electron-beam lithography (EBL), x-ray lithography, focused-ion beam lithography (FIBL), nanolithography, soft lithography, dip-pen nanolithography (DPNL), etc. Each method offers some advantages over other techniques but suffers from other limitations and drawbacks. The attention has been focused mainly on their basic principle. Lithography is the process of transferring a pattern into a reactive polymer film (i.e., resist), which will subsequently be used to replicate that pattern into an underlying thin film. Several lithographic techniques have been developed depending

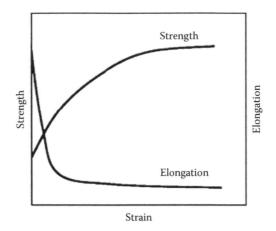

FIGURE 5.6
A typical general tendency of the change in strength and ductility during SPD. (Reprinted from *CIRP Ann. Manuf. Technol.*, 57, Azushima A. et al., Severe plastic deformation (SPD) processes for metals, 716–735, Copyright 2008, with permission from Elsevier.)

upon the source of lens systems and exposure radiations like photons, x-rays, electrons, ions, and neutral atoms. Lithography techniques can be divided into masked lithography such as photolithography, soft lithography and nanoimprint lithography (NIL), and mask-less lithography, which include EBL, FIBL, and scanning probe lithography. The mask-less lithography technique has low mass production. The lithography techniques have been used in the field of electronics and microsystems, medical and biotech, optics and photonics, and environment and energy harvesting. Photolithography is the most widely used technique in microelectronic fabrication, particularly for mass production of integrated circuit (IC).

Photolithography: Typical photolithographic process consists of producing a mask carrying the requisite pattern information and subsequently transferring that pattern, using some optical technique into a photoactive polymer or photoresist (or simply resist). In this, a resist material is applied as a thin coating (thickness ~1 μm) over an oxidized silicon wafer substrate and subsequently exposed to a radiation through a mask. This technique utilizes an exposure of a light-sensitive polymer (i.e., photoresist) to ultraviolet (UV) light to get a desired pattern. A UV light with suitable wavelengths is illuminated through a photomask to make an exposure on a photoresist which is coated on a substrate. In the exposed area, the photoresist polymer breaks down, resulting in more soluble photoresist polymer in a developer (or solvent) and subsequently, the exposed photoresist is removed in a developer. Figure 5.7 illustrates schematically the main steps in photolithography. An oxidized silicon wafer is coated with a thick photoresist layer, as shown in Figure 5.7a. A beam of UV light is incident on a mask through the mask apertures. This causes a photochemical reaction between the photon and the resist. The exposed parts of the resist are dissolved during the developing stage, as shown in Figure 5.7b, forming a replica of the mask pattern. This assembly is then placed in an acidic solution, which attacks the silica (Figure 5.7c). Once the silica has been removed, the resist is dissolved in a different acidic solution (Figure 5.7d). Further etching removes silicon from the exposed areas, creating channels (Figure 5.7e). The photolithography is a simple process but its implementation is complex and expensive. Masks need to be perfectly aligned with the pattern on the wafer. The silicon wafer has to be a perfect defect-free single crystal [23,24].

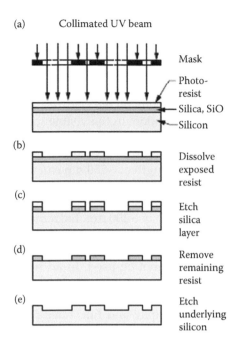

(a) Collimated UV beam

Mask

Photo-resist

Silica, SiO

Silicon

(b) Dissolve exposed resist

(c) Etch silica layer

(d) Remove remaining resist

(e) Etch underlying silicon

FIGURE 5.7
Schematic diagram of photolithography. (Reprinted from *Nanomaterials, Nanotechnologies and Design, an Introduction for Engineers and Architects,* Ashby M., Ferreira P., Schodek D., Elsevier Science Ltd. Copyright 2009, with permission from Elsevier.)

There are three types of photolithography: contact printing, proximity printing, and projection printing. Contact and proximity printings place the photomask in contact with or in a close proximity to the photoresist and they are capable of making patterns as small as a few micrometers. In contrast, a projection printing system utilizes an optical lens system to project a deep-UV pattern from an excimer laser (e.g., wavelength of 193 nm for ArF or 157 nm for F_2) on the photoresist enabling pattern size as small as a few tens of nanometers (37 nm) at a high throughput (60–80 wafers/h). However, due to sophisticated and expensive setup, the projection printing is used to manufacture advanced ICs and CPU chips. Photolithography is the main workhorse in the semiconductor and IC industry to manufacture ICs, microchips, and commercial MEMS devices. For decades, this technology has been used to produce ICs and microchips with a feature size ranging from a few nanometers up to tens of millimeters. The resolution limit (R) in conventional projection photolithography can be determined by Rayleigh's equation:

$$R = K_1 \cdot \lambda / NA = K_1 \cdot \lambda / (\eta \cdot \sin\theta) \qquad (5.1)$$

where λ is the wavelength of light, K_1 is the constant that depends on the resist material, process technology, and image-formation techniques used, and NA is the numerical aperture of the optical system. The η is the refractive index of image space ($\eta = 1$ for air or vacuum), and θ is the maximum cone angle of the exposure light beam. For hard contact printing, the resolution for 400 nm wavelength light and a 1 μm thick resist film is about 1 μm. Better resolution (or lower values of R) can be obtained by using shorter wavelengths of light and lens systems with larger numerical apertures (typically >0.5). However, such lens systems with high NA result in very small depth of focus and the exposure process becomes sensitive to slight variations in the thickness and absolute position of the resist layer.

The maximum possible resolution is restricted to $\sim\lambda/2$ because of diffraction limit, dust on substrates, and nonuniformity of the thickness of the photoresist and the substrate. Smaller the wavelength used, smaller will be the feature size. The depth of focus depends upon the penetration of incident radiation. In the visible range (700–400 nm), glass lenses and masks are used, while in the UV range, fused silica or calcium fluoride lenses are used. Hence, the conventional photolithography is capable of fabricating features of 200 nm and above only. The UV light with wavelength of $\sim$250 nm can reduce the feature size to about 100 nm. Deep UV lithography (DUVL) which includes KrF excimer laser, ArF excimer laser, and F_2 excimer lasers with a wavelength of 249 nm, 193 nm, and 157 nm, respectively, generates pattern with a minimal size of $\sim$100 nm. Extreme UV lithography (EUVL) with wavelengths in the range of 11–13 nm can produce features of 70 nm and below. However, EUVL has the drawback of the strong adsorption of light of such lower wavelengths, and therefore, refractive lens systems cannot be used. In addition, EUVL requires high-precision metrology system. Other advanced lithographic techniques including soft x-ray lithography, EBL, and FIBL can generate the features as small as a few nm [23–25].

X-ray lithography: The x-ray lithography and EBL have resolution better than the optical lithography. The shorter wavelength of the x-rays (0.1–10 nm) and electrons can produce features smaller than 100 nm, but these techniques are slow and expensive. In x-ray lithography, diffraction effects are minimized by the short wavelength, but conventional lenses are not capable of focusing x-rays, and the radiation damages the mask materials. Depending upon the wavelength of x-rays used, metals of suitable elements are chosen.

Electron-beam lithography: EBL utilizes an accelerated electron-beam focusing over a surface coated with a radiation-sensitive polymeric material. Subsequently, this electron-beam spot with a diameter as small as a few nanometers is scanned on the surface of resist in a dot-by-dot fashion to generate patterns in sequence. This causes a scission in the organic structure that is subsequently developed using a standard developing solution, whereby the exposed region dissolves completely in the developer. Subsequently, deposition of metal and/or dielectric layers is carried out following which the resist is dissolved to obtain the nanoscale patterns. EBL provides the highly desirable ability to create nanoscale features below the diffraction limit of the standard photolithography processes and does not require a physical mask to transfer the patterns [26]. Similarly, FIBL utilizes an accelerated ion beam (typically Ga^+) to directly hit a metallic film on the substrate. FIBL results in inevitable Ga^+-induced damage to the microstructure which changes the mechanical properties of the materials. Both EBL and FIBL processes are slow and significantly limit the sample throughput time. EBL is a powerful technique to achieve the features as small as 3–5 nm, but it is expensive. The resolution of EBL and FIBL techniques is of the order of 5–20 nm. In FIBL, the resists are more sensitive to ions as compared to electrons and it has low scattering in the resist and substrate. Most commonly used ions in FIBL are He^+, Ga^+, etc. with energy in the range of 100–300 keV.

The electron beam can be used to write patterns of very high resolution since the wavelength of electron is on the order of a few tenths of angstrom, and therefore their resolution is not limited by diffraction considerations. Resolution of EBL is limited by forward scattering of the electrons in the resist layer and back scattering from the underlying substrate. When an electron beam enters a polymer film or any solid material, it scatters and these scattering processes lead to a broadening of the beam, which in turn results in developed resist images wider than expected. The magnitude of electron scattering depends on the atomic number and density of both the resist and substrate as well as the velocity of the electrons or the accelerating voltage. EBL can generate a 40 nm pitch pillar grating after nickel liftoff when developing with ultrasonic agitation [24].

Scanning probe lithography: The scanning probe lithography-based nanofabrication involves two basic technologies: atomic force microscopy (AFM) and scanning tunneling microscopy (STM). An AFM tip is used as a pen and molecules are used as ink. Appropriate molecules picked up by the tip from the source of molecules can be transported and transferred at desired place on the substrate to pattern nanoscale features. The approach to deposit NPs or molecules selectively onto a substrate is called DPNL, which can be performed under ambient environment, for example, the transfer of thiol molecules to a freshly prepared gold surface. It involves immersing of an AFM cantilever in a solution of chemical molecules and subsequently its deposition onto a substrate surface. Traditional probe tips are made up of hard materials such as silicon, silicon nitride, and PDMS elastomer [27].

In STM lithography (STML), the STM probe with constant current mode is usually used as a lithographic source to produce nanoscale patterns of lines on an atomically flat surface of a substrate in vacuum or air. The STML has been used to perform indentations, local material deposition, local oxidation, and multiple tips or arrays of tips. It has extraordinary spatial and magnitude sensitivities. In contrast to EBL, the STML has advantages of easier to expose the high-resolution resists and the applied voltages are typically four orders of magnitude lower than that of EBL. The low voltage of STML produces less back scattering and minimizes the proximity effect. However, the low voltage of STML limits it to be used to expose very thin resist films. But, the *e*-beam formed in STML has a much higher resolution. In fact, in STML, the tip is so close to the target surface that the *e*-beam does not have a chance to diverge, while in direct writing EBL, various stages of electromagnetic lenses are used to deflect and accelerate electrons to form a relatively large *e*-beam. Furthermore, in some techniques, STML can be operated in ambient air, while EBL is always conducted under high-vacuum conditions.

For nanoscale patterning, the typical current, voltage, and scanning rate of the *e*-beam induced by STM are controlled on the order of 1 nA, 10 V, and 1 μm/s, respectively. Similar to EBL, the electrically conductive substrates are required in STML, which can put a constraint in the making of an optical device. When a voltage is applied, atoms or NPs can be transferred from the tip to the target surface, which is within the tunneling range. Although both positive and negative voltages have been used for the deposition, the tip-negative polarity is usually more stable with higher transfer probability and can yield smaller feature sizes. Si- and Ge-based nanostructures with lateral sizes of about 10 nm were successfully deposited on Si substrates using SiH_4, SiH_2Cl_2, and GeH_4 precursors.

The major difference between STML and AFML is that the STML is normally operated at air or vacuum environment under the constant (tunneling) current mode, while AFML is more in a constant voltage mode operated at an ambient air environment. Generally, STML can make nanostructures with higher resolution but at a lower speed, while AFML can have higher scan speeds but lower resolution. A resist exposure-based AFML technique can generate a line pattern with a width of 35 nm and periodicity of 68 nm. In FML, the contact mode is usually faster than the tapping mode. In contrast to AFM, the STM uses a sharpened conducting tip with a bias voltage between the tip and the target sample.

Dip-pen nanolithography: DPNL is a direct-write method based upon an AFM and works under ambient condition. Figure 5.8 shows the concepts of DPNL. When the AFM tip is

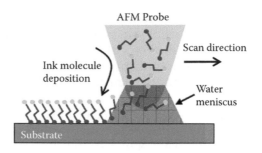

FIGURE 5.8
Illustrating the dip-pen nanolithography. (Reprinted from *Mater. Sci. Eng. R*, 54, Xie X. N. et al., Nanoscale materials patterning and engineering by atomic force microscopy nanolithography, 1–48, Copyright 2006, with permission from Elsevier.)

scanned across a substrate, the molecules are transferred from the tip to the substrate via the water-filled capillary due to chemisorption. It can develop nanostructures of about 15 nm dots spaced ~5 nm apart on an Au (111) substrate. A DPNL uses an AFM tip coated with a thin film of ink molecules (such as alkanethiols) that react with the substrate surface (such as gold) to write nanoscale patterns for lithography applications. As shown in Figure 5.8, when the tip is placed in a high-humidity atmosphere and is close to the Au substrate, a minute drop of water is naturally condensed between the AFM probe and the substrate. The drop of water acts as a bridge over which the ink molecules migrate from the tip to the gold surface where they are self-assembled. The capillary transport from the probe toward the tip apex provides a resupply of new molecules for a continuous writing. The molecules and substrates are chosen in order to have chemical affinity and favor adhesion of the deposited film. There are a number of ink materials including inorganics, organics, biomolecules, and conducting polymers, which are compatible with a variety of substrates such as metals, semiconductors, and functionalized surfaces. The molecular transport in DPNL depends on several variables including temperature, humidity, tip substrate contact force, writing speed, and the physicochemical properties of both the ink and the substrate. DPNL has capability to write letters with line thickness as small as 15 nm and distance between the letters is about 5 nm. DPNL has unique features of overwriting and erasing capability. For example, each line of about 2 μm long with less than 100 nm wide has been made of 1-octadecanethiol deposited on a gold substrate [28,29].

Nanoimprint lithography: There are four different types of NIL, which includes thermal (or hot) NIL, UV NIL, soft lithography, and roll-to-roll nanoimprint. Soft stamp imprinting is more popular than the hard one because it has an intimate contact with the surface, which can fulfill large-area imprinting. The most recent roll-to-roll NIL has gained popularity for volume fabrication. Although NIL has wide applications in nanodevices, biomedicine, organic devices, etc., there are many issues to overcome, such as mold fabrication and inspection, defects controlling, alignment, and overlay. It can provide nanoscale resolution of patterning, because it is not limited by the diffraction limit, the scattering effects, and the secondary electrons. NIL is a promising technology for producing 2-D or 3-D structures with sub-50-nm half-pitch features. The direct NIL process with a polydimethylsiloxane (PDMS) stamp has successfully produced silver line patterns in the range of 200–300 nm, and the combined NIL and liftoff process successfully produced silver line patterns in the range of 15–60 nm. The process has advantages of

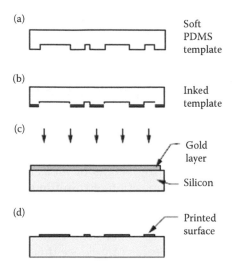

(a) Soft PDMS template

(b) Inked template

(c) Gold layer

Silicon

(d) Printed surface

FIGURE 5.9
Soft lithography processes. (Reprinted from *Nanomaterials, Nanotechnologies and Design, an Introduction for Engineers and Architects,* Ashby M., Ferreira P., Schodek D., Copyright 2009, with permission from Elsevier.)

low cost, high-replication reliability, and relatively high-throughput production. The NIL is suitable for the fabrication of electronic devices such as organic light emitting diodes (OLEDs), printed circuit boards, solar cells, active matrix OLEDs (AMOLEDs), printed capacitors, and chemical sensors [30,31].

Soft lithography: Soft lithography is developed as an alternative to photolithography and a replication technology for both micro- and nanofabrication. It includes contact printing, micromolding in capillaries, microtransfer molding, and replica molding. As shown in Figure 5.9, it involves fabrication of a patterned master, molding of master, and making replicas. A master is usually made using x-ray or EBL. A mold is usually made of polydimethylsiloxane (PDMS) due to its good thermal stability (~150°C), optical transparency, flexibility, and capability of cross-linking using radiation. The soft lithography is a useful alternative to obtain the resolution better than 100 nm at low cost. It can generate patterns and structures with feature sizes in the range from 30 nm to 100 μm. For this, a silicon or PDMS mold is taken and subsequently a precursor (or prepolymer) is poured on silicon-patterned substrate and cured. After proper heat treatment, the master is removed from the silicon pattern. The soft lithography does not require any special equipment. It is capable of producing wide range of nanostructures and can print or mold on curved as well as planar surfaces. It provides a convenient, effective, and low-cost method for the formation and manufacturing of micro- and nanostructures. Soft lithographic techniques require remarkably little capital investment and are procedurally simple. They can often be carried out in the ambient laboratory environment. They are not subject to the limitations set by optical diffraction and optical transparency. They provide alternative routes to structures that are <100 nm without the need for advanced lithographic techniques. They provide access to quasi-3-D structures and generate patterns and structures on nonplanar surfaces. The strength of soft lithography is in replicating rather than fabricating the master, but rapid prototyping and the ability to deform the elastomeric stamp or mold give it unique capabilities even in fabricating master patterns [23,25,32,33].

5.2 Bottom-Up Approaches

5.2.1 Physical Vapor Deposition (PVD)

Physical vapor deposition (PVD) is a process of transferring growth species from a source or target and depositing them on a substrate to form a film. The thickness of the deposits can vary from angstroms to millimeters. In general, PVD methods can be categorized into two groups, evaporation and sputtering, based on source of removing the species from the target. In evaporation, the growth species are removed from the source by thermal means, while in sputtering, atoms or molecules are dislodged from solid target through impact of plasma. Deposition of thin films by evaporation is carried out at a low pressure (10^{-3}–10^{-10} torr), and therefore, atoms and molecules in the vapor phase are not collided with each other prior to arrival at the substrate since the mean free path is very large as compared to the source-to-substrate distance. Therefore, the film quality is relatively poor. The pulsed laser beams can reduce this problem to a certain extent. In case of sputtering, atoms are ejected from the target surface by the impact of energetic ions and sputtering is capable of depositing high-melting point materials such as refractory metals and ceramics, which are difficult to convert to nanomaterials by evaporation. However, sputtered films are more prone to contamination than evaporated films due to the lower purity of the target materials.

Inert gas condensation (IGC) is the PVD method with additional feature of using the inert gas to reduce the mean free path of the species. It is commonly used to synthesize metallic and metal oxide nanopowders with a narrow particle size distribution. In this process, a metal is evaporated inside an ultrahigh-vacuum (UHV) chamber filled with inert gas. Typically, argon or helium is periodically admitted (Figure 5.10). The source of vapor can be an evaporation boat, a sputtering target, or a laser-ablation target. The vaporized species then loses energy via collisions with inert gas molecules. As collisions limit the mean free path, supersaturation is achieved above the vapor source. At high supersaturation, the vapors rapidly form large number of clusters that grow via coalescence and agglomeration. The clusters in the condensing gas are transported by convective flow to a vertical cold finger surface filled with liquid nitrogen. The size of NPs is in the range of 2–100 nm. The NPs can be collected by scraping them off the finger. They are collected via a funnel and transported to an *in situ* compaction device or coated with a surfactant agent that prevents them from agglomeration. The scraping and consolidation of particles are carried out under high vacuum to prevent the oxidation of the metallic NPs [34].

The size, morphology, and yield of the clusters are dependent on three fundamental parameters: (i) rate of supply of atoms to the region of supersaturation where condensation occurs, (ii) rate of energy removal from the hot atoms via the condensing gas medium, and (iii) rate of removal of clusters once nucleated from the supersaturated region. This method has several advantages like: higher-production rate ($\sim$60 g/h), high yield ($\sim$75%), high purity of final products, production of various compounds (metals, alloys, ceramics, intermetallic compounds, semiconductors, composites, etc.), flexible technique, and controlled particle size distribution. However, the method also has some disadvantages like formation of clusters and larger size particles, very expensive process (due to UHV and high-purity raw target), low yield per batch, and difficult to scale-up the process. In addition, it is not suitable for synthesizing semiconductors such as GaAs and InP because they easily decompose to the individual components [34,35].

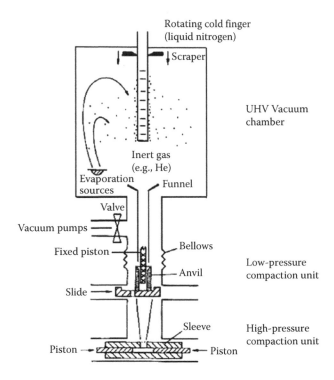

FIGURE 5.10
Schematic diagram of inert gas-condensation method. (Reprinted from *Mater. Sci. Eng. A*, 168, Siegel R. W., Synthesis and properties of nanophase materials, 189–197, Copyright 1993, with permission from Elsevier.)

5.2.2 Molecular-Beam Epitaxy

Molecular beam epitaxy (MBE) is a special technique for epitaxial growth of materials (films or nanowires) under ultra-high vacuum (UHV) conditions on the substrate surface. The growth occurs by the interaction of molecular beams from two or more beam sources at the substrate. It is generally used to deposit elemental or compound quantum dots, quantum well or quantum wires, and single-crystal film in a highly controlled evaporation of a variety of sources in UHV ($>10^{-8}$ Pa). As shown in Figure 5.11, a typical MBE setup consists of three vacuum chambers: the load chamber, transfer chamber, and the growth chamber. The growth chamber consists of evaporation cells, reflection high-energy electron diffraction (RHEED) gun and detector, mass spectrometer, and substrate. RHEED is used to monitor the growth of the product.

The substrate is inserted into the load chamber and subsequently placed inside the growth chamber. The pressures of load chamber, transfer chamber, and growth chambers are in the ranges of 10^{-4}, 10^{-8}, and 10^{-9} torr, respectively. The mean free path of the gas molecules at these low pressures (i.e., 10^{-9} torr) is larger than the dimensions of the growth chamber. Thus, the gas molecules emitted from the evaporation cells do not collide and react with other gaseous species until they strike the substrate. In contrast to PVD, this evaporated atoms or molecules do not interact with each other in the vapor phase. The source materials are resistively heated to the desired temperatures to obtain molecular beams of the constituent elements. The molecular or atomic beams are closely controlled, collimated, and directed onto the substrate surface in a high-vacuum chamber. The rate of deposition is kept very low and the substrate temperature is maintained high in order to achieve sufficient

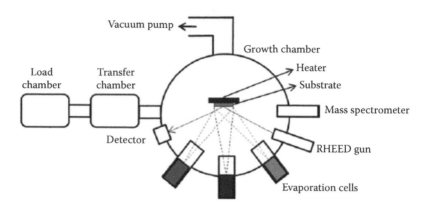

FIGURE 5.11
MBE method for the growth of AlGaAs on GaAs substrate. (From Meyyappan M., Sunkara M. (Eds.). 2009. *Inorganic Nanowires: Applications, Properties, and Characterization*. CRC Press, Taylor & Francis Group, New York. Reproduced with permission from Taylor & Francis Group.)

mobility of the elements on the substrate. The slow growth rate ensures sufficient surface diffusion and relaxation which results in a very smooth 2-D surface growth. Typical flux densities in the beams are of 10^{14}–10^{16} atoms/cm/s. The extremely clean environment, the slow growth rate, and independent control of the evaporation of individual sources enable the precise fabrication of nanostructures and nanomaterials at a single atomic layer. UHV environment ensures absence of impurity or contamination, and thus a highly pure epitaxy film can be obtained. For example, an AlGaAs layer can be grown on GaAs substrate by using atomic beams or fluxes of the elements Al, Ga, and As. In addition, if required, the beams of dopants, that is, Si for n-doping and Be for p-doping, can be fluxed. By controlling shutters for each beam, one can produce abrupt changes in crystal compositions and doping concentrations on the scale of one monolayer. MBE has also been used for the growth of III–V and II–VI semiconductor compound nanowires [24,32,36].

5.2.3 Chemical Vapor Deposition

Chemical vapor deposition (CVD) is the process of chemically reacting a volatile compound of a material to be deposited, with other gases, to produce a nonvolatile solid that deposits atomistically on a substrate. The deposit can be formed by a reaction between reacting (or precursor) gases in the vapor phase or by a reaction between a vapor and the surface of the substrate itself. It is widely used in industry because of relatively simple instrumentation, ease of processing, and possibility of depositing different types of materials at low cost. The CVD is performed in a reaction chamber (or reactor). In the reactor, a typical pressure of chemicals is ~10^4 Pa and heating is achieved by electric power. Basic CVD process involves a transport of reactant vapor or gas toward the substrate using some carrier gas (N_2 or Ar) as shown in Figure 5.12.

The substrate is generally kept at typical temperatures of 300–1200°C. The CVD mainly consists of two important functions: (i) transport of gas molecule to the reaction zone followed by reaction and (ii) deposition of the film onto the substrate surface. Deposition involves the adsorption of the reactant species onto the substrate surface sites followed by chemical reaction. Depending upon the type of materials to be grown, at an appropriate temperature the reactant cracks into different products/constituents which diffuse on the surface, undergo some chemical reaction at appropriate site, nucleate, and then

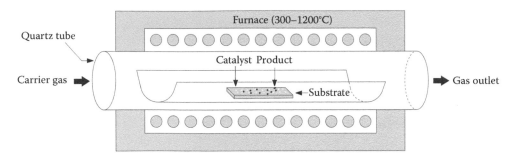

FIGURE 5.12
Basic CVD process setup.

grow to form the desired nanomaterial powder or film. The by-products created on the substrate have to be transported back to the gaseous phase. In this process, the reaction occurs at the substrate rather than in the gas phase. For the growth of Si layers, several sources of Si atoms like silicon tetrachloride ($SiCl_4$), silane (SiH_4), and dichlorosilane (SiH_2Cl_2) are used. The reaction of the $SiCl_4$ with hydrogen can be represented by the following equation:

$$SiCl_4 + 2H_2 \rightarrow Si + 4HCl \qquad (5.2)$$

The reaction of $SiCl_4$ with hydrogen can take place at temperatures in the range of 1150–1250°C. However, the reaction of SiH_4 and SiH_2Cl_2 with hydrogen can happen comparatively at lower temperatures, that is, 1000–1100°C. These reaction temperatures are well below the melting point of Si and provide efficient crystal growth onto the seed. In some cases, reaction temperatures can be reduced significantly by adding water to the system. However, the advantage of reduced reaction temperature is on the cost of having highly corrosive hydrochloric acid as a by-product in the system. This acid will corrode the inner side of the reactor walls. The growth of compounds of the elements belonging to group-III and group-V can take place at temperatures of ~600–700°C. Growth rate and the film quality depend upon the gas pressure and the substrate temperatures.

A variety of CVD methods and CVD reactors have been developed, depending on the types of precursors used, the deposition conditions applied, and the forms of energy introduced to the system to activate the chemical reactions required for the deposition of solid films on substrates. For example, when metalorganic compounds are used as precursors, the process is generally referred to as metalorganic CVD (MOCVD), and when plasma is used to promote the chemical reactions, this is a plasma-enhanced CVD (PECVD). There are many other modified CVD methods, such as low-pressure CVD, laser-assisted CVD, and aerosol-assisted CVD.

5.2.4 Colloidal or Wet Chemical Route

Colloids are considered as phase-separated small particles having shapes which may be spherical, rods, tubes, or plates. These particles may be made up of metal, alloy, and semiconductors synthesized in aqueous or nonaqueous media. These colloidal particles in liquids are to be stabilized by Coulombic repulsion or capping agents or passivation to prevent the excessive growth of NPs or their aggregates. Such capping or passivation of NPs provides sufficient stearic hindrance inhibiting the coalescence or aggregation.

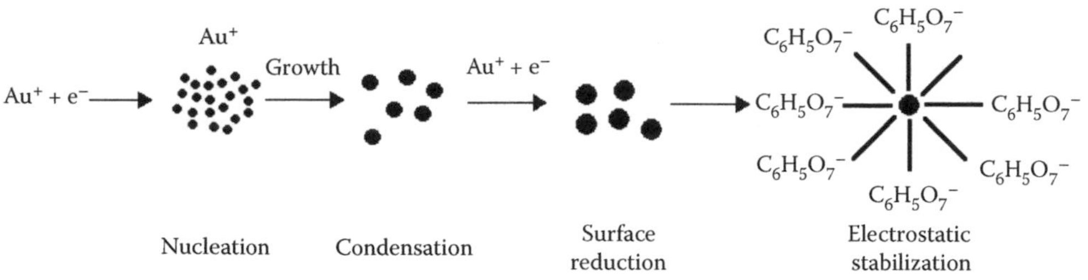

FIGURE 5.13
Stabilization of Au nanoparticles. (With kind permission from Springer Science+Business Media: *Nanotechnology: Principles and Practices*, 3rd ed., 2015, Kulkarni S. K.)

In the synthesis of metallic NPs, various types of precursors, reduction reagents, and other chemicals are used to control the reduction reactions, the initial nucleation, and the subsequent growth of initial nuclei. The most common precursors which include elemental metals, inorganic salts, and metal complexes are Ni, Co, $HAuCl_4$, H_2PtCl_6, $RhCl_3$, and $PdCI_2$. Reducing reagents commonly used are sodium citrate, hydrogen peroxide, hydroxylamine hydrochloride, citric acid, carbon monoxide, phosphorus, hydrogen, formaldehyde, aqueous methanol, sodium carbonate, and sodium hydroxide. Examples of polymeric stabilizers include polyvinyl alcohol (PVA) and sodium polyacrylate.

Synthesis of colloidal gold by wet chemical route has been reported by M. Faraday in 1857 and thereafter, a variety of methods have been developed for the synthesis of gold NPs. Chemical reactions for the synthesis of colloidal particles are carried out in a simple glass reactor (three-neck conical flask) of suitable size. Glass reactor should have a provision to introduce precursors, inert gas, thermocouple (to measure temperature), and sensor (to measure pH) during the reaction. Reaction is usually carried out under inert atmosphere (i.e., argon or nitrogen gas) to avoid any uncontrolled oxidation of the products. For the synthesis of gold NP, a dilute solution of chlorauric acid ($HAuCl_4$) in water is prepared. Then, few drops (or 1 mL) of 0.5% trisodium citrate ($Na_3C_6H_5O_7$) are added into the boiling solution followed by heating the solution at 100°C till color changes. The reduction of chlorauric acid by trisodium citrate at 100°C remains the most commonly used method. The reaction (Figure 5.13) can be represented by the following equation:

$$HAuCl_4 + NA_3C_6H_5O_7 \rightarrow Au^+ + C_6H_5O_7^- + HCl + 3\,NACl \tag{5.3}$$

The volume of the solution is maintained by adding water. This method results in a stable colloidal solution containing uniform particle size of $\sim$20 nm diameter. The final particle size depends upon the rate of nucleation and the growth of nuclei. Gold NPs with varying sizes have shown various colors such as intense red, magenta, blue, etc. Gold NPs can be stabilized by repulsive Coulombic interactions using thiol or some other capping molecules (Figure 5.13).

Similarly, amorphous silver NPs of $\sim$20 nm can be prepared by sonochemical reduction of an aqueous silver nitrate solution at a temperature of 10°C in an atmosphere of argon and hydrogen. The ultrasound resulted in decomposition of water into hydrogen and hydroxyl radicals. Hydrogen radicals would reduce silver ions into silver atoms, which subsequently nucleate and grow to silver nanoclusters. In addition, other metallic NP like palladium or copper or alloys can be synthesized by using appropriate precursors, temperature, pH, etc.

The types of reducing reagents used in the synthesis of NP affect their size and size distribution. In general, a strong reduction reaction provides a fast reaction rate and favors the formation of smaller NPs. A weak reduction reagent induces a slow reaction rate and favors relatively larger particles, which results in either wider or narrower size distribution. In addition, the use of solvent (acidic or basic) and polymer stabilizers play a very important role in controlling the size and shape of the metal NPs. The monosized metallic NPs are obtained by a combination of a low concentration of solute and polymeric monolayer adhered onto the growth surfaces. The combined effect of a low concentration and a polymeric monolayer would hinder the diffusion of the species from the surrounding solution to the growth surfaces and thus restrict the excessive growth of NPs. The NPs need to be surface passivated as colloids formed in liquids have a tendency to coagulate or grow due to attractive forces existing between them. The electrostatic and other repulsive forces may not be sufficient to keep them apart. As discussed above, stearic hindrance can be created by appropriately coating the NPs to keep them apart. This is often known as chemical capping. The capping of NPs has several advantages, including stable particles in solution or powder form, controlled size distribution, free from coagulation or aggregates, and chemically stable NPs (200–250°C) over a long time depending upon the capping molecules used. A variety of molecules such as thiophenol, mercaptoethanol, and sodium hexametaphosphate can be used to cap the NPs [24,32].

5.2.5 Reverse Micelle Method

Uniform and size-controlled NPs of metal, semiconductor, and metal oxides can be produced by the water-in-oil microemulsion (also called reverse micelle) method. In this, nanosized water droplets stabilized by a surfactant are dispersed in an oil phase. A schematic of a water-in-oil microemulsion is shown in Figure 5.14. Nanosized water droplets act as a microreactor, wherein particle formation occurs and helps to control the size of NPs. A unique feature of the reverse micelle process is that the particles are generally nanosized and monodispersed. This is because the surfactant molecules that stabilized the water

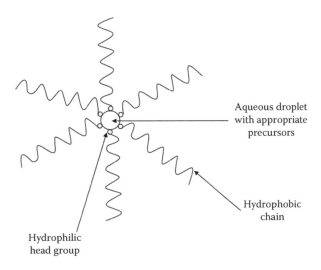

FIGURE 5.14
Schematic of a reverse micelle. (From Gogotsi Y. (Ed.). 2006. *Nanomaterials Handbook*. Taylor and Francis Group. Reproduced with permission from Taylor & Francis Group.)

droplets also adsorb on the surface of the NPs once the particle size approaches that of the water droplet. The reverse micelle method involves mixing two microemulsions that carry appropriate reactants. Water droplets of two microemulsions are allowed to collide with each other and the particle formation reaction takes place inside the water droplet. NP synthesis inside reverse micelles is accomplished by either hydrolysis of metal alkoxides or precipitation of metal salts with a base in case of metal oxide NPs. In case of metallic NPs, there is reduction of metal salts with a reducing agent (i.e., $NaBH_4$). The resultant particles are either filtered or centrifuged and then washed with acetone and water to remove any residual oil and surfactant molecules adsorbed on the surface of NPs. The major challenge of this process is to efficiently remove the NPs from the organic phase [37].

5.2.6 Green Chemistry Route

It is well known that the noble metallic NPs such as gold, silver, and platinum play an important role in the field of organic chemistry, bioelectronics, and pharmaceuticals. These noble metals are widely synthesized by wet chemical method using the reducing agents such as sodium borohydride, potassium bitartrate, methoxypolyethylene glycol, or hydrazine and the stabilizing agent such as sodium dodecyl benzyl sulfate or polyvinyl pyrrolidone. However, the noble metallic NPs synthesized by these methods must be free from the by-products and impurity which results in due to the use of reducing agent or stabilizing agents. These reducing agents are highly reactive chemicals and pose potential environmental and biological risks. However, the green chemistry method involves mainly three steps including the solvent medium, an environmentally benign reducing agent (e.g., sugar or β-D-glucose), and the nontoxic material for the stabilization of the NPs. Moreover, reducing agents are mild, renewable, inexpensive, and nontoxic. The green chemistry method uses starch as the protecting agent to protect or passivate the NPs surface. Moreover, environmental friendly multifunctional materials, the caffeine/polyphenols act as reducing as well as capping agent for Ag and Pd NPs. Caffeine, the most widely used drug, has high water solubility, low toxicity, and biodegradability.

Introduction: In general, NPs are synthesized by both bottom-up approach and top-down approach, which includes a number of physical and chemical methods. The most widely used method for synthesis of metallic NPs is wet chemical method, which involves reduction of metal ions in a liquid medium containing solution of metal salt and reducing agents. A stabilizing agent is also added to the reaction mixture to prevent the agglomeration of the NPs. Nevertheless, the chemical methods are low-cost process for high volume but they have drawbacks of contamination from precursors, toxic solvents, and hazardous by-products. Moreover, they have high-energy consumption and are difficult in: (i) scaling up the process of synthesis, (ii) purification, and (iii) separation of NPs. Hence, there is an increasing demand to develop high-yield, low-cost, nontoxic, and environmentally benign method/technique for the synthesis of metallic NPs. These drawbacks have been overcome by using the biological approach, which includes use of plants and plant extracts, algae, fungi, yeast, bacteria, and viruses for synthesis of NPs [38]. These biological approaches are also called green synthesis, which are simple and eco-friendly. They generate minimum wastes [39].

In recent years, biological synthesis (or green chemistry) has emerged as an attractive alternative to traditional physical or chemical methods for synthesizing NPs. Biological method involves using an environment-friendly green chemistry-based approach that employs unicellular and multicellular biological entities such as actinomycetes, fungus, bacteria, viruses, yeast, and plants. Biological method offers a clean, nontoxic, and environment-friendly method of synthesizing the NPs with a wide range of sizes, shapes, and compositions. These biological entities synthesize micro- to nanoscale structures. Compared to these biological methods (which use microorganisms), synthesizing the NPs using plants extracts is a relatively straightforward, much faster, and advantageous approach. Moreover, the plant approach does not need any special, complex, and multistep procedures such as isolation, culture preparation, and culture maintenance. Compared to biological methods which use bacteria and fungi, the plant-based biosynthesis methods have several advantages such as: (i) avoid the use of specific, well-conditioned culture preparation and expensive isolation techniques, (ii) safe, (iii) relatively short production times, (iv) a lower cultivation cost, and (v) a relatively simple process that can be easily scaled up for large-scale production of NPs. Each biological entity has varying degrees of biochemical processing capabilities to synthesize particular metallic or metallic oxide NPs. Due to the enzyme activities and intrinsic metabolic processes, all biological entities cannot synthesize NPs. Therefore, careful selection of the appropriate biological entity is necessary to produce NPs. Generally, biological entities with a potential to accumulate heavy metals have the best chance of synthesizing the metallic NPs. Recently, the biological synthesis of NPs using plants and plant extracts appears to be an attractive alternative to conventional chemical synthesis and the more complex culturing and isolation techniques needed for many microorganisms. Moreover, combinations of molecules found in plant extracts perform as both reducing and stabilizing (capping) agents during NP synthesis. These biological molecules are chemically complex, but have the advantage of being environment-friendly. The importance of developing environment-friendly sustainable metal NP producing technologies using the principles of green chemistry is discussed [37]. Recently, green process which uses natural extracts (of plants) as chelating agents has been found to be very effective for the synthesis of metals and oxides with average NPs size as small as 2.1 nm [39]. Several plants like alalfa, lemon grass, geranium leaves, *Cinnamomum camphora*, tamarind, neem, aloe vera, coriander, *Capsicum annuum*, *Hibiscus rosa sinensis* have been used to synthesize metal NPs. Plant extracts act both as reducing agents and stabilizing agents during the synthesis of NPs. The source of the plant extract influences the characteristics of the NPs because different extracts contain different concentrations and combinations of organic-reducing agents. These green processes are also called environmental friendly processes because they are free from toxic and dangerous by-products. This process is simple, low cost, eco-friendly, and safe for human therapeutic application. In addition, it is a single-step procedure suitable for large-scale production. The green synthesis method has several key merits like minimization of generated waste, utilization of nontoxic chemicals, environmentally benign solvents, and renewable materials. These features make this route a real alternative to traditional chemical methods. Plant extracts play a dual role in the process of NPs synthesis, that is, they reduce the metallic salts to NPs and act as stabilizing agents. Thus, it hinders the aggregation of the synthesized NPs [40–43]. Plant extracts containing bioactive alkaloids, proteins, phenolic acids, polyphenols, sugars, and terpenoids play an important role in first reducing the metallic ions and then stabilizing them. The variations in composition and concentration of these active biomolecules between different plants and their subsequent interaction with aqueous metal ions are the main factors contributing to different sizes and shapes of NPs (Table 5.1) [40].

The NPs produced by plants are more stable and the rate of synthesis is faster than that of microorganisms. They are suitable for large-scale synthesis of NPs with various shape and size [44].

TABLE 5.1

A Summary of Extracts of Plants, Metal Salts, Synthesized Nanoparticle, and Their Sizes

Sr. No.	Plants/Extract of Plants	Name of Metal Salt	Type of Nanoparticles Synthesized	Particle Size (nm)	Ref.
1	*Rosa rugosa*	Silver nitrate, auric acid	Ag, Au	11–12	[48]
2	*Chenopodium album* leaf	Silver nitrate, auric acid	Ag, Au	10–30	[44]
3	Apiin	Silver nitrate, chloroauric acid trihydrate ($HAuCl_4 \cdot 3H_2O$)	Ag, Au	7.5–23	[39]
4	*Hibiscus rosasinensis*	$AgNO_3$, $HAuCl_4 \cdot 3H_2O$	Ag, Au	14	[43]
5	*Euphorbia Jatropha* latex	Zinc nitrate	ZnO	50	[52]
6	*Punica granatum* peels	Copper acetate monohydrate $[Cu(CH_3COO)_2 \cdot H_2O]$	CuO	40	[49]
7	*Eucalyptus globulus*	Iron nitrate 9-hydrate	Fe_2O_3		[51]
8	Coffee, tea	Silver nitrate, $PdCl_2$	Ag, Pd	20–60	[46]
9	*Azadirachta indica*	$H_2PtCl_6 \cdot 6H_2O$ (aqueous)	Pt	5–50	[45]
10	*Aspalathus linearis*	$SnCl_4 \cdot 5H_2O$	SnO_2	2.1–19.3	[40]

Synthesis of NPs: Plant extracts containing bioactive alkaloids, proteins, phenolic acids, polyphenols, sugars, and terpenoids play an important role in first reducing the metallic ions and subsequently stabilize them. The variations in composition and concentration of these active biomolecules between different plants and their subsequent interaction with aqueous metal ions are the main factors contributing to different sizes and shapes of NPs as seen in Table 5.1 [39–40, 43–52]. This method involves mixing a sample of neat plant extract with a metal salt solution. Biochemical reduction of the salts starts immediately and the formation of NPs is indicated by a change in the color of the reaction mixture. During synthesis, there is an initial activation period when metal ions are converted from their mono or divalent oxidation states to zero-valent states and nucleation of the reduced metal atoms takes place. This is immediately followed by a growth of smaller neighboring particles to form thermodynamically stable larger NPs. Finally, the plant extracts stabilize the NPs to the most energetically favorable and stable morphology. The quality, size, and morphology of the NPs is influenced by concentration of the plants extract and metal salt, reaction time, reaction solution pH, and solution temperature [40].

5.2.6.1 Synthesis of Metallic NPs

- *Gold, Silver, Platinum, and Palladium NPs*: Owing to unique properties, Au NPs have potential applications in the fields of biosensors, hyperthermia therapy, therapeutic drugs, and antibacterial drugs. Similarly, due to excellent antimicrobial property, Ag is commonly used against pathogens. The high-surface area to volume ratio of Ag NPs enables them to closely interact with the bacterial cell membranes. Neem (*Azadirachta indica*) extract has been used as a reducing agent for the synthesis of silver, platinum, gold, and Au-core-Ag NPs.

- The synthesis of Ag NPs involves mixing of a solution of $AgNO_3$ and aqueous solution of soluble starch followed by addition of β-D-glucose as reducing agent. The heating of mixture to 40°C for about 20 h in presence of argon resulted in Ag NPs with size range of 1–8 nm. The hydroxyl groups present on starch act as passivation contacts for the stabilization of the NPs, and thus, the resultant dispersed Ag NPs in starch are highly stable and do not show signs of aggregation even after 2 months of storage [47]. Dwivedi and Gopal reported a synthesis of Ag and Au NPs using *Chenopodium album*, an obnoxious weed which is widely distributed in Asia, North America, and Europe. The leaf of *C. album* is thoroughly washed with distilled water to remove any dirt. The thoroughly washed leaf is boiled in distilled water for 30 min and suspension is filtered using Whatman 40 filter paper. The leaf extract is added to an aqueous solution of silver nitrate and auric acid at room temperature for 15 min. After some time, a change in solution color from a brownish-yellow and pinkish-red is observed indicating the formation of Ag NPs and Au NPs, respectively. The aqueous leaf extract of the herb reduces the metallic salt solutions to quasi-spherical-shaped metal NP with size range of 10–30 nm. The increase in the concentration of the metal ions increases their particle size, while the increase in leaf extract quantity decreases the particle size of the Ag and Au [44]. The use of leaves extracts of *Rosa rugosa* (ornamental plant), which is found naturally in Eastern Asia, Korea, northern parts of Japan and China, can also produce Ag and Au NPs. The fresh leaves of *R. rugosa* are washed with pure water and boiled thoroughly in pure water for 10 min. Then, an aqueous solution of appropriate concentration of silver nitrate and auric acid are prepared and mixed with leaves extract at room temperature for few minutes (i.e., 10 min). This results in a brown yellow and pink red solution indicating the formation of Ag NPs and Au NPs, respectively. This results in an average size of 10–12 nm, which are useful for biomedical and biotechnological applications [48]. The anisotropic Au and quasi-spherical Ag NPs have been synthesized by using apiin (from henna leaves) as the reducing and stabilizing agent. The size and shape of the NPs have been controlled by varying the ratio of metal salts to apiin compound in the reaction medium. The average size of the Au and Ag NPs obtained by this method is in the range of 21–39 nm [39]. The Au and Ag NPs were also synthesized using the hibiscus leaf extract at ambient conditions. Hibiscus leaf extract contains proteins, vitamin C, organic acids (i.e., malic acid), flavonoids, and anthocyanin. The ability of hibiscus extract as a natural starch and sucrose blocker is found to lower starch and sucrose absorption when injected at reasonable doses by inhibiting amylase which in turn influences the glycemic load favorably. For the synthesis of Au or Ag NPs, a hibiscus extract is added to a vigorously stirred aqueous solution of $HAuCl_4 \cdot 3H_2O$ or $AgNO_3$ and stirred continuously for 1 min. The presence of malic acid molecules in the hibiscus leaf extract reduces and stabilizes the NPs. The reduction of Ag ions starts, when the pH of the solution is adjusted to about 6.8 using NaOH. The size and shape of Au NPs are controlled by varying the ratio of metal salt and extract in the reaction medium. Variation of pH of the reaction medium gives Ag NPs of different shapes. The stabilization of Au and Ag NPs occurs through amine groups and carboxylate ion, respectively [43].

- For the synthesis of platinum NPs, the neem plant, which is easily available in Asian continent, is added to a solution of aqueous $H_2PtCl_6 \cdot 6H_2O$ and the mixture is maintained at 100°C for 1 h in a sealed flask to avoid the evaporation. The increased temperature catalyzes the rate of reduction process. The protein biomolecule is responsible for the reduction of chloroplatinic ion into platinum NPs. The reduced platinum solution is sonicated for 30 min to break the particle agglomerates. This results in primary particles with size in the range of 5–50 nm indicating a very simple and economically feasible method for the synthesis of platinum NPs, which have medicinal and catalytic applications [45].

- The use of coffee and tea extract can generate bulk quantities of nanocrystals of noble metals such as silver (Ag) and palladium (Pd) at room temperature. The NPs with size range of 20–60 nm and face centered cubic (FCC) symmetry are obtained. Their size and shape depends upon the source of coffee or tea extract used for reducing the metal salts [46]. The use of barberry fruit extract synthesizes Pd NPs supported on reduced graphene oxide under the mild conditions. This fruit extract reduces and stabilizes the graphene oxide and Pd^{2+} ions. The Pd NPs/reduced graphene oxide can be used as an efficient and heterogeneous catalyst [50].

- Bimetallic NP synthesis involves the competitive reduction between two aqueous solutions each containing a different metallic ion precursor that is mixed with a plant extract. In the case of an Au–Ag bimetallic NP, Au having the larger reduction potential will form first to create the core of a resulting core-shell structure. Subsequent reduction of Ag ions results in Ag coalescing on the core to form the shell [40].

5.2.6.2 Synthesis of Oxide NPs

- The SnO_2 is a transparent, n-type semiconductor with a bandgap of ~3.6 eV. After doping, it can be made useful in several applications such as transparent conducting oxides, varistors, gas sensors, thick film resistors, and electrochemical devices as well as effective photocatalysts. The use of *Aspalathus linearis'* natural extract as chelating agent produces SnO_2 NPs. The addition of extract into a salt solution $SnCl_4 \cdot 5H_2O$ generates a white deposit at the bottom after ~10 min. The resultant solution is centrifuged several times at 1000 rpm for 10 min each. The white deposit is dried at about 80°C, which results in an average size of 2–3 nm. The SnO_2 NPs exhibit effective photocatalytic responses to methylene blue, Congo red, and Eosin Y [40].

- Copper (Cu) and copper oxide (CuO) NPs exhibit antibacterial activity against a pathogen. CuO is an important semiconductor with a narrow bandgap of 1.7 eV. It is used in pesticides formulation and antibacterial agents. Cu and CuO NPs have been synthesized by a variety of plant extracts. Cu NPs have been biologically synthesized using magnolia leaf extract to produce stable NPs with size in the range of 40–100 nm [3]. For synthesizing CuO NPs, a typical reaction mixture of copper acetate monohydrate is dissolved in deionized water and stirred magnetically at room temperature for 5 min. Afterward, *Punica granatum* peels aqueous extract is added dropwise under stirring. As soon as the peels extract comes in contact with Cu ions, spontaneous change in the blue color of Cu ions to green color occurs. When the resultant green mixture is left under stirring at room temperature for about 10 min, the green mixture changed to a brown suspended mixture, indicating the formation of water-soluble monodispersed CuO NPs. This process results in crystalline NPs with an average size of 40 nm. The morphology and size of the CuO NPs could be controlled by tuning the concentration of *P. granatum* peels aqueous extract and Cu ions [49].

- A green synthesis of iron oxide (i.e., maghemite) NPs is obtained using eucalyptus extract as reducing agent. The controlled conditions lead to the formation of the encapsulation of the iron oxide NPs in chitosan beads (or matrix), also called magnetic organic/inorganic hybrids. These hybrid nanomaterials can be used as sorbent to remove completely arsenic ions [51]. Zinc oxide (ZnO) NPs are an important biomedical and cosmetic product. The use of latex of *Euphorbia jatropha* as reducing agent synthesizes ZnO NPs. The concentration of plant latex plays an important role in controlling the size and morphology of the NPs. The average particle size of ∼15 nm is obtained [52]. A list of metal salts and their reducing agent is summarized in Table 5.1.

5.2.6.3 Factors Affecting Size and Morphology of NPs

The green chemistry synthesis method of metallic NPs is affected by several factors such as pH, metal salt concentrations, reaction time, and temperature. The variation in pH of the reaction medium tends to generate NPs with different shape and size. For example, lower pH (i.e., 2) values produce larger particles and relatively smaller particles at pH 3 and 4. This is due to the fact that at lower pH fewer functional groups are available within the extract for particle nucleation; hence, lower pH results in particle aggregation to form larger NPs. In contrast, more functional groups at higher pH values nucleate more number of nuclei and hence result in comparatively smaller particles. For a given concentration of metal salt solution, the increasing concentrations of plant extracts change the particle shape from low symmetry to higher symmetry (i.e., spherical). The reaction time may vary from few minutes to several hours for the complete synthesis of NPs. On increasing the reaction time after a few minutes, the rate of synthesis of NPs decreases and the NPs grow in size at the expense of smaller-sized particles. The particles nucleation rate becomes faster with increasing reaction temperature, which leads to decrease in the average particle size with steady increase in the particle conversion rate with increasing temperature [40].

5.2.7 Sol-gel Method

The sol-gel method is one of the most attractive and versatile methods to synthesize nanomaterials at low temperature and low cost. A "sol" is defined as a stable dispersion of colloidal particles or polymers in a solvent, whereas colloid is defined as suspension of dispersed solid particles (1–1000 nm). The gel consists of a 3-D continuous network, which encloses liquid phase. Sol-gel processes are categorized into aqueous based that involved water during reaction and alcohol based that excludes water. Sol-gel processes can be categorized into alkoxides and nonalkoxide processes according to volatility of the precursor material. Metal alkoxides are widely used because metals are capable of forming the metal alkoxides with general chemical formula $M(RO)_x$, where M and R represent the metal and alkyl group and x is the valence state of the metal. To get a homogeneous gel without precipitation, some additives are added to increase the time of gelation to increase the stability of sol-gel-derived product. The basic sol-gel reaction involves dispersing colloidal particles having very small size such that gravitational force on them is negligible and exhibits random Brownian motion within the fluid matrix. The interaction among them is dominated by short-range forces like van der Walls attraction and surface charges. This interaction extends only for a few nanometer distances.

Sol-gel process involves hydrolysis and polycondensation reactions using water as a medium. In some cases, catalysts are required to increase the reaction kinetics that promotes the hydrolysis of the precursors. As represented in Figure 5.15, the hydrolysis (or polymerization) establishes metal–OH–metal or metal–O–metal bridges between metallic atoms of the precursors. As a result, colloidal particles are formed. The next step involves the linking of these colloidal particles to build a 3-D network structure within solvent.

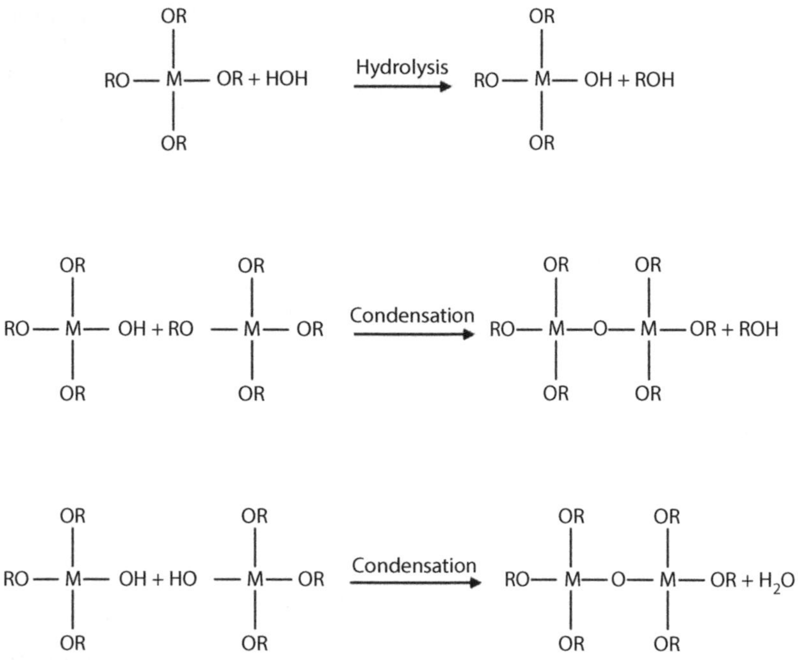

where R = –CH$_3$, –C$_2$H$_5$, –C$_3$H$_7$, etc.

M = Metal cation (e.g., Ti^{4+}, Si^{4+})

FIGURE 5.15
A typical schematic diagram of hydrolysis and condensation reactions in the sol-gel process.

The step of gelation arises when linking process enhances to an extent such that a giant spanning cluster is formed. At this point, the gel possesses high viscosity. The aging plays a crucial role when a large number of sol particles are to be reacted. This gel has the ability to take the shape of the container or mold in which it is prepared. Then, the gel structure is allowed to ripen in order to strengthen the interparticle interactions prior to drying. The drying is usually done for several hours to several days. During this (or aging) process, the gel continues to transform. If the gel is dried by simple evaporation, then the capillary forces shrink the gel and a xerogel is formed. In contrast, if it is dried under supercritical conditions, the 3-D network structure having high porosity forms, also called as aerogel. For example, a supercritical drying by cold CO$_2$ method produces materials with higher-specific surface areas and porosity. Aerogels are disordered NP networks with interconnected pore structures. The aerogels possess porosities up to 99.8% and densities as low as 0.004 g/cm^3. The high-surface area and relatively large and interconnected pores makes these aerogel excellent platforms for catalysis and sensing applications. Their architecture contributes to extraordinary low-thermal conductivity and a high degree of elasticity. The sol-gel process is a very simple and cost-effective for synthesis of NPs. Due to the use of small molecules with high purity, this method produces final product with high purity.

The silicon-based sol-gel process is probably the one that has been most investigated. Principally, R$_{4-n}$SiX$_n$ compounds ($n = 1–4$, X = alkoxide or halogen) are used as molecular precursors, in which the Si–X bond is labile toward hydrolysis reactions forming unstable silanols (Si–OH) that condensate leading to Si–O–Si bonds. The basic steps of the sol-gel synthesis of oxide NPs are shown in the flowchart presented in Figure 5.16. The metal

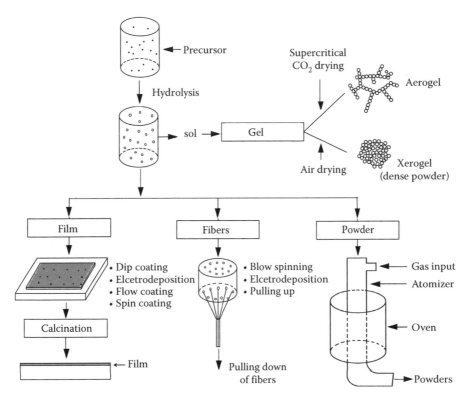

FIGURE 5.16
A schematic diagram of synthesis of oxide nanoparticles by a sol-gel process.

alkoxide and water are dissolved in a common alcoholic solvent in order to carry out the reaction. The alkoxide group, being highly electronegative, creates a positive charge on the central metal atom. The water molecule has a partial negative charge on the oxygen atom, which attacks the metal atom from the alkoxide and results in the hydrolysis of the alkoxide. The result is formation of the M–O–M (where M is metal) bond within the solution. The kinetics of hydrolysis and condensation reactions are governed mainly by the ratio of molar concentrations water to alkoxide (R). In general, low R values (<3) are suitable for fiber and thin-film formation, while large R values (>3) generate powder particles. The size, size distribution, and surface morphology of oxide NP are affected by the value of R. Sol-gel method can produce ultrafine particles, thin films, nanofibers, and nanoporous membranes. For the synthesis of TiO_2 and SiO_2 NPs, $Ti(OC_4H_9)_4$ and $Si(OC_4H_9)_4$ precursors are used, respectively [35].

5.2.8 Combustion Method

Combustion synthesis (CS) or self-propagating high-temperature synthesis (SHS) is an effective, and low-cost method for production of various industrially useful materials. The conventional solid state SHS being a gasless combustion process typically yields much coarser particles than solution combustion synthesis (SCS) method which has initial reaction in aqueous solution. One of CS method is of gas-phase synthesis which has the ability of producing nonagglomerated fine particles. However, this is the less effective method due to high cost of the final products. It is difficult task to produce the nanomaterials by conventional SHS, where the typical size of initial solid reactants is on the order of 10–100 μm. The SHS method coupled with postsynthesis treatment like intensive milling or chemical dispersion has been used to synthesize nanomaterials. In alkali metal-molten-salt-assisted combustion (additive-modified SHS) method, the reducing metal (e.g., Mg) reacts with transition metal oxide (Me_2O_x) in the alkali metal-molten-salt (e.g., NaCl). A large amount of heat is generated during the combustion reaction, which melts the salt at 1083 K and further nucleation of metal (Me) particles occurs in the molten NaCl, which protects them from agglomeration and growth. The by-product (i.e., MgO) can be removed by washing powder in acid solution. This method can also be used to synthesize other pure metals (i.e., Ti, Mo, W), carbides (i.e., TiC or WC), and composites (i.e., WC-Co). The nanograined WC-Co composites with grain size of 50–200 nm exhibit extremely higher hardness. This method has relatively low product yield due to the presence of MgO. If metal is replaced by carbon as the reaction fuel, a high rate of CO_2 release facilitates synthesis of highly porous (∼70%) powders having particle size in the range of 50–800 nm. This method can produce advanced oxide nanopowders of ferroelectrics ($BaTiO_3$, $SrTiO_3$, $LiNbO_3$), multiferoics ($HoMnO_3$, $BiFeO_3$), fuel cell components ($LaGaO_3$, $La_{0.6}Sr_{0.4}MnO_3$), battery electrode material ($LiMn_2O_4$), hard/soft ferrites ($BaFe_{12}O_{19}$, $CoFe_2O_4$, Ni–Zn, Mn–Zn–ferrites, $Y_3Fe_5O_{12}$), and catalysts ($LaCrO_3$, $LiCrO_3$). This approach can also synthesize the oxides of $CaSnO_3$ and $LaGaO_3$, which cannot be produced by conventional SHS due to the pyrophoric nature of metals (La, Li) or metals with low-melting point (e.g. Ga, Hg, Cs).

SCS method has been primarily used to produce a variety of metals, alloys, and complex nanooxides for the applications in catalysts, fuel cells, and biotechnology. It involves uniform reaction solution preheating followed by self-ignition, which leads to steady-state self-propagation of SCS of nanopowders. SCS method, due to its versatility and simplicity, is a unique and cost-effective method for the production of nanomaterials. Typical SCS method involves a self-sustained reaction in a homogeneous solution of a metal precursor (oxidizer) and a fuel (e.g., urea, glycine, hydrazine, etc.). The reaction between fuel and oxygen-containing species, formed during the decomposition of the nitrite species, provides high-temperature rapid interaction. In a typical scheme, an initial liquid solution of desired reagents after preheating to a moderate temperature (150–200°C) self-ignites leading to the synthesis of nanomaterials with tailored composition. The initial aqueous solution allows molecular level mixing of the reactants, thus permitting the precise and uniform composition on the nanoscale. For example, when nickel nitrate and glycine react together and decompose, an exothermic reaction occurs, which makes SCS self-sustained. Further, increasing the glycine content in the initial solution leads to a hydrogen-rich reducing environment in the combustion wave that in turn results in the formation of pure nickel nanopowder. The high-reaction temperature (∼1000–2000°C) ensures high-product purity and crystallinity. Hence, this method does not need high-temperature calcination process. Interestingly, short process duration (few seconds) and the formation of various gases during combustion inhibit the particle size growth and favor the synthesis of nanomaterial with high-specific surface area. The SCS of ferric nitrite and glycine system synthesizes iron oxide. The SCS method can synthesize Cu nanopowders with the size of 150–200 nm. Its specific surface area is ∼15 m^2/g, which is higher relative to an oxide prepared by precipitation and calcination method [53–56].

5.2.9 Atomic Layer Deposition

Atomic layer deposition (ALD) is a surface-controlled method for the deposition of films from gas phase. In ALD, the gaseous reactants are alternately pulsed to the substrates, and between the reactant pulses the reactor is purged with an inert gas. The film growth proceeds via self-limiting saturative surface reactions, which controls the film thickness and results in an excellent conformality and large area uniformity. The separate pulsing of precursors allows the use of highly reactive precursors. Figure 5.17 shows a schematic representation of a basic ALD cycle using two precursors. In this example, ZnS film is deposited with use of $ZnCl_2$ and H_2S as precursors. First, $ZnCl_2$ is pulsed into the reaction chamber and a monolayer of this precursor is chemisorbed on the substrate. Its excess amount is purged away with an inert gas. The second precursor H_2S is pulsed into the reactor which reacts with the chemisorbed $ZnCl_2$, forming a ZnS layer on the substrate and HCl gas as a by-product. The HCl gas and excess precursor (i.e., H_2S) are then purged away with an inert gas. The second purge completes one deposition cycle. The desired film thickness is obtained by repeating the deposition cycle an appropriate number of times. A complete monolayer is formed in every deposition cycle if the process is ideal and the film is free from the impurities. The good quality and excellent conformality of the films deposited by ALD make it a promising technique for a variety of microelectronic applications. Compared to CVD and PVD methods, the ALD method is a low-temperature method producing films with less impurities, high-aspect ratio structures, and excellent conformality, as shown in Figure 5.18.

The ALD has the advantage of precise thickness control at the Ångstrom or monolayer level. No other thin film technique can approach the conformality achieved by ALD on high-aspect structures. ALD has emerged as an important technique for depositing thin films for a variety of applications in semiconductor industries for high-dielectric constant gate oxides in the MOSFET structure, copper diffusion barriers in backend interconnects, and deposition of high-quality dielectrics to fabricate trench capacitors for DRAM. Miniaturization in the semiconductor industry has led to the requirement for atomic level control of thin film deposition [56]. ALD produces a thin film of Al_2O_3 using trimethylaluminum and H_2O at ~177°C, which is smooth and extremely conformal to the underlying high-aspect ratio trench substrates (Figure 5.18). It was also found that the growth of Al_2O_3 film per cycle decreases progressively with temperature between 177°C and 300°C [57].

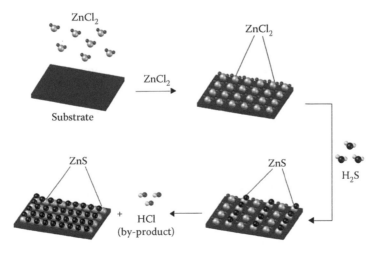

FIGURE 5.17
A schematic representation of a formation of ZnS film using the basic ALD method.

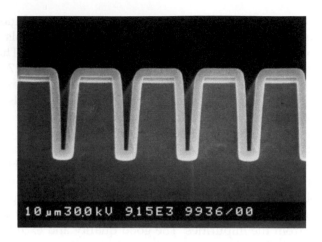

FIGURE 5.18
Cross-sectional SEM image of an Al_2O_3 film (thickness: ~300 nm) on a Si wafer with a trench structure prepared by ALD. (Ritala M. et al.: Perfectly conformal TiN and Al_2O_3 films deposited by atomic layer deposition. *Chem. Vap. Depos.* 1999. 5. 7–9. Copyright Wiley-VCH Verlag GmbH & Co. KGaA. Reproduced with permission.)

References

1. El-Eskandarany M. S. 2001. *Mechanical Alloying for Fabrication of Advanced Engineering Materials.* Noyes Publications, New York. pp. 2–16.
2. Suryanarayan C. 2008. Recent developments in mechanical alloying. *Rev. Adv. Mater. Sci.* 18: 203–211.
3. Zhang G., Schlarb A. K., Tria S., Elkedim O. 2008. Tensile and tribological behaviours of PEEK/nSiO$_2$ composites compounded using a ball milling technique. *Compos. Sci. Technol.* 68: 3073–3080.
4. Tadayyon G., Zebarjad S. M., Sajjadi S. A. 2011. Effect of both nano-size alumina particles and severe deformation on polyethylene crystallinity index. *J. Thermoplast. Compos. Mater.* 25: 479–490.
5. Suryanarayana C. 2001. Mechanical alloying and milling. *Prog. Mater. Sci.* 46: 1–184.
6. Prakash T. 2012. Review on nanostructured semiconductors for dye sensitized solar cells. *Electron. Mater. Lett.* 8(3): 231–243.
7. Othman A. R., Sardarinejad A., Masrom A. K. 2015. Effect of milling parameters on mechanical alloying of aluminum powders. *Int. J. Adv. Manuf. Technol.* 76: 1319–1332.
8. Can M. M., Ozcan S., Ceylan A., Firat T. 2010. Effect of milling time on the synthesis of magnetite nanoparticles by wet milling. *Mater. Sci. Eng. B* 172: 72–75.
9. Goyal R. K., Dhoka P., Bora A., Gaikwad P. 2016. Synthesis of iron nanopowder and its nanocomposites with high electrical conductivity and thermal stability. *J. Mater. Sci. Surf. Eng.* 4(1): 339–341.
10. Yadav T. P., Yadav R. M., Singh D. P. 2012. Mechanical milling: A top down approach for the synthesis of nanomaterials and nanocomposites. *Nanosci. Nanotechnol.* 2(3): 22–48.
11. Suryanarayana C., Ivanov E., Boldyrev V. V. 2001. The science and technology of mechanical alloying. *Mater. Sci. Eng. A* 304–306: 151–158.
12. Tiwary C. S., Kashyap S., Biswas K., Chattopadhyay K. 2013. Synthesis of pure iron magnetic nanoparticles in large quantity. *J. Phys. D: Appl. Phys.* 46: 385001.
13. Malek A., Basu P. 2010. Mechanochemical synthesis of nano-structured TiC from TiO$_2$ powders. *J. Alloys Compd.* 500: 220–223.
14. Karimbeigi A., Zakeri1 A., Sadighzadeh A. 2013. Effect of composition and milling time on the synthesis of nanostructured Ni-Cu alloys by mechanical alloying method. *Iranian J. Mater. Sci. Eng.* 10(3): 27–31.

15. Zakeri M., Rahimipour M. R., Abbasi B. J. 2013. Synthesis of nanostructure tetragonal ZrO$_2$ by high energy ball milling. *Mater. Technol.* 28: 181–186.
16. Valiev R. 2004. Nanostructuring of metals by severe plastic deformation for advanced properties. *Nat. Mater.* 3: 511–516.
17. Halfa H. 2014. Recent trends in producing ultrafine grained steels. *J. Miner. Mater. Charact. Eng.* 2: 428–469.
18. Valiev R. Z., Alexandrov I. V. 1999. Nanostructured materials from severe plastic deformation. *Nanostruct. Mater.* 12: 35–40.
19. Wang Y. M., Ma E., Chen M. W. 2002. Enhanced tensile ductility and toughness in nanostructured Cu. *Appl. Phys. Lett.* 80(13): 2395–2397.
20. Swaminathan S., Shankar M. R., Rao B. C., Compton W. D., Chandrasekar S., King A. H., Trumble K. P. 2007. Severe plastic deformation (SPD) and nanostructured materials by machining. *J. Mater. Sci.* 42: 1529–1541.
21. Sabirov I., Enikeev N. A., Murashkin M. Y., Valiev R. Z. 2015. *Bulk Nanostructured Materials with Multifunctional Properties*. Springer, Switzerland.
22. Azushima A., Kopp R., Korhonen A., Yang D. Y., Micari F., Lahoti G. D., Groche P., Yanagimoto J., Tsuji N., Rosochowski A., Yanagida A. 2008. Severe plastic deformation (SPD) processes for metals. *CIRP Ann. Manuf. Technol.* 57: 716–735.
23. Ashby M., Ferreira P., Schodek D. 2009. *Nanomaterials, Nanotechnologies and Design, an Introduction for Engineers and Architects*. Elsevier Science Ltd. USA.
24. Cao G. 2004. *Nanostructures and Nanomaterials: Synthesis, Properties and Applications*. Imperial College Press, London.
25. Xia Y., Whitesides G. M. 1998. Soft Lithography. *Annu. Rev. Mater. Sci.* 28: 153–184.
26. Walia S., Shah C. M., Gutruf P., Nili H., Chowdhury D. R., Withayachumnankul W., Bhaskaran M., Sriram S. 2015. Flexible metasurfaces and metamaterials: A review of materials and fabrication processes at micro- and nano-scales. *Appl. Phys. Rev.* 2: 011303.
27. Pimpin A., Srituravanich W. 2011. Reviews on micro- and nanolithography techniques and their applications. *Eng. J.* 16(1): 37–55.
28. Tsenga A. A., Notargiacomob A., Chen T. P. 2005. Nanofabrication by scanning probe microscope lithography: A review. *J. Vac. Sci. Technol.* 23(3): 877–894.
29. Xie X. N., Chung H. J., Sow C. H., Wee A.T.S. 2006. Nanoscale materials patterning and engineering by atomic force microscopy nanolithography. *Mater. Sci. Eng. R* 54: 1–48.
30. Kim Y. J., Kim G., Lee J. 2010. Fabrication of a conductive nanoscale electrode for functional devices using nanoimprint lithography with printable metallic nanoink. *Microelectron. Eng.* 87: 839–842.
31. Zhou W. 2013. *Nanoimprint Lithography: An Enabling Process for Nanofabrication*. Springer-Verlag Berlin, Heidelberg.
32. Kulkarni S. K. 2015. *Nanotechnology: Principles and Practices*. 3rd ed. Springer, New Delhi.
33. Xia Y., Whitesides G. M. 1998. Soft Lithography. *Angew. Chem. Int. Ed.* 37: 550–575.
34. Siegel R. W. 1993. Synthesis and properties of nanophase materials. *Mater. Sci. Eng. A* 168: 189–197.
35. Koch C. C., Ovid'ko I. A., Seal S., Veprek S. 2007. *Structural Nanocrystalline Materials: Fundamentals and Applications*. Cambridge University Press, New York.
36. Meyyappan M., Sunkara M. (Eds.). 2009. *Inorganic Nanowires: Applications, Properties, and Characterization*. CRC Press, Taylor & Francis Group, New York.
37. Gogotsi Y. (Ed.). 2006. *Nanomaterials Handbook*. Taylor and Francis Group, Boca Raton, FL.
38. Thakkar K. N., Mhatre S. S., Parikh R. Y. 2010. Biological synthesis of metallic nanoparticles. *Nanomed. Nanotechnol. Biol. Med.* 6: 257–262.
39. Diallo A., Manikandan E., Rajendran V., Maaza M. 2016. Physical and enhanced photocatalytic properties of green synthesized SnO$_2$ nanoparticles via *Aspalathus linearis*. *J. Alloy. Compd.* 681: 561–570.
40. Shah M., Fawcett D., Sharma S., Tripathy S. K., Poinern G. E. J. 2015. Green synthesis of metallic nanoparticles via biological entities. *Materials* 8: 7278–7308.

41. Dwivedi A. D., Gopal K. 2010. Biosynthesis of silver and gold nanoparticles using *Chenopodium album* leaf extract. *Colloid. Surf. A: Physicochem. Eng. Aspects* 369: 27–33.
42. Kasthuri J., Veerapandian S., Rajendiran N. 2009. Biological synthesis of silver and gold nanoparticles using apiin as reducing agent. *Colloid. Surf. B: Biointerfaces* 68: 55–60.
43. Philip D. 2010. Green synthesis of gold and silver nanoparticles using *Hibiscus rosa sinensis*. *Physica E* 42: 1417–1424.
44. Iravani S. 2011. Green synthesis of metal nanoparticles using plants. *Green Chem.* 13: 2638–2650.
45. Raveendran P., Fu J., Wallen S. L. 2003. Completely "Green" synthesis and stabilization of metal nanoparticles. *J. Am. Chem. Soc.* 125: 13940–13941.
46. Dubey S. P., Lahtinen M., Sillanpaa M. 2010. Green synthesis and characterizations of silver and gold nanoparticles using leaf extract of *Rosa rugosa*. *Colloid. Surf. A: Physicochem. Eng. Aspects* 364: 34–41.
47. Thirumurugan A., Aswitha P., Kiruthika C., Nagarajan S., NancyChristy A. 2016. Green synthesis of platinum nanoparticles using Azadirachta indica—An ecofriendly approach A. *Mater. Lett.* 170: 175–178.
48. Nadagouda M. N., Varma R. S. 2008. Green synthesis of silver and palladium nanoparticles at room temperature using coffee and tea extract. *Green Chem.* 10: 859–862.
49. Nasrollahzadeh M., Sajadi S. M., Rostami-Vartooni A., Alizadeh M., Bagherzadeh M. 2016. Green synthesis of Pd nanoparticles supported on reduced graphene oxide using barberry fruit extract and its application as a recyclable and heterogeneous catalyst for the reduction of nitroarenes. *J. Colloid Interf. Sci.* 466: 360–368.
50. Ghidan A. Y., Al-Antary T. M., Awwad A. M. 2016. Green synthesis of copper oxide nanoparticles using *Punica granatum* peels extract: Effect on green peach aphid. *Environ. Nanotechnol. Monit. Manage.* 6: 95–98.
51. Geetha M. S., Nagabhushana H., Shivananjaiah H. N. 2016. Green mediated synthesis and characterization of ZnO nano particles using *Euphorbia Jatropa* latex as reducing agent. *J. Sci. Adv. Mater. Devices* 1(3): 301–310.
52. Martínez-Cabanas M., López-García M., Barriada J. L., Herrero R., de Vicente M. E. S. 2016. Green synthesis of iron oxide nanoparticles. Development of magnetic hybrid materials for efficient As (V) removal. *Chem. Eng. J.* 301: 83–91.
53. Kumar A., Wolf E. E., Mukasyan A. S. 2011. Solution combustion synthesis of metal nanopowders: Copper and copper/nickel alloys. *AIChE J.* 57(12): 3473–3479.
54. Aruna S. T., Mukasyan A. S. 2008. Combustion synthesis and nanomaterials. *Curr. Opin. Sol. State Mater. Sci.* 12: 44–50.
55. Mukasyan A. S., Epstein P., Dinka P. 2007. Solution combustion synthesis of nanomaterials. *Proc. Combust. Inst.* 31: 1789–1795.
56. George S. M. 2010. Atomic layer deposition: An overview. *Chem. Rev.* 110: 111–131.
57. Ritala M., Leskela M., Dekker J. P., Mutsaers C., Soininen P. J., Skarp J. 1999. Perfectly conformal TiN and Al_2O_3 films deposited by atomic layer deposition. *Chem. Vap. Depos.* 5(1): 7–9.

6

Synthesis, Properties, and Applications of Carbon Nanotubes

Carbon nanotubes (CNTs) are unique tubular structures of carbon with a few nanometer diameter and large aspect ratio. The nanotubes may consist of one to few tens of concentric shells of carbons with adjacent shells separation of ~0.34 nm. CNTs made of one hollow seamless graphitic shell are called single wall nanotubes (SWNTs) and have diameters typically 0.6–3 nm. CNTs made of two or more seamless concentric shells are called multi-walled nanotubes (MWNTs). Their high Young's modulus and tensile strength makes them preferable reinforcement for composites based on polymer, metal, and ceramic matrices with improved thermal, mechanical, and electrical properties. The CNTs can be metallic or semiconducting depending on their structure. Due to their unique electronic properties, CNTs find application in electronic devices such as field effect transistors (FETs), single electron transistors, and rectifying diodes. The huge surface to volume ratio makes them very useful as hydrogen storage materials. This chapter is intended to summarize some of the important synthesis methods including arc discharge, laser ablation, catalytic growth, or chemical vapor decomposition (CVD), and high-pressure carbon monoxide (HiPCO) decomposition. Moreover, the optical, electrical, mechanical and thermal properties, and applications of CNTs are also discussed.

6.1 Introduction

In 1991, for the first time Iijima [1] investigated multiwalled carbon nanotubes (MWNTs) with adjacent shell separation of ~0.34 nm. After 2 years, Bethune et al. [2] and Iijima et al. [3] reported synthesis of single-walled carbon nanotubes (SWNTs). Nowadays, MWNTs and SWNTs are synthesized mainly by four techniques such as arc discharge, laser ablation, catalytic growth, and HiPCO decomposition. There are several other methods for the synthesis of CNTs but those methods are not discussed here. The CNTs have been suggested for use in nanocomposites, FETs, single electron transistors and rectifying diodes, logic circuits, flat panel displays, sports goods, water purification, nuclear waste management, and hydrogen storage media.

6.2 Synthesis of CNTs

After the discovery of MWNTs, the SWNTs were discovered in the soot of the arc discharge method in 1993 [1]. The arc discharge method has been used since long time in the production of carbon fibers and fullerenes. Thereafter, laser ablation, CVD, and HiPCO decomposition methods have been employed to synthesize CNTs, as discussed below.

6.2.1 Arc Discharge Method

Arc discharge is the easiest and most commonly used method for producing CNTs in an inert gas at ~100 torr. In this method, when an electric arc discharge is generated between two graphite electrodes under inert atmosphere of helium or argon, a very high temperature of about 4000 K is obtained which allows the sublimation of the carbon from the anode. As shown in Figure 6.1a, arc discharge reactor consists of a cylinder of about 30 cm in diameter and about 1.0 m in length. The sapphire windows are located opposite to plasma zone to observe the arc. The reactor chamber is first evacuated to a pressure of 0.1 Pa and then it is filled with an inert gas up to the desired working pressure. Under a desired inert gas pressure, when the anode is moved toward the cathode to maintain a distance of about 1 mm, a direct current of 50–120 A passes through the electrodes and creates plasma between them. The temperature of the plasma typically reaches about 4000 K and the carbon is vaporized from the anode and reorganized at the cathode, forming a cylindrical deposit. This method can easily produce straight and near-perfect MWNTs. The diameter of the cathode is usually larger than that of anode and both electrodes are water-cooled. The anode must be continuously moved to ensure a constant distance (~1 mm) between the electrodes. The diameter of produced MWNTs ranges from 4 to 30 nm and length up to 1 μm. Each MWNT comprises 2–50 coaxial tubes of graphene sheets separated by a distance of 0.34 nm. On each of the tubes, the carbon-atom hexagons are arranged in a helical fashion about the MWNT axis. The tips of the tubes are usually closed by pentagons and heptagons of carbon atoms. Large-scale synthesis of MWNTs can be carried out at a potential of ~18 V DC between two electrodes in helium atmosphere at ~500 torr. The yield of nanotubes is ~75% relative to the starting graphitic material. This is a common method to produce MWNT without a catalyst. The important process parameters of this method are gaps between electrodes (>1 mm), high current (50–120 A), plasma between the electrodes, and voltage range (30–35 V) under the specified electrode dimensions [4,5].

In contrast to MWNTs, the synthesis of SWNTs requires metallic powder as catalyst. In 1993, the SWNTs were synthesized by using arc discharge chamber filled with a gas mixture of methane and argon in the presence of an ultrafine metal powder such as Ni, Fe, Co, Cr, or rare earth elements in the anode. When a DC current is applied between the electrodes, the arc discharge is generated which results in SWNTs bundles. These bundles consist of SWNTs with diameters varying between 0.7 and 1.65 nm and an average diameter of ~1 nm. The diameters can be controlled by varying the gas composition and magnetic field. Their diameters can be reduced to <1 nm by applying a magnetic field of 0.2–2.0 kG. For example, a diameter of <1 nm was obtained at 2 kG magnetic field [6]. In this method, the yield of SWNTs is quite low and it can be increased by using more than one or mixture of Ni, Co, Y, Pt, etc. The quality and quantity of SWNTs are affected by the catalyst concentration, inert gas pressure, nature of gas, the current, and system geometry.

The structure and properties of CNTs are highly sensitive to the production method and synthesis parameters such as temperature, reactor size, gas flow and pressure, precursors, etc. For example, arc-discharge-grown MWNTs are highly crystalline and straight, and have few defects whereas CVD-grown MWNTs are larger in length and diameters, highly defective, and are not straight. The presence of defects affects the electronic structure, mechanical properties, and thermal properties of CNTs. Due to the defects, Young's modulus of CVD-grown MWNTs is orders of magnitude less than Young's modulus

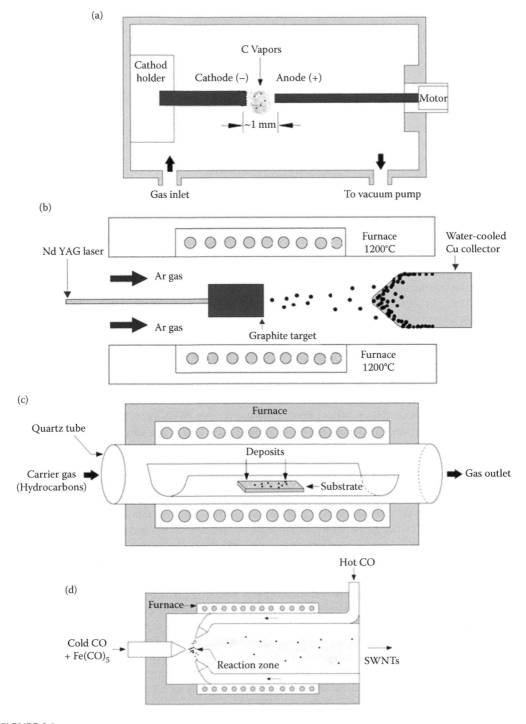

FIGURE 6.1
Schematic drawing of reactors used for CNTs synthesis: (a) arc discharge, (c) laser ablation, (b) chemical vapor deposition, and (d) high-pressure carbon monoxide.

of arc-discharge-grown MWNTs [7]. The arc discharge technique can produce a large quantity of nanotubes, but this method has relatively little control over the alignment or chirality of the CNTs. Additionally, it needs purification of the CNTs from the metallic catalysts [8].

6.2.2 Laser Ablation

Guo et al. [9] were the first to synthesize CNTs by the laser ablation method in 1995, and the apparatus used by them is shown in Figure 6.1b. In 1996, Smalley and coworkers [10] reported synthesis of SWNTs by this method using graphite rods with small amounts of Ni and Co. This process is known to produce CNTs with the highest quality and high purity of SWNTs. In this process, a graphite pellet containing the catalyst is put in the middle of an inert gas-filled quartz tube placed in an oven. The oven temperature is maintained at about 1200°C. This method involves an intense laser beam (pulsed or continuous laser) to vaporize a target consisting of a mixture of graphite and metal catalyst under an inert gas atmosphere (helium or argon) at a pressure of ~500 torr (or 66.66 kPa). A very hot vapor plume forms, then expands, and cools rapidly. As the vaporized species cool, small carbon molecules and atoms quickly condense to form larger clusters. The catalysts vapor also vaporizes and attaches to carbon clusters and prevents their closing into cage structures. These clusters finally produce SWNTs until the catalyst particles become too large, or the temperature is too low for carbon's diffusion through or over the catalyst's surface. A constant flow of inert gas (Ar or He) is passed through the tube in order to transfer the soot generated to a water-cooled copper collector, the quartz tube walls, and the backside of the pellet. The metal atom must have a sufficiently high electronegativity to avoid the formation of fullerenes [10]. Both MWNTs and SWNTs can be synthesized using this process. However, this method favors the growth of SWNTs with controlled diameter depending on reaction temperature [11]. The produced CNTs are uniform in diameters but they are in the form of ropes of 5–20 nm diameter and tens to hundreds of micrometers long. However, this method involves evaporation of carbon atoms from solid target at temperatures exceeding 3000°C, and thus, the CNTs are entangled. The pulsed laser demands a much higher light intensity (100 kW/cm^2) compared to continuous laser (12 kW/cm^2). The scaling-up of both arc discharge and laser ablation reactor is a difficult task [10]. Both laser ablation and arc discharge techniques operate at similar conditions and use the condensation of carbon atoms generated from the vaporization of graphite targets. The SWNTs usually condense as ropes or bundles consisting of several individual SWNTs. The CNTs also consist of amorphous carbon and/or encapsulated metal catalyst particles. The laser ablation technique favors the growth of SWNTs. The quality, length, diameter, and chirality distribution of the SWNTs produced by laser ablation are believed to be comparable with those of arc discharge SWNTs [9]. The arc discharge and laser ablation methods have the advantage of high yields (i.e., >70%) and annealed CNTs. The laser ablation produces CNTs with relatively low-metallic impurities, but the obtained CNTs are not necessarily uniformly straight. This method is costly because of involvement of high-purity graphite rods and the costly laser powers. In addition, the quantity of CNTs synthesized per day is lower than that of arc discharge technique [8].

6.2.3 Chemical Vapor Decomposition

The thermal CVD method is the most commonly used method which consists of a quartz tube, a substrate, and a tubular furnace. Figure 6.1c shows a schematic setup for CVD reactor. A substrate with metal catalysts such as Ni, Fe, Co, etc. are placed in the quartz tube and heated to temperatures of 500–1000°C. Hydrocarbons such as methane, ethane, ethylene, acetylene are slowly fed in the quartz tube. At an appropriate temperature and pressure, the hydrocarbon molecules decompose into carbon atoms. The metal (or catalyst) nanoparticles assist the hydrocarbon gas to decompose and the resultant carbon atoms dissolve in the metal nanoparticles and precipitate out when the carbon–metal solution becomes supersaturated. The precipitation of carbon from the metal nanoparticles leads to the formation of MWNTs or SWNTs, which have tubular structure. These CNTs consist of carbon with sp^2 structure on the sidewall and a hemispherical end cap. The SWNTs are found to be produced at higher temperatures with a well-dispersed and -supported metal catalyst. In contrast, the MWNTs are formed at lower temperatures. The CVDs generate large quantities of CNTs. Nowadays, CVD technique is most successfully commercialized method. The controlling parameters for the growth of CNTs are pressure of the vaporized carbon source, the purity and flow rate of the carrier gas, the residence time for thermal decomposition, and the temperature of the furnace. This method has drawback of high-defect density in the CNTs owing to low synthesis temperatures compared to arc discharge and laser ablation. Hence, the tensile strength of the CVD-grown CNTs is only one-tenth of those synthesized by arc discharge. The by-products of this method are usually aromatic carbon, amorphous carbon, polyhedral carbon, metal particles, etc.

In the presence of transition metal catalysts such as Fe, Ni, Co, etc., CVD generates MWNTs and SWNTs. MWNTs are mainly produced at lower temperatures (300–800°C) in an inert gas atmosphere, whereas SWNTs require higher temperatures (600–1150°C) and a mixture of H_2 and an inert gas. CVD and plasma-enhanced CVD (PECVD) or plasma-assisted CVD (PACVD) are commonly used to grow aligned MWNTs [9]. Compared to laser ablation method, CVD is a cheaper process for producing relatively pure CNTs in large scale with better control of the reaction conditions [8].

HiPCO method is the advanced technique of CVD method. Figure 6.1d shows a schematic diagram of the HiPCO reactor. This reactor consists of a quartz tube surrounded by electrical heated elements inside a thick-walled aluminum cylinder. Carbon monoxide (CO) and $Fe(CO)_5$ flow through the quartz tube. The space between the quartz tube and the aluminum cylinder's inner wall is filled with an argon gas atmosphere. The pressure of argon gas is maintained slightly higher than the pressure of the CO inside the quartz tube. Mixtures of $Fe(CO)_5$ and CO, which are source of catalyst particles (i.e., Fe) and carbon, respectively, are injected into the reactor. At a pressure of ∼30–50 atm and temperature of ∼900–1100°C, the $Fe(CO)_5$ decomposes and the iron (Fe) atoms condense into clusters. These clusters serve as catalytic particles upon which SWNTs nucleate and grow in the gas phase via CO disproportionation (i.e., a redox reaction which transforms a molecule into two or more dissimilar products).

$$CO + CO \rightarrow CO_2 + C_{(SWNTs)}$$

The yield and quality of CNTs depend on the temperature, pressure, and catalyst concentration. This method can produce a ∼10 g/day of high purity (up to 97 mol%) SWNTs. The showerhead CO preheater heats the cold flow emerging from the injector nozzle. As shown in Figure 6.1d, CO for the showerhead is flowed through these channels and heated by contact with graphite rods. The heated CO passes from the channels into the showerhead orifice circle and into the mixing zone. There it collides and mixes with the cold injector flow of CO and Fe(CO)$_5$ emerging from the copper nozzle. The temperature of the heated CO passing through the showerhead is controlled by varying the current passed through the heating rods. Inside the inner bore of the graphite heater, the CO/Fe(CO)$_5$ mixture is rapidly heated and mixed with the showerhead CO. Fe(CO)$_5$ decomposes and forms Fe clusters, which help in nucleating and growing of SWNTs. The SWNTs and Fe particles are collected into the product collection apparatus. The maximum production of SWNTs is obtained at 1050°C. This method produces CNTs at rates up to 10.8 g/day. This method has drawback of termination of CNT growth due to accumulation of Fe atoms onto growing CNTs [12]. The Fe content in pristine SWNTs has been reduced from 26% Fe (w/w) to 0.4% (w/w) for the functionalized SWNTs [13]. The diameters of SWNTs produced by HiPCO process are 0.79–1.2 nm and increase to about 1.47 nm after heat treatment at 1800°C. The Fe content can be decreased from about 30% in the SWNTs to about 2% after the heat treatment at 1800°C [14]. The CNTs synthesized by any of the above method may consist of one or more impurities including graphite, amorphous carbon, fullerenes, coal, and metal nanoparticles. The type of impurities present in CNTs is analyzed by a suitable characterization technique before purification.

6.3 Characterization Techniques to Analyze the Purity

Scanning electron microscopy (SEM) is used to determine the nature of the CNTs and obtain a rough idea about material quality. High-resolution transmission electron microscopy (HRTEM) is mostly used to monitor the surface texture of individual ropes, qualitative determination of impurities, defects in CNTs, and to distinguish between the SWNTs and MWNTs. However, electron microscope cannot provide a quantitative evaluation of the purity of CNTs. Raman spectroscopy is a fast, convenient, and non-destructive analysis technique to estimate the extent of amorphous carbon and damage to the tube surfaces. The characteristic peaks of MWNTs appear due to disordered carbon (D-band), and out-of-phase graphene sheet-like vibrations (G-band). These peaks occur at ~1290 and 1600/cm, respectively. These peaks may also be used to qualitatively describe the purity of the MWNTs. An increase in the ratio of the area of the primary "G" peak to that of the "D" peak (G:D ratio) is correlated to increasing graphite crystallite size and a decrease in amorphous carbon in graphitic materials. However, it cannot provide direct information on the nature of metal impurities. TGA is useful to find out metal impurities in CNTs simply by burning CNT sample in air. Oxidation temperature of the sample in TGA serves as a measure of thermal stability of CNTs in air. The presence of defects in CNT walls generally lowers the thermal stability. Active metal particles present in the CNT samples catalyze carbon oxidation. A higher oxidation temperature (>500°C) is always associated with purer and less defective CNT samples. It is impossible to distinguish type of metal catalysts and carbonaceous impurities. UV–Vis–NIR spectrometer is used to get absorption spectra and optical density with time of nanotube solutions. The optical density versus time serves as a measure of dispersibility of CNTs. Atomic-force microscopy measures the bundle heights of nanotubes deposited onto suitable substrate. X-ray photoelectron spectroscopy (XPS) is often used to characterize the functional groups on the walls of CNTs, and energy-dispersive spectroscopy (EDS) is used to semiquantitatively identify the catalyst particles in CNTs [11].

6.4 Purification of CNTs

As discussed above, the synthesized CNTs contain carbonaceous impurities and metal catalyst particles. The amount of the impurities commonly increases with the decrease of CNT diameter. Typical carbonaceous impurities are amorphous carbon, fullerenes, graphite pieces, and carbon nanoparticles (CNPs). These impurities come from the vaporization of graphite rods in arc discharge and laser ablation. The graphitic pieces may come from the breaking of the graphite rods during evaporation. Moreover, graphitic polyhedrons with enclosed metal particles also coexist with CNTs synthesized by arc discharge, laser ablation, and the high-temperature CVD. Metal impurities usually remain from the transition metal catalysts and sometimes they are encapsulated by carbon layers; hence, their dissolution in acid during purification becomes difficult. Moreover, the presence of broad particle size distributions of these impurities and different defects or curvature makes it difficult to develop a single purification method to obtain high-purity CNT materials [11]. Depending on technique of CNTs synthesis, there are several methods for purification, which include removal of large graphite particles and aggregates with filtration, dissolution of metal catalysts in appropriate solvents (concentrated acids) and dissolution of fullerenes in organic solvents, and microfiltrations and chromatography to size separation [8].

Purification methods of CNTs can be mainly classified into three categories: chemical, physical, and a combination of both. The chemical method purifies CNTs based on the idea of selective oxidation, wherein carbonaceous impurities are oxidized at a faster rate than CNTs, and the dissolution of metallic impurities by acids. This method can effectively remove amorphous carbon and metal particles. The physical method separates CNTs from impurities based on the differences in their physical size, aspect ratio, gravity, and magnetic properties, etc. In general, the physical method is used to remove the graphitic sheets, CNPs, aggregates, or separate CNTs with different diameter/length ratios. Since this method does not require oxidation, severe damage of CNTs can be prevented. However, the physical method is complicated, time-consuming, and less effective method. A combination of chemical method and physical method of purification (or multistep purification) combines the merits of both purification methods. This method can lead to high yield and high-quality CNT products. Owing to the mixture of as-prepared CNTs, CNT entanglements, CNT morphology and structure, metal catalysts, amorphous carbon, graphite, etc., the multistep purification method needs a skillful selection of various purification techniques to obtain high-purity CNTs.

6.4.1 Chemical Oxidation Method

Compared to CNTs, the carbonaceous impurities usually have higher oxidation reactivity due to the presence of more dangling bonds and structural defects. In addition, the high reactivity of the CNPs is due to their large curvature and pentagonal carbon rings. This method involves oxidation of impurities by gases using air, O_2, Cl_2, H_2O, etc. and liquids using acids and refluxing. The carbonaceous impurities are oxidized at a faster rate than CNTs. This method often opens the end of CNTs and reduces their aspect ratio, damages surface structure, and introduces oxygenated functional groups such as –OH, –C=O, and –COOH on the surfaces of CNTs. In gas-phase oxidative purification, CNTs are purified by oxidizing carbonaceous impurities at a temperature ranging from 225 to 760°C. High-temperature oxidation in air is a simple method to purify the CNTs free from metal and have fewer defects on tube walls. The purification yield can be increased by uniform exposure of the as-synthesized CNTs and by enhancing the difference in oxidation resistance to air between CNTs and carbonaceous particles. However, air oxidation of SWNTs consumes a large fraction of SWNTs due to the larger amount of curvature by the graphene sheet of SWNTs. The presence of metal impurities catalyzes the low-temperature oxidation of carbon; hence, it is better to remove metal particles before gas-phase oxidation of SWNTs. Gas-phase oxidation is a simple method to remove the carbonaceous impurities and open the caps of CNTs. This method cannot directly remove the metal catalyst and large graphite particles. The gas-phase purification method is better to purify CNTs free from metal catalyst. The liquid-phase purification method can remove both metal catalyst and amorphous carbon by using HNO_3, H_2O_2, or a mixture of H_2O_2 and HCl, a mixture of H_2SO_4, HNO_3, $KMnO_4$ and NaOH, and $KMnO_4$. In this, oxidative ions and acid ions dissolved in a solution attack the network of raw CNTs and dissolve metal catalysts. The resultant solution is filtered and washed to obtain a high-purity CNTs with a high yield. However, this process causes by-products on the surface of CNTs, adds functional groups, and decreases aspect ratio of CNTs. HNO_3 is the most commonly used inexpensive and nontoxic reagent for CNT purification.

Chen et al. treated CNTs with a diluted HNO_3 solution at 90°C for 3 h to dissolve iron or its oxide particles. The treatment of MWNTs with a concentrated HNO_3 oxidizes amorphous carbon layers and particles present on the surface of MWNTs. The resultant suspension was filtered and washed with deionized water. Further, the washed suspension was filtrated using 0.20 μm copolymer membrane to obtain a layer of pure MWNTs. This purification treatment reduced the Fe content significantly. Fullerene can be easily removed owing to its solubility in certain organic solvents including toluene. The liquid-phase oxidation method often provides surface modification to CNTs, which increases the chemical activity and the solubility of CNTs in most organic and aqueous solvents. The resultant modified CNTs act as efficient reinforcement for the composites. This method has drawbacks of damage to CNTs, inability to remove the large graphite particles, and the loss of a large amount of SWNTs with small diameter. This method results in shortened and defective tubes. It is very difficult to obtain purified SWNTs with high-purity and high yield without damage [15].

CNTs can be extracted by filtering a stable colloidal dispersion of the CNTs in a water/surfactant solution, while leaving the nanoparticles in the filtrate. This provides CNTs without any damage to tube tips or tube walls [16]. Two-step method, that is, acid washing followed by hydrogen treatment (700–1000°C), removes both metallic as well as amorphous carbon, thus producing high-quality CNTs. The acid washing dissolves the metal particles, while the hydrogen treatment removes amorphous carbon as well as the carbon coating on the metal nanoparticles. Hydrogen treatment converts carbon into CH_4 whereas in other routine method, air oxidizes the amorphous carbon into CO_2 [17]. The oxidation of arc-discharge-grown MWNTs reduces the yield by 99% and the acid treatment or surfactants may damage or functionalize the nanotube surface, which may also alter their thermal, electrical, and mechanical properties [9].

6.4.2 Physical Method

Different morphology, aspect ratio, size, solubility, gravity, and magnetism of CNTs from those of impurities enable one to separate CNTs from impurities by using physical techniques such as filtration, chromatography, centrifugation, electrophoresis, and high-temperature annealing. Separation by filtration is based on the differences in physical size, aspect ratio, and solubility of CNTs, CNSs, metal particles, and fullerenes. The physical method is a nondestructive and nonoxidizing treatment. This is time-consuming process.

It involves suspension of as-received CNTs in a solvent containing surfactants followed by sonication and then filtration. Small-size particles or soluble impurities in solution are filtered out, and CNTs with large aspect ratio remain on filter paper. The purification by microfiltration gives CNTs with purity of >90% by weight [18].

Centrifugation is based on the effect of gravity on particles in suspension. In this, two particles of different masses settle in a tube at different rates in response to gravity. The centrifugation can separate amorphous carbon and CNPs based on the different stabilities in dispersions consisting of amorphous carbon, CNTs, and CNPs in aqueous media. The requirement of this process is that CNTs need to be first treated with nitric acid, which introduces functional groups on their surface [18]. Microwave-assisted purification is the most promising technology for large-scale purification. This gives significant efficiency improvements over traditional acid reflux with minimal CNTs damage under optimized conditions. This reduces the long-processing times or avoids multiple stages [19].

High-temperature annealing is one of the most efficient methods for the removal of metal particles from the CNTs. It also removes structural defects from the CNTs. In this, the raw CNTs sample is heated at high temperature (>1400°C) under inert atmosphere or high vacuum. At such high temperatures, the graphite or CNT is stable and the structural defects are eliminated. Moreover, the metal catalyst evaporates above their evaporation point, thus making the resultant CNTs almost free from metal. This method increases the mechanical strength and thermal stability of the CNTs. However, this cannot remove the carbonaceous impurities, which is more difficult to remove after graphitization [11]. Ultrahigh purity MWNTs are obtained by annealing the raw CVD-grown MWNTs under a vacuum at temperatures between 1500°C and 2150°C. The high-temperature vacuum process efficiently removes the residual metal catalysts and metal oxide carriers and enhances the graphitization of the MWNTs. This results in MWNTs with purity about 99.9%. This method is a nondestructive and commercially viable purification method for CNTs. Figure 6.2a and b shows TEM image of the MWNTs before and after purification treatment, respectively. The

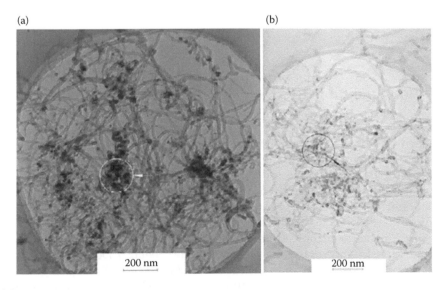

(a) (b)

200 nm 200 nm

FIGURE 6.2
Typical HRTEM of (a) as-grown MWNTs and (b) MWNTs annealed at 2000°C for 5 h. (Reprinted from *Carbon*, 41, Huang W. et al., 99.9% purity multi-walled carbon nanotubes by vacuum high-temperature annealing, 2585–2590, Copyright 2003, with permission from Elsevier.)

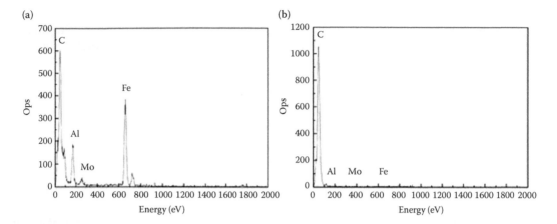

FIGURE 6.3
Typical EDS of (a) as-grown MWNTs and (b) MWNTs annealed at 2000°C for 5 h. (Reprinted from *Carbon*, 41, Huang W. et al., 99.9% purity multi-walled carbon nanotubes by vacuum high-temperature annealing, 2585–2590, Copyright 2003, with permission from Elsevier.)

typical impurity in the raw sample of MWNTs is catalyst and amorphous carbon which is distributed throughout the sample. Figure 6.2a clearly shows metal catalyst particles at the nanotube tips and in the tube cores. The black areas are probably due to aggregates of metal catalysts and Al_2O_3 carrier. The EDS spectra of a marked area (in a white circle) are shown in Figure 6.3a indicating the presence of impurities including Fe, Mo, Al, and O in the MWNTs. Figure 6.3b shows the TEM image of the sample annealed at 2000°C for 5 h indicating the disappearance of the impurities [20]. Separating CNTs with different chirality is more complex. The electrically conducting nanotubes can be separated from semiconducting by the application of current between the metal electrodes, which causes conducting nanotubes to burn, leaving the semiconducting ones intact.

6.5 Structures and Properties of CNTs

6.5.1 Structures

CNTs are formed when graphene sheets (or hexagonal lattice) are coiled into seamless hollow cylinders. Figure 6.4a shows the unrolled hexagonal lattice of an SWNTs with the chiral vector (*C*), the translational vector (*T*), the unit vectors (a_1 and a_2), and the chiral angle (θ). The chiral vector is called the circumferential vector, and it is always perpendicular to the translational vector. The circumference of an SWNTs is determined by the chiral vector. The circumference vector or chiral vector is defined by $(C) = na_1 + ma_2$, where *n* and *m* are a pair of integers (or indices), and a_1 and a_2 are two unit vectors. The chiral vector defines the relative location of two sites in a graphene sheet [21]. Depending on the orientation of the graphene sheet relative to the tube axis (*T*), the SWNTs may be obtained in the form of armchair, chiral, and zigzag nanotubes. The SWNTs is said to be armchair and zigzag when the translational vector (or tube axis) is perpendicular and parallel, respectively, to the C–C bonds, located on opposite sides of each hexagon in the honeycomb lattice. The rest of the SWNTs, where the translational vector lies at an angle with respect to

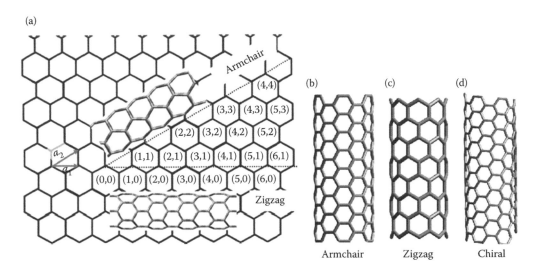

FIGURE 6.4
(a) The graphene sheet of an SWNTs, with the different (*m, n*) values and possible chiral vectors. The three different configurations of SWNTs are: (b) armchair, (c) zigzag, and (d) chiral. (From Dervishi E. et al. 2009. *Particul. Sci. Technol.* 27: 107–125. Reproduced with permission from Taylor & Francis Group.)

the C–C bonds, are known as chiral SWNTs. Figure 6.4b–d shows all three chiralities of SWNTs. Thus, depending upon the pair of integers and chiral angle (θ), the SWNTs may be armchair, zigzag, and chiral tubes. The armchair SWNTs are formed when $m = n$ and the chiral angle is 30°, whereas zigzag nanotubes are formed when either n or m is 0 and the chiral angle is 0°. All other tubes wrapped with a finite angle θ with $0° < \theta < 30°$ are said to be chiral or helical. Therefore, depending on the values of the indices, m and n, SWNTs are metallic when $(2n + m) = 3p$, where p is an integer, and otherwise are semiconducting. Since all armchair CNTs satisfy this equation, they are metallic, while the CNTs with other configurations (chiral or zigzag) can be either metallic or semiconducting. Depending on the chirality and diameter, one-thirds of the SWNTs are metallic and two-thirds are semiconducting [21].

When the graphite sheet is rolled into a cylinder, the π orbital becomes more delocalized outside the CNT, resulting in the σ bonds being slightly out of plane. Due to this σ–π rehybridization, the nanotubes become mechanically stronger and electrically and thermally more conductive than graphite. Compared to graphite sheets, the conductivity of CNTs increases as the delocalization of the π orbital increases. Furthermore, the σ–π rehybridization leads to change in the electronic properties of CNTs, that is, the electronic states of the CNTs split into 1-D subbands, instead of a single wide electronic energy band. If a gap exists between the valence band and the conduction band, the material behaves as a semiconductor; otherwise, it behaves as a metal. CNTs possess higher electrical conductivity than copper due to their low resistance and very few defects. The electrical resistivity of CNTs is as low as $10^{-6}\ \Omega \cdot m$. This value can be altered by modifying the structure of the nanotube lattice. The Fermi level of CNTs is very sensitive to the type of dopant that comes into contact with the nanotubes. There are mainly two types of doping methods for CNTs: interstitial doping and substitutional doping. In interstitial doping, the CNT lattice remains the same and the dopants are adsorbed at the surface. In substitutional doping, the dopant atoms replace the carbon atoms and form sp^2 bonding in the CNT structure. The doping of MWNTs or SWNTs with boron (B) or nitrogen (N) increases the electrical conductivity of the CNTs by an order of magnitude compared to pure CNTs. The B and N dopants make the CNTs either p- or n-types, respectively, through substitutional doping. It is the intercalation of different metals in between the tubes and chemical doping that significantly enhances the electrical conductivity of CNTs [21].

Depending on the number of layers, the inner diameter of MWNTs varies from 0.4 nm to few nanometers and outer diameter varies from 2 nm to 20–30 nm. Both tips of MWNT are usually closed and the ends are capped by half-fullerene molecules (or pentagonal defects). On the other hand, SWNTs diameters vary from 0.4 to 2–3 nm, and their length is typically of the micrometer range. SWNTs are usually found in bundles or ropes, where SWNTs are hexagonally arranged to form a crystal-like construction. The chirality of the SWNTs affects electrical properties of nanotubes. Due to sp^2 bonds between the individual carbon atoms, CNTs have a higher tensile strength than steel and Kevlar. This bond is even stronger than the sp^3 bond in diamond [8].

Raman spectroscopy is one of the most powerful tools for characterization of CNTs. All allotropic forms of carbon are active in Raman spectroscopy. This is a fast and nondestructive technique which does not need sample preparation. Raman spectroscopy provides unique information about vibrational and electronic properties of the CNTs. The most important Raman spectrum includes mainly radial breathing modes (RBM) at 100–250/cm, the higher frequency D (disordered) at 1300–1360/cm, G (graphite) at 1500–1700/cm, and G' (second-order Raman scattering from D band variation) modes at 2500–2900/cm [22]. The D band originates from a hybridized vibrational mode associated with graphene edges and it indicates the presence of some disorder to the graphene structure. This band is often referred to as the disorder band or the defect band. Its intensity relative to that of the G band is often used as a measure of the quality of CNTs. The Raman spectra for CNTs are generally obtained in the range 1000–2000/cm. Typical Raman spectra of MWNTs show a G band at 1588/cm and a D band at 1357/cm. The G band is wide and its intensity is lower than twice that of the D band. The width of the G band gives size distribution of the CNTs whereas the intensity of the D band decreases with the degree of graphitization of the tubes (Figure 6.5) [23]. The RBM bands are unique to SWNTs and correspond to the

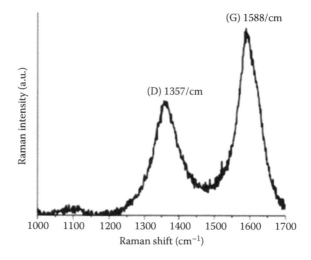

FIGURE 6.5
Raman spectrum of CVD–MWNT. (Reprinted from *Compos. Sci. Technol.*, 62, Allaoui A. et al., Mechanical and electrical properties of a MWNT/epoxy composite, 1993–1998. Copyright 2002, with permission from Elsevier.)

expansion and contraction of the tubes. The frequency of these bands can be correlated to the diameter of SWNTs. In contrast to SWNTs, the MWNTs show lack of RBM modes and a much more prominent D band. In MWNTs, the RBM modes are not present because the outer tubes restrict the breathing mode. The more prominent D band in MWNTs indicates more disorder in the structure. The RBM mode completely disappears for double walled carbon nanotubes (DWNTs) and MWNTs. However, the D bands and G' bands get proportionately larger as the number of layers adds to the walls of the nanotubes. In other words, the ratio of the intensities of D and G bands is a good indicator of the quality of CNT samples. Similar intensities of these bands indicate a high quantity of structural defects. Raman spectrum of MWNTs shows the lowest ratio, consequently higher quantity of structural defects due to its multiple graphite layers. However, SWNT's structure is associated with smaller structural distortions and therefore the intensity of D band in their Raman spectra is typically weak compared to that of G band intensity.

6.5.2 Electrical Properties of CNTs

In addition to the conduction and the valence band of the graphene which touch on the six corners (K points) of the Brillouin zone, SWNTs also have 1-D subbands. If one of these subbands passes through one of the K points, the nanotube will be metallic, otherwise it will be semiconductor with a gap inversely proportional to the diameter. For a tube with a diameter of 3 nm, the gap falls in the range of the thermal energy at room temperature. The electronic structure of SWNTs depends both on the orientation of the honeycomb lattice with respect to the tube axis and the radius of curvature imposed to the bent graphene sheet. In MWNTs, the conduction occurs essentially through the external tube (for c-MWNTs), but the interactions with the internal coaxial tubes may make the electronic properties vary. The presence of several coaxial tubes applies stresses to the tube, which change its electronic behavior as a consequence of the variation brought to the position of the subbands with respect to the K points of the Brillouin zone [24]. As discussed in previous section, depending on indices one-thirds of the SWNTs are metallic and two-thirds are semiconducting. Armchair SWNTs are always metallic. In contrast, zigzag SWNTs are semiconducting with a band gap (E_g) in the range of 0.2–0.9 eV. The band gap of SWNTs is inversely proportional ($E_g = {\sim}0.7$ eV/d [nm]) to their diameter [25]. Bockrath et al. [26] investigated electrical transport through bundles of SWNTs bridging contacts separated by 200–500 nm. The I–V characteristic exhibits strong suppression of the conductance near $V = 0$ for $T < 10$ K. The linear-response conductance (G) of the bundle as a function of the gate voltage V_g consists of a series of peaks separated by regions of very low conductance. CNTs can produce electric current-carrying capacity 1000 times higher than copper wires [27,28].

In bundles of metallic SWNTs, van der Waals interaction between individual nanotubes can induce a pseudo-gap of ${\sim}0.1$ eV, which mimics semiconducting behavior. For semiconducting SWNTs, in the absence of impurities or defects (or doping), the Fermi energy (E_F) is 0. However, a *real* nanotube, due to the presence of adsorbates from the ambient, shows the E_F either <0 or >0 for the p-type or n-type nanotube, respectively. Due to high contact/Ohmic resistances, it is difficult to measure the low resistance of CNTs. The smallest possible resistance of SWNTs is ${\sim}6.5$ kΩ. The SWNTs outperform Cu interconnect for the lengths larger than 1 μm and the current-carrying capacity is 10^9 A/cm^2, which is higher than that of Cu (i.e., 10^6 A/cm^2). In addition, CNTs do not have problem of electromigration. Thus, CNTs are useful in high power applications particularly in *long* interconnect due to their high-current-carrying capabilities [29].

The resistivity of CNT buckypaper measured, in a magnetic field at 7 T, along parallel to the alignment direction is lower than that of perpendicular direction throughout the temperature range of 1.5–295 K. At 295 K, the resistance measured along parallel direction is 0.91 m$\Omega \cdot$ cm. The resistivity of pure nanotube fibers can be further decreased by stretching the nanofibers [30]. The maximum current density of the metallic SWNTs is 10^9 A/cm^2, which is about four orders of magnitude higher than the value of 10^5 A/cm^2 for normal metals [31]. Individual MWNT exhibits resistivities (at 300 K) of about 1.2×10^{-4} to 5.1×10^{-6} $\Omega \cdot$ cm, depending upon the defects, chirality, diameter variations, and degree of crystallinity or hexagonal lattice perfection. Bundles of SWNTs have shown resistivities in the range of 0.34×10^{-4} and 1.0×10^{-4} $\Omega \cdot$ cm [32].

6.5.3 Mechanical Properties

The CNTs have high stiffness and axial strength due to the carbon–carbon sp^2 bonding [33]. The average Young's modulus of isolated SWNTs is 1.25 ± 0.40 TPa [34], significantly higher than that of carbon fibers (i.e., 680 GPa) [35]. The estimated Young's modulus of MWNTs [36] is 1.26 TPa and nanotube buckle elastically at large deflection angles of ~10° for length of 1 μm. The average bending strength is 14.2 ± 8.0 GPa. The fracture strength of MWNTs is between 11 and 63 GPa [37]. The theoretical average Young's and shear moduli of SWNTs ropes are about 1.0 and 0.5 TPa, respectively [38].

As discussed above, CNTs have an elastic modulus of >1 TPa which is comparable to that of a diamond (i.e., 1.2 TPa) and strengths 10–100 times higher than the strongest steel [39]. The tensile strength is in the range from 11 to 63 GPa depending on the outer shell diameter. The nanotubes are very flexible. It can be elongated, twisted, flattened, or bent into circles before fracturing. The nanotubes have breaking strain of up to 5.3%. The stress–strain curves of SWNTs rope show that the load is carried primarily by SWNTs on the periphery of the ropes, which results in breaking strengths from 13 to 52 GPa [40]. This is far less than that reported for a single MWNT [41]. The elastic modulus, Poisson's ratio, and bulk modulus are found to be dependent on the tube radius. The mechanical properties of the CNTs have been shown in Table 6.1, which shows that properties also depend on the technique of measurement.

The modulus as well as strength is highly dependent on the nanotube growth method and subsequent processing which may introduce uncontrolled defects. For example, the Young's modulus values of 3–4 GPa were obtained in MWNT produced by pyrolysis of organic precursors [44]. MWNTs and SWNTs bundles may be stiffer in bending but are expected to be weaker in tension due to pullout of individual tubes. The specific modulus and specific strength are ~19 and 56 times better than the steel, respectively. The untwisted yarns are very weak but twisted single-strand yarns exhibited strengths in the range 150–300 MPa and increased further to 250–460 MPa in the two-ply yarns. The nanotube films or buckypapers have shown useful applications in sensors and actuators. However, the tensile modulus, strength, and elongation-to-break values for a solution cast film with random SWNTs orientations are 8 GPa (at 0.2% strain), 30 MPa, and 0.5%, respectively. These values are much lower than those of fibers. This is due to the failure via interfibrillar slippage in the films rather than fracture within a fibril [45]. Isolated SWNTs have a modulus as high as 1 TPa; however, the modulus (measured by AFM) of SWNTs decreases dramatically as the size of the SWNTs bundle increases (Figure 6.6). The low modulus of larger ropes at low strains indicates that slippage occurs between the individual nanotubes within the rope. Individual strands can slip until they are tightly packed, thus there is poor load transfer between the SWNTs within the rope and they are needed to separate SWNTs from the ropes before use in nanocomposites [46].

TABLE 6.1

Mechanical Properties of CNTs

Material	Tensile Strength (GPa)	Young's Modulus (TPa)	Measurement Technique	Ref.
SWNTs	–	2.8–3.6	Micro-Raman spectroscopy	[42]
SWNTs rope	45 ± 7	0.32–1.47	Micro-Raman spectroscopy	[42]
SWNTs rope	13–52 (breaking strength)	–	Tensile-loading experiment (fracture strain <5.3%)	[30]
Buckypaper of SWNTs	0.030	0.008 (at 0.2% strain)	Tensile-loading experiment (elongation to break 0.5%)	[30]
SWNTs	–	1.2	Atomic force microscopy	[42]
SWNTs	–	0.9–1.7	TEM/vibrational theory	[42]
MWNT	–	1.7–2.4	Micro-Raman spectroscopy	[42]
MWNT	–	0.27–0.95	Atomic force microscopy	[42]
MWNT rope	1.72 ± 0.64	–	–	[43]
Twisted single-strand yarn of MWNT	0.15–0.30	–	–	[30]
Twisted two-ply yarns of MWNT	0.25–0.46	–	–	[30]

6.5.4 Thermal Properties

In materials, the thermal conductivity is dominated by both phonons and electrons. In graphite, it is generally dominated by the phonons which in turn dominate the specific heat above ~20 K. In CNTs, the phonon contribution dominates at all temperatures, while the electronic contribution is only ~1% of the total. At low temperatures, specific heat and thermal conductivity depend linearly with temperature from 1 K to room temperature.

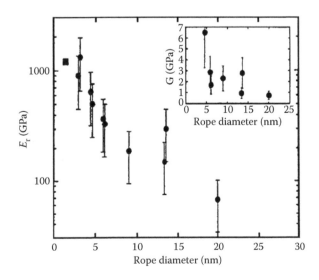

FIGURE 6.6

A plot of modulus versus SWNTs rope diameter showing the decrease in the modulus of SWNTs ropes as a function of rope diameter. (Adapted from Ajayan P. M., Schadler L. S., Braun P. V. (Eds): *Nanocomposite Science and Technology*. 2003. Copyright Wiley-VCH Verlag GmbH & Co. KGaA. Reproduced with permission.)

In a metallic SWNTs, the phonon specific heat is about 100 times that of electronic specific heat. The specific heat of a bulk sample of highly purified SWNTs increases from 0.5 to 100 mJ/g · K over the temperature range from 2 to 100 K. The specific heat of SWNTs increases linearly with temperature from 2 to 8 K and upturn above 8 K, due to 1-D quantized nature [47]. The specific heat of MWNTs [48] increases linearly with temperature from 10 to 300 K. At higher temperatures, the specific heat will have a mixed 2-D–3-D behavior due to the weak intertube interactions in both MWNTs and bundles of SWNTs.

At low temperatures, Umklapp scattering disappears and inelastic phonon scattering is generally due to sample boundaries or defects. Thus, at low temperature ($T <$ Debye temperature), the temperature-dependent phonon thermal conductivity is similar to that of the specific heat. Individual tubes into ropes or MWNTs may introduce intertube scattering, thus change in the thermal conductivity with increasing temperatures is observed. Figure 6.7a shows the thermal conductivity of CVD-grown MWNTs as a function of temperature from 4 to 300 K. At low temperature ($T < 100$ K), the thermal conductivity increases as $\sim T^2$. In contrast to graphite, the room temperature thermal conductivity of MWNTs is comparatively small and the MWNTs do not show a maximum in thermal conductivity versus temperature curve due to Umklapp scattering. However, the thermal conductivity of single MWNT increases as T^2 up to ~ 100 K, peaks near 300 K, and decreases above 300 K. The room temperature thermal conductivity value is over 3000 W/mK [49]. The thermal conductivity of an individual MWNTs measured by 3-ω method is 650 and 830 W/mK for the MWNTs with diameters of 46 and 42 nm, respectively [50].

Figure 6.7b shows the thermal conductivity of a bulk-aligned SWNTs. It can be seen that the thermal conductivity of bulk-aligned SWNTs increases with increasing temperature to 300 K. At high temperature, there is a decreasing slope indicating the onset of Umklapp scattering. For an SWNT mat, which has disordered tangled SWNTs, the room temperature thermal conductivity is ~ 35 W/mK. In contrast, the thermal conductivity of the well-aligned SWNTs is >200 W/mK, which is smaller by an order of magnitude than that of highly crystalline graphite. A single tube or a rope of continuous tubes may have

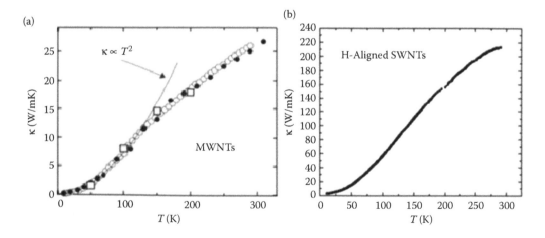

FIGURE 6.7
Temperature-dependent thermal conductivity of (a) MWNTs (From Yi W. et al., 1999. *Phys. Rev. B Rapid Comm.* 59: R9015. Reproduced with permission from American Physical Society) and (b) bulk sample of aligned SWNTs (Reprinted with permission from Hone J. et al., 2000. Electrical and thermal transport properties of magnetically aligned single wall carbon nanotube films. *Appl. Phys. Lett.* 77: 666–668. Copyright 1999 by the American Institute of Physics).

significantly higher thermal conductivity than the mats or ropes due to the absence of lower rope–rope junctions, respectively. Similarly, sintered sample of CNTs shows room temperature thermal conductivity value of 2.3 W/mK, which is much lower than the values of pure metals or high-quality graphite at room temperature. This discrepancy might be attributed to several reasons including: additional phonon modes in graphite, which are lacking in an isolated nanotube, different scattering processes in graphite and tubes, and quantization of the transverse component of the phonon wave vector due to the periodic boundary conditions imposed by the cylindrical geometry of CNTs [51].

The room temperature thermal conductivity of an individual MWNT is >3000 W/mK which is two orders of magnitude higher than the values determined for mat samples (~35 W/mK). At low temperatures ($8 < T < 50$ K), it increases following a power law with an exponent 2.5. In the intermediate temperature range ($50 < T < 150$), it increases quadratically in T. Above this temperature range, it exhibits a peak at 320 K. As the diameter of the MWNT increases, the thermal conductivity versus temperature (T) becomes similar to the bulk. The $T^{2.5}$ dependence is due to the interlayer phonon modes giving rise to slight 3-D nature. The T^2 dependence is due to the higher frequency phonon modes of the separate layers. As T increases, the strong phonon–phonon Umklapp processes become more effective and the Umklapp path decreases. Once the static and Umklapp paths become equal, a peak in thermal conductivity versus temperature appears. The simulated thermal conductivity [52] of about 6600 W/mK has been reported for an isolated (10, 10) nanotube at room temperature, which is comparable to the thermal conductivity of graphite. These high values are associated with the large phonon mean free paths in these systems. The experimental thermal conductivity of SWNTs bundle is in the range of 30–35 W/mK, which is two orders of magnitude lower than the theoretical value. This lower value results from the boundary phonon scattering between individual nanotubes in the bundle and the uncertainty of the thermal resistance in the tube–tube junctions and the filling factor of CNTs in the bundles [50]. In contrast, the experimental thermal conductivity of the pure graphite is ~6000 W/mK at the peak and 2000 W/mK at room temperature [51].

6.5.5 Thermoelectric Power

The thermoelectric power (TEP) can provide experimental confirmation of band structure calculations and the type of dominant current carrier in SWNTs, which varies linearly with the T. TEP of a semiconductor has inverse relationship ($1/T$). Therefore, the low-temperature measurements can confirm whether there is an opening of an energy gap at the Fermi level or not [53]. At room temperature, the TEP is positive and of moderate amplitude of about 50 μV/K, while at low temperatures it is linear with temperature and approaches zero as $T \to 0$. The large positive TEP at high temperatures is usually attributed to hole-like carriers.

For MWNT bundles, the TEP is positive over the temperature range of 4.2–300 K [54], that is, the TEP has a value of about 22 μV/K at 300 K and increases with the increase of temperature. A two-band model indicates that the TEP is due to both metallic band carriers and semiconducting band carriers. Below 30 K, the TEP deviates from the linear temperature dependence due to the changes in the transport mechanism. The TEP of SWNTs-bundles prepared by using catalysts (Fe, Co, Ni) is also found positive over the temperature range of 10–400 K indicating the dominant contribution of hole carriers [55]. The TEP exhibits linear behavior at low temperatures and a peak in the temperature range of 80–100 K. The presence of such a peak has been attributed to magnetic impurities in the bundles indicating that the interaction of the Fe, Co, and Ni nanoparticles may play a role of enhancing the thermoelectric efficiency.

6.6 Applications of CNTs

Due to huge aspect ratio (>1000), and unique thermal, mechanical and electrical properties, the CNTs have been suggested for a wide variety of applications including nanocomposites, chemical sensors, field emission materials, electronic devices, nanotweezers, batteries, supercapacitors, and hydrogen storage. The CVD-grown large MWNTs have been commercialized for the applications in lithium-ion batteries and in plastics for electrostatic discharge applications. However, SWNTs have not been commercialized yet but its potential application in the field emission display has been discussed [56].

6.6.1 Nanocomposites

For structural or load-bearing applications, the addition of CNTs to the polymer matrices can increase mechanical properties like stiffness, strength, and toughness significantly. The mechanical properties of CNT-reinforced polymer composites strongly depend on the extent of load transfer between the matrix and CNTs. The addition of CNTs into polymers increases significantly the modulus of elasticity, strength, Vickers hardness, and breaking strength of the resultant nanocomposites [57]. The ultimate strength and shear modulus increases from 30% to 70% with the addition of 1–4 wt% SWNTs. The strain to failure also increases indicating an increase in toughness [58]. The addition of 1 wt% amino-functionalized CVD-grown MWNTs with an aspect ratio of >1000 into the epoxy matrix doubles the flexural strength, while the flexural modulus changes slightly. This is due to the fact that the CNTs are embedded and tightly held to the matrix, which results in broken MWNTs instead of being pulled out during flexural test. This also indicates the existence of strong interfacial bonding between amino-functionalized MWNTs and the epoxy matrix [59]. The addition of CNTs also improves skid resistance and reduces abrasion of the tire [60].

The addition of ∼1 wt% MWNT to epoxy resin enhances stiffness and fracture toughness by 6% and 23%, respectively, without compromising other mechanical properties and thus making them useful for the load-bearing applications. These enhancements depend on CNT diameter, aspect ratio, alignment, dispersion, and interfacial interaction with the matrix. Moreover, CNTs addition into polymer increases the damping property, which is good for the sporting goods including tennis racquets, baseball bats, and bicycle frames. It can also be added to polymer–carbon fiber composites to improve further the mechanical properties of wind turbine blades and hulls for maritime security boats. Moreover, the addition of CNTs into 1 μm diameter carbon fibers enhances ∼35% the strength (4.5 GPa) and stiffness (463 GPa) compared to the fiber without CNTs. Furthermore, the addition of small amounts of CNTs to polymers or metals provides increase in tensile strength and modulus, and decrease in coefficient of thermal expansion (CTE), thus making the nanocomposites useful for the applications in aerospace and automotive structures. For example, Al–MWNT nanocomposites show strengths comparable to stainless steel (0.7 to 1 GPa) at a one-thirds of the density of the steel. The addition of MWNTs, as electrically conductive fillers, into the polymers matrices results in a percolation network at loading as low as 0.01 wt%, and the highest electrical conductivity for a disordered MWNT-filled polymer nanocomposites can reach as high as 100 S/cm at 10 wt% loading. Such CNT–polymer nanocomposites can be useful in the automotive industry for the electrostatic-assisted painting of mirror housings, fuel lines, and filters. Moreover, these nanocomposites are useful to dissipate electrostatic charge and to use in electromagnetic interference (EMI) shielding packages and wafer carriers for the microelectronics industry. As a replacement for haloge-nated flame retardants, MWNTs can be used as a flame retardant additive to polymers [61].

6.6.2 Hydrogen Storage

Nanotubes have been reported as potential candidates useful for hydrogen storage in fuel cells that can power electric vehicles, laptop, computers, etc. As the specific surface area of CNTs increases, the hydrogen storage increases and then gets saturated above a spe-cific surface area of 850 m^2/g. The maximum hydrogen storage capacity of SWNTs and DWNTs at room temperatures and 10 MPa is 0.5 wt%. Purification of the CNTs samples by oxidative acid treatment or high-temperature annealing decreases the hydrogen stor-age capacity. The presence of metal impurity or open CNTs has little or no influence on the amount of hydrogen storage values. Probably, the functionalization of CNTs blocks potential adsorption [62]. Hydrogen can become an ideal energy carrier and truly useful for the end user particularly in high-energy density rechargeable batteries and fuel cells, if its storage capacity is increased. For these applications, metal hydridation has limitations of high density and high cost. In addition, liquid nitrogen temperature and high pressure are required to hold the physically adsorbed H_2. SWNTs soots could absorb about 5–10 wt% of H_2 at 133 K and 300 torr while high-purity SWNTs store 8.25 wt% of H_2 at 80 K and 100 atm. All these systems of H_2 uptake need high pressure and/or subambient tempera-ture. Lithium- or potassium-doped CNTs (MWNT with diameter ~25–35 nm and specific surface area ~130 m^2/g) can absorb ~20 or 14 wt% of hydrogen at moderate (200–400°C) or room temperatures, respectively, under ambient pressure. These values are greater than those of metal hydride and graphite systems. For example, H_2 uptake of Li- and K-doped CNTs is ~40% and 180%, respectively, greater than those of graphite powder (specific sur-face area = ~8.6 m^2/g). The high uptake of alkali-doped CNTs is due to the hollow cylin-drical tube of CNTs with much more open edge and greater interplanar distance (0.347 nm for CNT vs. 0.335 nm for graphite). FTIR shows gradual appearance and broadening in C–H band with increasing H_2 uptake, while a decrease and finally vanish of C–H band upon desorption of H_2. Ultraviolet photoelectron spectroscopy (UPS) shows that the Li doping creates an extra half-filled electron density of state containing the Fermi edge and reduces the energy barrier for H_2 dissociation, and thus it results in high H_2 uptake from the Li-doped CNTs. Compared to Li-doped CNT, K-doped CNT absorbs H_2 at lower tem-perature. However, the Li-doped CNT materials are thermally and chemically more stable than those of K-doped CNT materials [63]. Due to cylindrical, hollow geometry, and nano-meter-sized diameters of CNTs, the hydrogen is stored in the inner cores through a capil-lary effect and the hydrogen is also absorbed by physisorption (through the van der Waal forces) on CNTs surfaces. The hydrogen does not form any chemical bonds with the CNTs.

6.6.3 Energy Storage Devices

CNTs have been studied for the applications in lithium-ion batteries and electrochemical devices. MWNTs are widely used in lithium-ion batteries for notebook computers and mobile phones. In lithium-ion batteries, small amounts of MWNT powder are blended with active materials and a polymer binder, such as 1 wt% CNT loading in $LiCoO_2$ cath-odes and graphite anodes. CNTs provide increased electrical connectivity and mechani-cal integrity, which enhances rate capability and cycle life. The supercapacitor made up of binder and additive-free forest-grown SWNTs has an energy density of 16 Wh/kg and a power density of 10 kW/kg. However, high cost of SWNTs is a major challenge [61].

Due to the high-surface area, high-electrical conductivity, and excellent mechanical properties, CNTs are attractive as electrodes for electrochemical devices. The separation between the charge on the electrode and the countercharge in the electrolyte is about a nanometer (vs. micrometer for ordinary capacitors) for nanotubes in electrodes, which results in very large capacitances, typically between 15 and 200 F/g, depending on the surface area of the nanotube array. Thus, large amounts of charge can be injected by applying only a few volts. This charge injection is used for energy storage in nanotube supercapacitors and to provide the electrode expansions and contractions that can do mechanical work in electromechanical actuators. Nanotube electromechanical actuators function at a few volts, compared to the ~100 V used for piezoelectric stacks and the ~1000 V used for electrostrictive actuators. Nanotube actuators have been operated at temperatures up to 350°C. The maximum developed isometric actuator stress of SWNTs actuators is 26 MPa. This is ~10 times the stress initially reported for these actuators and ~100 times that of the stress generation capability of natural muscle. It approaches the stress generation capability of high-modulus commercial ferroelectrics (~40 MPa) [64].

6.6.4 Field Emission Emitters

Field emission is a phenomenon of emitting electrons at a very high rate from the ends of the tube by applying a small electric field parallel to the axis of a CNT. It involves extracting of electrons from a conducting solid using an electric field. Both SWNTs and MWNTs have been proved as field emitters. The vertically aligned CNTs have the largest potential. The most important application of field emission effect is in the development of flat panel displays [21]. Nanometer-sized radius of curvature of CNTs makes them useful for the field emitter arrays in displays. As an electron emitter, CNTs provide a large enhancement factor (β) in the field emission, at a lower threshold voltage. The use of CNTs in cold cathode devices allows for instantaneous turn-on, high power, and low-control voltage operation, along with long lifetimes and miniaturization. Compared to thermionic emission, CNT-induced field emission shows low-power consumption at room temperature. For a metal with a flat surface, the threshold electric field for field emission is as high as 10^4 V/μm, which is three orders of magnitude higher than that of CNTs (<10 V/μm). CNTs exhibit larger current of up to 1 μA from a single MWNT and current density of about 200 mA/cm^2. Vertically orientated CNTs have the best emission characteristics. In general, SWNTs films, due to a higher degree of structural perfection, has the highest β (i.e., >3000), while catalytically grown MWNTs have a β ~800. Compared to SWNTs, the low value of β for MWNTs has been attributed to their larger tip radius [29].

SWNTs and MWNTs can be used as field emission electron sources for flat panel displays, lamps, gas discharge tubes providing surge protection, and x-ray and microwave generators. A potential applied between a CNT-coated surface and an anode produces high-local fields, as a result of the small radius of the nanofiber tip and the length of the nanofiber. These local fields cause electrons to tunnel from the nanotube tip into the vacuum. Electric fields direct the field-emitted electrons toward the anode, where a phosphor produces light for the flat panel display application. Nanotube field-emitting surfaces are relatively easy to manufacture by screen printing nanotube pastes and do not deteriorate in moderate vacuum (10^{-8} torr). In contrast, tungsten and molybdenum tip arrays require a vacuum of 10^{-10} torr and are difficult to fabricate. CNTs provide stable emission, long lifetimes, and low-emission threshold potentials. They have current density as high as 4 A/cm^2, which are much higher than the value of 10 mA/cm^2 needed for flat panel field emission displays and the values >0.5 A/cm^2 required for microwave power amplifier tubes. Furthermore, compared to liquid crystal displays, CNTs have the advantages of low-power consumption, a wide viewing angle, high brightness, a fast response rate, and a wide operating temperature range. Recently, Samsung has demonstrated a 9 in. red–blue–green color display which can reproduce moving images [64–66].

In order to minimize the electron emission threshold field, it is desirable to have emitters with a low-work function and a large field enhancement factor. The field enhancement factor is inversely proportional to the radius of the emitter tip. CNTs have the unique combination of structural integrity, nanometer size diameter, high-electrical conductivity, and chemical stability, which make them good electron emitters. Hence, CNTs exhibit a lower threshold electric field compared to conventional emitters, as shown in Table 6.2. For a current density of 10 mA/cm^2, the threshold field for random SWNTs films is in the range of 2–3 V/μm, while for random and aligned MWNTs it is typically in the range of 3–5 V/μm. These threshold field values are significantly lower than those of conventional field emitters such as the Mo and Si tips (i.e., 50–100 V/μm). The aligned MWNT films do not perform better than the random films due to the electrical screening effect arising from closely packed nanotubes. The low threshold field for electron emission by the CNTs is due to large field enhancement factor. The reduced electron work function for SWNTs is about 4.8 eV. Compared to MWNTs, SWNTs generally have a higher degree of structural perfection and thus, SWNTs have higher current densities and a longer lifetime. The current densities of the CNTs are significantly higher than those of conventional emitters such as nanodiamonds (i.e., <30 mA/cm^2). CNTs emitters are also attractive for microwave amplifiers. However, the emission site density of CNTs is about 10^3–10^4/cm^2 at the turn-on field which is too low for high-resolution display applications. High-resolution display devices typically require emission site densities of about 10^6/cm^2 [65–68,74].

However, the electrical properties are affected significantly by the surrounding environment which influences the removal of heat from current-carrying nanotubes. For example, suspended nanotubes display drastically different electron transport from that of those on

TABLE 6.2

Threshold Electrical Field Values for Materials for a Current Density of 10 mA/cm^2

Material	Threshold Electrical Field (V/μm)	Field Enhancement Factor (Approx.)	Ref.
Flat metal surface	10×10^3	–	[67]
Mo or Si tips	50–100	–	[68]
Amorphous diamond	20–40		[68]
CVD-grown MWNTs on Inconel 718 substrate	5	2500	[69]
Catalytic MWNT films	4.8	830	[70]
SWNTs films	3.9	3400	[70]
Opened MWNT films	3.7	1100	[70]
CVD-grown MWNTs on Si substrate	3	2500	[69]
12.5 wt% boron-doped MWNT/PS films	2.6	–	[71]
Closed MWNT films	2.2	1600	[70]
CNTs printed on nonpatterned KOVAR substrate	2.02	1892	[71]
12.5 wt% MWNT/PS films	1.6	–	[71]
CNTs printed on a 740 nm (pitch size) line-patterned KOVAR substrate	0.58	8710	[72]
MWNT arrays grown on TiN substrate	~1.6	–	[73]

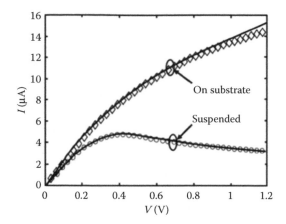

FIGURE 6.8
Current–voltage characteristics of CNTs on a substrate and suspended over a trench. (Reprinted with permission from Pop E., Mann D. 2005. Negative differential conductance and hot phonons in suspended nanotube molecular wires. *Phys. Rev. Lett.*, 95: 155–505. Copyright 2005 by the American Physical Society.)

substrates. Figure 6.8 illustrates the current–voltage characteristics of the SWNTs situated on the nitride-based substrate and suspended over a 0.5 μm deep trench. The measurements were conducted at room temperature in vacuum. One can see that the substrate-supported SWNTs displays a monotonic increase in the current, approaching 20 μA under increasing voltage, while the current in the suspended SWNTs reaches a peak of ~5 μA only followed by a noticeable current drop. This different behavior is due to significant self-heating effects of the SWNTs carrying ultrahigh current densities. In the case of the suspended tube, the Joule heating cannot be removed effectively because heat dissipation is possible only through nanotube contacts. As a result, the temperature of the tube increases, especially in its central region [65,66].

6.6.5 Microelectronics

Owing to low electron scattering and band gap, the SWNTs are compatible with field effect transistor (FET) architectures and high-k dielectrics. SWNTs-FETs with sub-10 nm channel lengths have shown a normalized current density (2.41 mA/μm at 0.5 V) greater than those obtained for silicon devices. Transistors based on horizontally aligned CNT arrays have shown mobility (~80 cm²/V/s) much higher than that of amorphous silicon (~1 cm² V/s). Additionally, these transistors can be deposited by low temperature and non-vacuum methods hence CNT thin-film transistors (TFTs) are attractive for driving organic light-emitting diode (OLED) displays. The vertical CNT-FETs have also shown sufficient current output to drive OLEDs at low voltage, enabling red-green-blue emission by the OLED through a transparent CNT network. In future, copper may be replaced by CNTs in microelectronic interconnects, owing to their low scattering, high current-carrying capacity, and resistance to electromigration [61].

6.6.6 Electrical Brush Contacts

MWNT-based electrical brush contacts have been suggested to be useful in electric motors and generators because of their super-compressibility, high-electrical and thermal conductivity, mechanical strength, tunable friction behavior, and environmental insensitivity. MWNT brush has much lower density (i.e., 0.3 g/cm^3) than the commercial carbon–copper composite brush (~2.84 g/cm^3). They are soft, durable, stable, and less noisy as compared to the commercial electric brushes used in ordinary electric motor [74].

6.6.7 Sensors and Probes

The nonmetallic nanotubes-based sensors are very sensitive to a small amount of species and hence, they respond. CNTs-based scanning probe tips for atomic probe microscopes are sold in the market by Seiko Instruments. The mechanical robustness and low-buckling force of CNTs (in probes) significantly increase the probe life and minimize the sample damage during repeated hard crashes into substrates. Moreover, the cylindrical shape and small tube diameter enable imaging in narrow, deep crevices, and improve resolution in comparison to conventional nanoprobes. The probe with modified CNTs enables the mapping of chemical and biological functions. Nanotweezers based on CNTs are driven by the electrostatic interaction between two nanotubes on a probe tip. They are good to manipulate the properties of materials on the nanometer scale [31]. SWNTs biosensors exhibit large changes in electrical impedance and optical properties in response to the surrounding species. This is due to adsorption of a target on the CNT surface. The efficiency of CNT sensors for low-detection limits and high selectivity depends upon the functional groups, diameter and length of CNTs, and design of the sensors. CNT sensors have been used for gas and toxin detection in the food industry, military, and environmental applications [61,67].

6.6.8 Artificial Implants

MWNTs or SWNTs can be used as implants in the form of artificial joints and other implants without host rejection response. High-tensile strength as well as compressive strength can make the CNTs or modified CNTs useful in bone substitutes and implants. Ultrahigh-surface area of CNTs makes them useful in tissue engineering as a substitute for damaged or diseased tissue and as promising potential for delivery of drugs, peptides, and nucleic acids. The specific drug or gene can be integrated to walls and tips of CNTs and recognize cancer-specific receptors on the cell surface, and carry therapeutic drugs or genes more safely and efficiently in the cells that are previously inaccessible [8].

6.6.9 Water Filters

Owing to the lack of fresh and clean water, water crisis has become a global problem around the world. Pollution makes the situation worse for water shortage problems due to various contaminants such as heavy metals, distillates, microcystins, and antibiotics. The conventional water treatment methods are simply unable to efficiently remove some of these pollutants from water. The mechanical flexibility and robustness, thermal stability and resistance to harsh environments enable CNTs potential adsorbents to remove both organic and inorganic contaminants from water systems. CNTs are also superior sorbents for dioxin removal. The stronger chemical–nanotube interactions, tailored surface chemistry, and high sorption capacity of CNTs have made CNTs sponge as superior sorbents for a wide range of organic chemicals and inorganic contaminants than the conventional activated carbons. The CNT sponge is a randomly intertwined three-dimensional (3-D) structure exhibiting very low density. It can float on oil-contaminated water and remove oil with a large adsorption capacity, that is, 80–180 times its own weight for a wide range of solvents and oils. The sponge has a tendency to move to the oil film area due to its high hydrophobicity, which leads to the unique "floating-and-cleaning" capability. In addition to serving as direct adsorbents, CNTs can also be utilized as excellent scaffolds for macromolecules or metal oxides with intrinsic adsorption ability.

6.6.10 Coatings

MWNTs are emerging as important ingredients and/or fillers for paints to reduce biofouling of ship hulls by discouraging the attachment of algae and barnacles. They can also be used as a possible alternative to environmentally hazardous biocide-containing paints. Moreover, the addition of CNTs in anticorrosion coatings for metals can enhance coating stiffness and strength [61].

References

1. Iijima S. 1991. Helical microtubules of graphitic carbon. *Nature* 354: 56.
2. Bethune D. S., Kiang C. H., de Vries M. S., Gorman G., Savoy R., Vazquez J., Beyers R. 1993. Cobalt-catalysed growth of carbon nanotubes with single-atomic-layer walls. *Nature* 363: 605.
3. Iijima S., Ichihashi T. 1993. Single-shell carbon nanotubes of 1-nm diameter. *Nature* 363: 603.
4. Ebbesen T. W., Ajayan P. M. 1992. Large-scale synthesis of carbon nanotubes. *Nature* 358: 220.
5. Paradise M., Goswami T. T. 2007. Carbon nanotubes—Production and industrial applications. *Mater. Des.* 28: 1477–1489.
6. Tessonnier J. P., Su D. S. 2011. Recent progress on the growth mechanism of carbon nanotubes: A review. *Chem. Sus. Chem.* 4: 824–847.
7. Grobert N. 2007. Carbon nanotubes—Becoming clean. *Mater. Today* 10: 28–35.
8. Eatemadi A., Daraee H., Karimkhanloo H., Kouhi M., Zarghami N., Akbarzadeh A., Abasi M., Hanifehpour Y., Joo S. W. 2014. Carbon nanotubes: Properties, synthesis, purification, and medical applications. *Nanoscale Res. Lett.* 9: 393.
9. Guo T. 1995. Catalytic growth of single-walled nanotubes by laser vaporization. *Chem. Phys. Lett.* 243: 49–54.
10. Thess A., Lee R., Nikolaev P., Dai H., Petit P., Robert J., Xu C., Lee Y. H., Kim S. G., Rinzler A. G., Colbert D. T., Scuseria G. E., Tomanék D., Fischer J. E., Smalley R. E. 1996. Crystalline ropes of metallic carbon nanotubes. *Science* 273: 483.
11. Hou P. X., Liu C., Cheng H. M. 2008. Purification of carbon nanotubes. *Carbon* 46: 2003–2025.
12. Bronikowski M. J., Willis P. A., Colbert D. T., Smith K. A., Smalley R. E. 2001. Gas-phase production of carbon single-walled nanotubes from carbon monoxide via the HiPco process: A parametric study. *J. Vac. Sci. Technol. A.* 19(4): 1800–1805.
13. Georgakilas V., Voulgaris D., Vázquez E., Prato M., Gudi D. M., Kukovecz A., Kuzmany H. 2002. Purification of HiPCO carbon nanotubes via organic functionalization. *J. Am. Chem. Soc.* 124: 14318–14319
14. Yudasaka M., Kataura H., Ichihashi T., Qin L. C., Kar S., Iijima S. 2001. Diameter enlargement of HiPCO single-wall carbon nanotubes by heat treatment. *Nano Lett.* 1(9): 487–489.
15. Chen H., Chen Y., Williams J. S. *Proceedings of the 1st Nanomaterials Conference*, Coombs S., Dicks A. (Eds.). ARC Centre for Functional Nanomaterials, Brisbane, Australia. pp. 7–9.
16. Bonard J. M., Stora T., Salvetat J. P., Maier F., Stöckli T., Duschl C., Forró L., de Heer W. A., Châtelain A. 1997. Purification and size-selection of carbon nanotubes. *Adv. Mater.* 9(10): 827.

17. Vivekchand S. R. C., Govindaraj A., Seikh M. M., Rao C. N. R. 2004. New method of purification of carbon nanotubes based on hydrogen treatment. *J. Phys. Chem. B*. 108: 6935–6937.

18. Bandow S., Rao A. M., Williams K. A., Thess A., Smalley R. E., Eklund P. C. 1997. Purification of single-wall carbon nanotubes by microfiltration. *J. Phys. Chem. B*. 101: 8839–8842.

19. MacKenzie K., Dunens O., Harris A. T. 2009. A review of carbon nanotube purification by microwave assisted acid digestion. *Sep. Purif. Technol.* 66: 209–222.

20. Huang W., Wang Y., Luo G., Wei F. 2003. 99.9% purity multi-walled carbon nanotubes by vacuum high-temperature annealing. *Carbon* 41: 2585–2590.

21. Dervishi E., Li Z., Xu Y., Saini V., Biris A. R., Lupu D., Biris A. S. 2009. Carbon nanotubes: Synthesis, properties, and applications. *Particul. Sci. Technol.* 27: 107–125.

22. Kar K. K. 2011. *Carbon Nanotubes: Synthesis, Characterization, and Applications*. Research Publishing, Singapore.

23. Allaoui A., Bai S., Cheng H. M., Bai J. B. 2002. Mechanical and electrical properties of a MWNT/epoxy composite. *Compos. Sci. Technol.* 62: 1993–1998.

24. Bhushan B. (Ed.). 2004. *Springer Handbook of Nanotechnology*. Spinger-Verlag, Berlin.

25. Javey A., Kong J. (Eds.). 2009. *Carbon Nanotube Electronics*. Springer, New York.

26. Bockrath M., Cobden D. H., McEuen P. L., Chopra N. G., Zettle A., Thess A., Smalley R. E. 1997. Single-electron transport in ropes of carbon nanotubes. *Science* 275: 1922.

27. Maiti A., Brabec C. J., Roland C. M., Bernholc J. 1995. Theory of carbon nanotube growth. *Phys. Rev. B*. 52: 14850.

28. Lee Y. H., Kim S. G., Jund P., Tomanek D. 1997. Catalytic growth of single-wall carbon nanotubes: An ab initio study. *Phys. Rev. Lett.* 78: 2393.

29. Bandaru P. R. 2007. Electrical properties and applications of carbon nanotube structures. *J. Nanosci. Nanotechnol.* 7: 1–29.

30. Gogotsi Y. (Ed.). 2006. *Nanomaterials Handbook*. CRC Press, USA.

31. Baughman R. H., Zakhidov A. A., de Heer W. A. 2002. Carbon nanotubes—The route toward applications. *Science* 297: 787–792.

32. Terrones M. 2003. Science and technology of the twenty-first century: Synthesis, properties, and applications of carbon nanotubes. *Annu. Rev. Mater. Res.* 33: 419–501.

33. Robertson D. H., Brenner D. W., Mintmire J. W. 1992. Energetics of nanoscale graphitic tubules. *Phys. Rev. B*. 45: 12592.

34. Krishnan A., Dujardin E., Ebbesen T. W., Yianilos P. N., Treacy M. M. J. 1998. Young's modulus of single-walled nanotubes. *Phys. Rev. B*. 58: 14013.

35. Jacobsen R. L., Tritt T. M., Guth J. R., Ehrlich A. C., Gillespie D. J. 1995. Mechanical properties of vapor-grown carbon fiber. *Carbon* 33: 1217.

36. Wong E. W., Sheehan P. E., Lieber C. M. 1997. Nanobeam mechanics: Elasticity, strength and toughness of nanorods and nanotubes. *Science* 277: 1971.

37. Yu M. F., Lourie O., Dyer M. J., Moloni K., Kelly T. F., Ruoff R. S. 2000. Strength and breaking mechanism of multi-walled carbon nanotubes under tensile load. *Science* 287: 637.

38. Lu J. P. 1997. Elastic properties of carbon nanotubes and nanoropes. *Phys. Rev. Lett.* 79: 1297.

39. Charlier J. C., Blasé X., de Vita A., Car R. 1999. Electronic structure at carbon nanotube tips. *Appl. Phys. A*. 68: 267.

40. Yu M. F., Files B. S., Arepalli S., Ruoff R. S. 2000. Tensile loading of ropes of single-wall carbon nanotubes and their mechanical properties. *Phys. Rev. Lett.* 84: 5552.

41. Wang Z. L., Gao R. P., Poncharal P., De Heer W. A., Dai Z. R., Pan Z. W. 2001. Mechanical and electrostatic properties of carbon nanotubes and nanowires. *Mater. Sci. Eng. C*. 16: 3–10.

42. Wernik J. M., Meguid S. A. 2010. Recent developments in multifunctional nanocomposites using carbon nanotubes. *Appl. Mech. Rev.* 63: 050801.

43. Ruoff R. S., Qian D., Liu W. K. 2003. Mechanical properties of carbon nanotubes: Theoretical predictions and experimental measurements. *C. R. Phys.* 4: 993–1008.

44. Li F., Cheng H. M., Bai S., Su G., Dresselhaus M. S. 2000. Tensile strength of single-walled carbon nanotubes directly measured from their macroscopic ropes. *Appl. Phys. Lett.* 20: 3161–3163.

45. Gogotsi Y. (Ed.). 2006. *Nanomaterials Handbook*. CRC Press, Taylor & Francis Group, Boca Raton, FL, USA.

46. Ajayan P. M., Schadler L. S., Braun P. V. (Eds.). 2003. *Nanocomposite Science and Technology*. Wiley-VCH Verlag GmbH Co. KGaA, Weinheim.

47. Hone J. 2004. Carbon nanotubes: Thermal properties. In *Dekker Encyclopedia of Nanoscience and Nanotechnology*. J. A. Schwarz, C. I. Contescu, K. Putyera (Eds.) Marcel Dekker, New York, pp. 603–610.

48. Yi W., Lu L., Zhang D. L., Pan Z. W., Xie S. S. 1999. Linear specific heat of carbon nanotubes. *Phys. Rev. B.* 59: 9015.

49. Kim P., Shi L., Majumdar A., McEuen P. L. 2001. Thermal transport measurements of individual multiwalled nanotubes. *Phys. Rev. Lett.* 8721: 215502.

50. Choi T. Y., Poulikakos D., Tharian J., Sennhauser U. 2005. Measurement of thermal conductivity of individual multiwalled carbon nanotubes by the 3-ω method. *Appl. Phys. Lett.* 87: 013108.

51. Hone J., Whitney M., Piskoti C., Zettl A. 1999. Thermal conductivity of single-walled carbon nanotubes. *Phys. Rev. B.* 59(4): R2514.

52. Berber S., Kwon Y. K., Tománek D. 2000. Unusually high thermal conductivity of carbon nanotubes. *Phys. Rev. Lett.* 84: 4613.

53. Hone J., Ellwood I., Muno M., Mizal A., Cohen M. L., Zettl A., Rinzler A. G., Smalley R. E. 1998. Thermoelectric power of single-walled carbon nanotubes. *Phys. Rev. Lett.* 80: 1042.

54. Tian M., Li F., Chen L., Mao Z. 1998. Thermoelectric power behavior in carbon nanotubule bundles from 4.2 to 300K. *Phys. Rev. B.* 58: 1166.

55. Grigorian L., Sumanasekera G. U., Loper A. L., Fang S. L., Allen J. L., Eklund P. C. 1999. Giant thermopower in carbon nanotubes: A one-dimensional Kondo system. *Phys. Rev. B.* 60: R11309.

56. Jorio A., Dresselhaus G., Dresselhaus M. S. (Eds.) 2008. Carbon nanotubes. *Topics Appl. Phys.* 111: 13–62.

57. Hone J., Llaguno M. C., Nemes N. M., Johnson A. T., Fischer J. E., Walters D. A., Casavant M. J., Schmidt J., Smalley R. E. 2000. Electrical and thermal transport properties of magnetically aligned single wall carbon nanotube films. *Appl. Phys. Lett.* 77: 666–668.

58. Kociak M., Henrard L., Stéphan O., Suenaga K., Colliex C. 2000. Plasmons in layered nanospheres and nanotubes investigated by spatially resolved electron energy-loss spectroscopy. *Phys. Rev. B.* 61: 13963.

59. Shen J., Huang W., Wu L., Hu Y., Ye M. 2007. Thermo-physical properties of epoxy nanocomposites reinforced with amino-functionalized multi-walled carbon nanotubes. *Compos. Part A: Appl. Sci. Manuf.* 38(5): 1331–1336.

60. Langer L., Bayot V., Grivei E., Issi J. P., Heremans J. P., Olk C. H., Stockman L., Van Haesendonck C., Bruynseraede Y. 1996. Quantum transport in a multi-walled carbon nanotube. *Phys. Rev. Lett.* 76: 479.

61. De Volder M. F. L., Tawfick S. H., Baughman R. H., Hart A. J. 2013. Carbon nanotubes: Present and future commercial applications. *Science* 339: 535–539.

62. Bacsa R., Laurent C., Morishima R., Suzuki H., Le Lay M. 2004. Hydrogen storage in high surface area carbon nanotubes produced by catalytic chemical vapor deposition. *J. Phys. Chem. B.* 108: 12718–12723.

63. Chen P., Wu X., Lin J., Tan K. L. 1999. High H_2 uptake by alkali-doped carbon nanotubes under ambient pressure and moderate temperatures. *Science* 285: 91–93.

64. Baughman R. H., Zakhidov A. A., de Heer W. A. 2002. Carbon nanotubes—The route toward applications. *Science* 297: 787–792.

65. Mitin V. V., Kochelap V. A., Stroscio M. A. 2008. *Introduction to Nanoelectronics: Science, Nanotechnology, Engineering, and Applications*. Cambridge University Press, India.

66. Pop E., Mann D. 2005. Negative differential conductance and hot phonons in suspended nanotube molecular wires. *Phys. Rev. Lett.* 95: 155–505.

67. Bandaru P. R. 2007. Electrical properties and applications of carbon nanotube structures. *J. Nanosci. Nanotechnol.* 7: 1–29.

68. Dresselhaus M. S., Dresselhaus G., Avouris P. (Eds.). 2001. Carbon nanotubes. *Topics Appl. Phys.* 80: 391–425.
69. Sridhar S., Ge L., Tiwary C. S., Hart A. C., Ozden S., Kalaga K., Lei S., Sridhar S. V., Sinha R. K., Harsh H., Kordas K., Ajayan P. M., Vajtai R. 2014. Enhanced field emission properties from CNT arrays synthesized on inconel superalloy. *ACS Appl. Mater. Interfaces* 6: 1986–1991.
70. Bonard J. M., Salvetat J. P., Stöckli T., Forfo L., Châtelain A. 1999. Field emission from carbon nanotubes: Perspectives for applications and clues to the emission mechanism. *Appl. Phys. A.* 69: 245–254.
71. Poa C. H., Silva S. R. P., Watts P. C. P., Hsu W. K., Kroto H. W., Walton D. R. M. 2002. Field emission from nonaligned carbon nanotubes embedded in a polystyrene matrix. *Appl. Phys. Lett.* 80: 3189.
72. Kim S. J., Park S. A., Kim Y. C., Ju B. K. 2017. Enhanced field emission properties from carbon nanotube emitters on the nanopatterned substrate. *J. Vac. Sci. Technol. B.* 35: 011802.
73. Rao A. M., Jacques D., Haddon R. C., Zhu W., Bower C., Jin S. 2000. In situ-grown carbon nanotube array with excellent field emission characteristics. *Appl. Phys. Lett.* 76: 3813.
74. Toth G., Mäklin J., Halonen N., Palosaari J., Juuti J., Jantunen H., Kordas K., Sawyer W. G., Vajtai R., Ajayan P. M. 2009. Carbon-nanotube-based electrical brush contacts. *Adv. Mater.* 21: 1–5.

7

Synthesis, Properties, and Applications of Nanowires

This chapter provides an overview of recent research on synthesis, properties, and applications of inorganic nanowires (NWs), particularly metallic and semiconducting NWs. NWs are one-dimensional (1-D) anisotropic structures which are small in diameter (10–100 nm) and large in length (several micrometers), and hence, a very high-surface surface-to-volume ratio. In other words, NWs are 1-D structural material having two quantum-confined directions and one unconfined direction for electrical conduction. This facilitates them to be used in applications where electrical conduction, rather than tunneling transport, is required. These NWs exhibit unique electrical, electronic, optical, thermoelectric, magnetic, *and* chemical properties, which are different from that of their bulk counterpart. The physical properties of NWs are affected by the morphology, diameter, carrier density of states, etc. In the last two decades, significant progress has been made for the synthesis of metallic and semiconductor NWs of controlled size and composition. The synthesis methods for NWs can be divided broadly into two categories: vapor-phase techniques and liquid-phase techniques. The most commonly used vapor-phase technique is the vapor–liquid–solid (VLS) technique which will be discussed in detail in this chapter. Other vapor-phase methods are chemical vapor deposition (CVD) using catalyst metals, chemical vapor transport, reactive vapor transport, laser ablation, carbothermal reduction, chemical beam epitaxy (CBE), thermal evaporation and thermal decomposition, and plasma- and current-induced methods. Liquid-phase methods are sol-gel synthesis, hydrothermal, and electrodeposition, which may or may not utilize templates for producing NWs. Liquid-phase techniques can also be divided into template-based and template-free approaches. Metallic NWs such as bismuth (Bi), copper (Cu), silver (Ag), gold (Au), nickel (Ni), platinum (Pt), iron (Fe), indium, cobalt, palladium, and zinc have been grown successfully using the template method.

7.1 VLS Method

The VLS method was originally developed by Wagner and Ellis to produce micrometer-sized whiskers during the 1960s. A typical VLS process starts with the dissolution of gaseous reactants into nanosized liquid droplets of a catalyst metal followed by nucleation and growth of single crystalline wires. The VLS process has been widely used for synthesizing semiconductor NWs. In this process, Au, Cu, Ni, and tin are used as typical metal catalysts. For example, silicon (Si) NWs are grown by the absorption of a gaseous source material (e.g., silane) into an Au liquid droplet. After supersaturation of the liquid alloy, a nucleation event generates a solid precipitate of the source material. This seed serves as a preferred site for further deposition of material at the interface of the liquid droplet, promoting the elongation of the seed into a NW or a whisker. When the gas flow of a source material is maintained, the source material diffuses through the molten Au–Si droplet and grows epitaxially at the liquid–solid interface. The NWs thus obtained are of high purity, except for the end containing the solidified catalyst as an alloy particle. The diameter of the NWs can be controlled by controlling the nanosized catalyst droplets. Laser-assisted catalytic VLS growth method generates NWs under nonequilibrium conditions.

The VLS growth method consists of three main stages: metal alloying, crystal nucleation (or precipitation), and crystal growth. In the metal alloying stage, nanosized metallic (or catalyst) particles are formed on a substrate by laser ablation or by annealing a very thin metallic film above the eutectic temperature. The size of catalyst particles is around 10–20 nm. Then the source of desired material, that is, silane (SiH_4 or $SiCl_4$) in the case of silicon NW growth, is introduced into a tube furnace maintained above the eutectic temperature. As shown in Figure 7.1a, Au–Si alloy has a eutectic temperature and composition of 363°C and 3 wt% (i.e., ∼18.6 at%) Si. The only phases present are the liquid, the Si

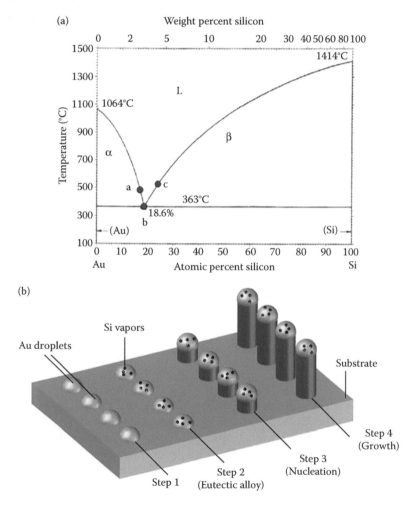

FIGURE 7.1
(a) The phase diagram of an Au–Si binary system. (Reprinted from *J. Mater. Process. Tech.*, 206, Lin J. S. et al., Fabrication and characterization of eutectic gold-silicon (Au-Si) nanowires, 425–430. Copyright 2008, with permission from Elsevier.) (b) A schematic diagram illustrating VLS mechanism for growth of Si nanowires.

and Au solid phases with low mutual solid solubility. A schematic diagram is shown in Figure 7.1b, which explains all the steps of the synthesis of Si NWs. When, the Si-gas source decomposes on the surface of the Au (i.e., catalyst), and Si diffuses through the catalyst droplets, an Au–Si alloy is formed. Once the supersaturation is reached, Si precipitates at the liquid–solid interface. On continuously feeding Si vapors, growth of NWs takes place and the liquid alloy drop remains on the NWs as it grows in length. The pressure and temperature of the tube furnace control the catalyst size and its state. The initial size of the catalyst and, to some extent, the growth conditions affect the diameter of the NWs. For the growth of Si NWs, Au, aluminum (Al), Ag, and Fe can be used as catalyst. The carrier gas reacts with catalyst particles in the furnace and forms liquid eutectic droplets. The eutectic temperature for Au–Si, Al–Si, Ag–Si, and Fe–Si alloys is 363°C, 577°C, 837°C, and >1200°C, respectively. Higher eutectic temperatures are more prone to oxidation of the catalyst particle. In contrast to Au–Si alloy, Al–Si alloy has a eutectic temperature of 577°C and a eutectic composition of 12.6 wt% Si. Moreover, compared to Au, Al is highly reactive in the ambient air and hence, it is easily oxidized to form aluminum oxide (ca. 2–5 nm thick). This oxide layer prevents the diffusion of Si vapor through the eutectic alloy. Hence, there is a challenge for using Al as catalyst for the growth of Si NWs by VLS method. Si NWs with uniform diameters can also be synthesized using gallium (Ga) as catalyst below 400°C in hydrogen plasma [1].

Similar to growth of Si NWs, the growth of germanium NWs can be obtained from Au catalysts using Ge as a source of vapor. In this case, Au–Ge eutectic alloy (i.e., ~28 at % Si) is formed when the temperature is higher than the eutectic point (363°C). With increasing amount of Ge vapor, the liquid alloy becomes supersaturated with Ge and its nucleation (or precipitation) in the form of NW occurs at the solid–liquid interface. Further supplying of the Ge vapor pushes the interface forward or backward to form NWs. The diameter of the NWs grown by the VLS method is determined by the diameter of the catalyst particles. The growth temperature can be set in between the eutectic point and the melting point of the material. Physical methods such as laser ablation or thermal evaporation as well as CVD are used to generate the reactant species in vapor form. Major advantage of this route is that patterned NWs can be obtained by using patterned deposition of catalyst particles. This is the easiest method. The NWs grown by the VLS method consist of a crystalline core coated by an amorphous oxide layer of around 2–3 nm. This oxide layer may be removed by hydrofluoric (HF) acid. This method has been used successfully to obtain the NWs of elements, oxides, carbides, phosphides, etc. with good control over the diameter and diameter distribution [2].

It is to be noted that Au droplets have been reported as catalyst in several books as well as in this chapter. However, the VLS growth is not really a catalytic process. The Au (or seed metal) droplet just receives the source material and once the supersaturation is reached, the excess material precipitates out of the droplet in the form of a NW. Thus, the seed metal acts as a soft template to collect the material and facilitate NW precipitation and its growth along its axial direction. In chemistry, a catalyst is the material which increases the rate of a chemical reaction while remaining intact in the process. However, practically, the activation energy for the Si NW growth using Au seeds and Si thin film growth in microelectronics industry (without seeds) is almost similar, that is, ~130 kJ/mol. Similarly, the activation energy for the growth of GaAs NWs was found in the range of 67–75 kJ/mol using both MOCVD and thin-film epitaxy method indicating that Au does not aid in increasing the reaction rate. In spite of this evidence, we have used the commonly used term "catalyst" throughout this chapter [3].

7.2 Template Method

Template method is a convenient and versatile method for generating 1-D nanostructures by using templates. The templates are made of anodic or anodized alumina membranes (AAMs), diblock copolymer, glass, silica, and polymers such as polycarbonate, polyester, and polyethylene terephthalate (PET). Polymer and alumina membranes are prepared by track etching and anodization, respectively. The typical pore sizes of the membranes range from 20 to 200 nm, thickness of few micrometers and pore density from 10^4 to 10^9 pores/ cm^2. Plastic membranes are preferred over mica and glass because they are easy to dissolve. In this method, a template or a membrane is used to deposit the material of interest inside the hollow 1-D pores, as shown in Figure 7.2. The produced NWs are released from the templates by removing the host matrix using etchants. They have been used to fabricate NWs of inorganic materials such as Au, Ag, Pt, TiO_2, MnO_2, ZnO, SnO_2, In_2O_3, CdS, CdSe, CdTe, electronically conducting polymers such as polypyrole, poly(3-methylthiophene), and polyaniline. In contrast to VLS method, it is difficult to obtain single-crystalline NWs by using template method [2]. In this section, electrochemical deposition, electroless deposition, and pressure injection technique methods will be discussed.

7.2.1 Electrochemical Deposition

This method has been widely used to prepare a variety of metal nanowires such as Au, Ag, Cu, Pt, and Ni. Whitney et al. [4] reported the synthesis of nickel and cobalt NWs using electrochemical deposition of the metals into polycarbonate membranes or templates. One side of the membrane is sputter coated by a metallic film (e.g., Cu), which acts as working electrode in an electrochemical cell (Figure 7.3a). The electrodeposition

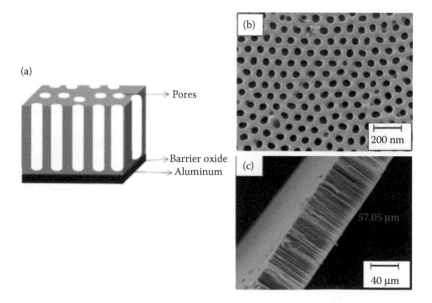

FIGURE 7.2
A nanoporous alumina membrane. (Reproduced from Meyyappan M., Sunkara M. (Eds.). 2009. *Inorganic Nanowires: Applications, Properties, and Characterization*, CRC Press, Florida. With permission from Taylor & Francis Group.)

solution is confined to the bare side of the membrane. The nickel sulfamate solution and cobalt sulfate solution are used to get the arrays of Ni and Co NWs, respectively. The deposition starts onto the Cu film from within the pores. The electrodeposition is carried out at constant potential.

For a constant potential and under the pseudo-steady-state conditions, the electrodeposition current is directly proportional to the area of the electrodeposit. After an initial transient, the current–time curves exhibit three distinct regions (Figure 7.3b). Region I corresponds to the electrodeposition of metal into the pores in the membrane. Region II indicates that the pores are completely filled with the deposited metal and the electrodeposit begins to form hemispherical caps over the end of each NW. In region III, on further increasing the deposition time, the caps coalesce into a planar metallic top layer for which the current is constant. Thus, an array of NWs can be formed by stopping the electrodeposition process at the transition point between region I and region II. The resultant arrays of Co and Ni NWs showed coercivities in the range from 150 to 680 oersteds, which is significantly higher than the values of (i.e., a few tens of oersteds) bulk Ni and Co. The enhancement of coercivities increases with decreasing wire diameter. This is due to the fact that larger wire diameter facilitates the formation of multidomains, hence degrades the coercivity [4]. The lengths of NWs can be controlled by varying the amount of metal to

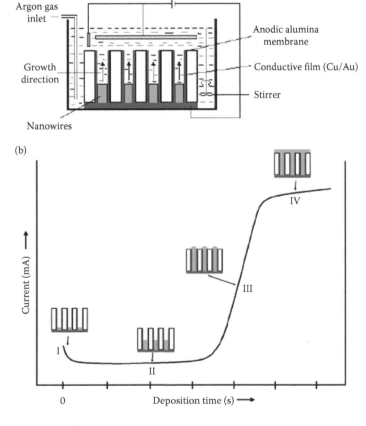

FIGURE 7.3
(a) Schematic illustration of electrode arrangement setup for deposition of NWs and (b) current–time curve for electrodeposition of Ni NWs into a membrane. The inset shows three stages of the electrodeposition.

be deposited. This method can also be used to synthesize the conductive polymers within the pores of these templates.

7.2.2 Electroless Deposition

Electroless metal deposition involves the use of a chemical reducing agent to plate a metal from solution onto a surface. This method involves applying a sensitizer (typically Sn^{2+}) to the inside surfaces of the membrane. The sensitizer binds to the surfaces via complexation with surface amine, carbonyl, and hydroxyl groups. This sensitized membrane is then activated by an exposure to Ag^+, resulting in the formation of discrete nanoscopic Ag particles on the membrane's surfaces. Finally, the Ag-coated membrane is immersed into the plating bath containing Au and a reducing agent, which results in Au plating on the membrane faces and pore walls. The key feature of this process is that metal deposition starts in the pores at the pore wall. Therefore, hollow metal tubules and solid metal NWs are obtained after short deposition time and long deposition time, respectively. In contrast to the electrochemical deposition method where the length of the metal NW can be controlled at will, the electroless deposition method provides structures that run the complete thickness of the template membrane. Thus, internal diameter and the outer diameter of the tubules/NWs can be controlled by varying the deposition time and the diameter of the pores in the template. In addition, this method has the advantage that the surface to be coated need not be electronically conductive [5].

7.2.3 Pressure Injection Technique

In this method, the NWs are formed by pressure injecting the low-melting point material (i.e., metal or semiconductor) in liquid form into the evacuated pores of the template. The templates made up of anodic alumina and glass must be chemically and structurally stable to withstand high temperatures and high pressures. This method can synthesize both metallic (Bi, In, Sn, and Al) or semiconducting (Se, Te, GaSb, and Bi_2Te_3) NWs using anodic alumina templates. It can yield highly crystalline NWs. Some surfactants (e.g., Cu in Bi melt) are added into liquid to decrease the surface tension of the liquid and the contact angle between the liquid and the template. This reduces the required pressure and to maximize the filling factor. This method produces the NWs with high crystallinity and a preferred crystal orientation along the wire axis [6].

7.3 Properties of NWs

Some of the important thermal, optical, and electrical properties of the NWs are discussed below.

7.3.1 Thermal Properties

7.3.1.1 Melting Point

Due to the high surface-to-volume ratio and low-coordination number of the surface atoms, the NWs show lower melting point compared to their bulk counterpart. The Bi NWs synthesized by template method exhibit a size-dependent melting point, that is, the melting point decreases from 544 K for the bulk Bi to 536 K for 12 nm diameter wires. However, the melting point was found independent of their orientation for a fixed diameter wire. The depression in melting point for Bi NWs was much smaller than those of Bi particles [7], Ge NWs [8], and GeTe. The melting of GeTe NW synthesized by VLS method commences at 390°C which shows 46% decrease compared to bulk melting point of 725°C. Its outside oxide layer, which has a higher-melting point of 1115°C, is left behind as nanotube. Similarly, the In_2Se_3 NWs melt at 680°C compared to bulk melting point of 890°C [9]. In contrast, the melting point of the embedded NWs in the matrix is found to increase, compared to free-standing metallic NWs. This may be due to the higher cohesive force between the outer surface of NWs and the template's surface. Moreover, the thermal stability of nanomaterials is controlled by the surface and diffusion processes and influenced by the material characteristics, temperature, and geometrical parameters. In particular, it was predicted that NWs of Cu, Au, and Ni may fragment into a chain of nanospheres above a temperature that is much lower than the corresponding bulk melting point. For example, the fragmentation of Ni NWs takes place at temperature of about 900°C (bulk melting point = 1453°C). The thermal stability is also influenced by the NW structure. The single crystalline Au NWs oriented along the <110> direction were found to be more stable than their polycrystalline counterparts [10].

7.3.1.2 Coefficient of Thermal Expansion

The Zn NWs with an average diameter of about 40 nm and prepared by an electrochemical method possess smaller thermal expansion coefficient compared to the conventional bulk Zn. The thermal expansion coefficient of the Zn NWs along c-axis direction is far smaller (i.e., 5.2×10^{-6} /K) than that of bulk Zn (i.e., 18.4×10^{-6} /K) and along the a-axis direction, it is slightly smaller (i.e., 5.4×10^{-6} /K) than that of bulk Zn (i.e., 5.7×10^{-6} /K). During the increasing temperature, the NW vacancies and vacancy clusters move easily along the grain boundaries onto the surfaces of the NWs and disappear fast. As a result, atoms in the grain boundaries rearrange rapidly, resulting in the increase of the order degree of grain boundary structure. This makes nonlinear vibration contribution in grain boundaries to thermal expansion coefficient decrease fast and at the same time, the NW grain boundary is the main contribution to thermal expansion coefficient. Therefore, the thermal expansion coefficient of Zn NWs becomes smaller than that of conventional Zn bulks [11]. In contrast, the linear thermal expansion coefficient of nano-Cu bulk [12], nano-Fe bulk [13], and nanocrystalline Se [14] samples is significantly higher than that of their bulk counterparts. The high-thermal expansion coefficient of nanobulk materials was attributed to the high-volume fraction of the grain boundaries.

7.3.1.3 Thermal Conductivity

The thermal conductivity of the VLS-grown Si NWs is strongly diameter dependent at the nanoscale. Figure 7.4 shows thermal conductivity values for the Si NWs of various diameters varying between 22 and 115 nm. These NWs were found to have a crystalline structure. At a given temperature, a reduction in thermal conductivity was found with a decrease in NW diameter and its value for a NW with 22 nm diameter was smaller by two orders of magnitude than that of bulk Si. This was attributed to an increase in phonon scattering at the nanoscale. For the NWs with diameters of 37, 56, and 115 nm, the thermal conductivities reach their peak values at around 210, 160, and 130 K, respectively. In contrast, the peak value of bulk Si occurs at ~25 K. The shifting of the peak values from lower

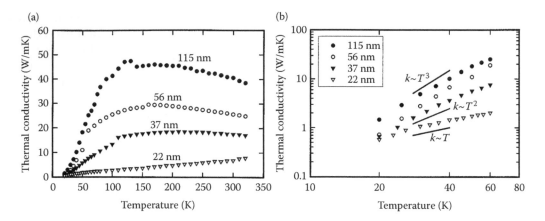

FIGURE 7.4
Thermal conductivity of Si NWs (of different diameters) as a function of temperature; (a) 30–320 K and (b) 20–60 K. The number with symbols indicates corresponding wire diameter. (Reprinted with permission from Li D. et al. 2003. Thermal conductivity of individual silicon nanowires. *Appl. Phys. Lett.* 83: 2934. Copyright 2003, American Institute of Physics.)

temperatures to higher temperatures with decreasing wire diameter indicates that the phonon boundary scattering dominates over phonon–phonon umklapp scattering. This results of the thermal conductivity with an increase in the temperature. In addition, the thermal conductivity of the NW with a diameter of 22 nm did not exhibit a peak within the experimental temperature range [15].

Figure 7.4b shows temperature-dependent thermal conductivity of the NWs between 20 and 60 K, in log–log scale. NWs with diameters of 115 and 56 nm fit Debye T^3 law quite well with the temperature range of 20–60 K. However, for the smaller diameter wires (i.e., 37 and 22 nm), the power exponent gets smaller as the diameter decreases indicating that there are some other effects which play important role besides phonon boundary scattering [15].

7.3.2 Optical Properties

Quantum size effect has an impact on electronics and photonics applications of NW. Quantum confinement occurs when the nanomaterial dimensions approach the size of an exciton Bohr radius. A decrease in diameter (particularly below 10 nm) of NWs increases its bandgap. For example, the bandgap for Si NWs with 7 nm diameter is close to that of bulk Si (i.e., 1.1 eV). However, as the diameters of NWs decrease below 7 nm, the bandgap increases to 1.5 eV (for 2.5 nm) and to 3.5 eV (for 1.3 nm) [16].

Photoluminescence (PL) or fluorescence spectroscopy is a common method to study quantum confinement effects in the NWs. The PL studies of Si NWs showed a blueshift and a reduction in PL lifetime with decreasing diameter. Moreover, doping of the Si NWs by p-type (i.e., B and P) dopant and n-type (i.e., Bi) dopant is desirable to tune the electronic properties suitable for fabricating diodes and transistors. Figure 7.5 shows the PL of InP NWs as a function of wire diameter, thus providing the effective bandgap. It can be seen that as the wire diameter of an InP NW becomes smaller than the bulk exciton diameter of 19 nm, quantum confinement effects set in, and the bandgap is increased. This results in an increase in the PL peak energy because of the stronger electron-hole Coulomb-binding

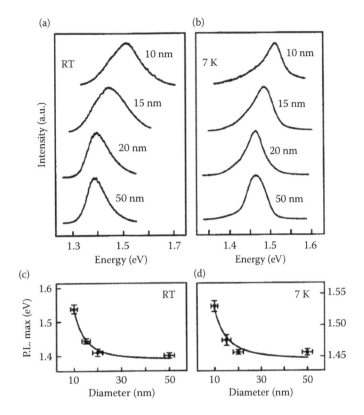

FIGURE 7.5
Photoluminescence (PL) spectra from single InP nanowires of varying diameters at (a) room temperature and
(b) 7 K. The emission energy maxima versus nanowire diameter at (c) room temperature and (d) 7 K. PL spectra
are shifted upward for clarity. (Reprinted with permission from Gudiksen M. S., Wang J., Lieber C. M. 2002.
Size-dependent photoluminescence from single indium phosphide nanowires. *J. Phys. Chem. B* 106(16): 4036–
4039. Copyright 2002, American Chemical Society.)

energy within the quantum-confined NWs as the wire radius gets smaller than the effective Bohr radius for the exciton of bulk InP. The smaller the effective mass, the larger is
the quantum confinement effects. The line widths of the PL peak for the small diameter
NWs (10 nm) are smaller at low temperature (7 K). The observation of strong quantum
confinement and bandgap tunability effects at room temperature is significant for photonics applications of NWs [3,6,17].

Metallic NWs exhibit interesting plasmon absorption effect. The energy of the surface
plasmon band is sensitive to particle size, shape, composition, surrounding media, and
interparticle interactions. Spherical metallic nanoparticles show single absorption band.
However, NWs or nanorods of Ag split original single absorption band to two absorption
bands. Their absorption intensity changes with increasing aspect ratio. The first peak and
second peak arise due to the transverse plasmon resonance and the longitudinal plasmon
resonance, respectively. When NWs are exposed to NH_3 gas, water vapor, or other environment, the electrical resistance of the HF-etched NWs decreases relative to the nonetched
NWs. Thus, NWs can serve for a sensor. Figure 7.6 shows the experimental absorption
spectra for a variety of Au nanoparticle–alumina composites. The Au particle aspect ratio
(length/diameter) varies from 7.7 to 0.38, and the diameter of each Au particle is ∼52 nm.
The reduction in absorption intensity with decreasing aspect ratio is expected due to the

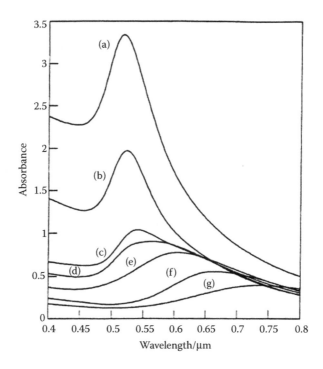

FIGURE 7.6
Experimental absorption spectra of the Au nanoparticle-containing membranes. The spectrum with the highest absorbance maximum is for the membrane containing the aspect ratio of 7.7 (a), followed by 2.7 (b), 1.3 (c), 0.77 (d), 0.54 (e), 0.46 (f), and 0.38 (g), respectively. (Hulteen J. C., Martin C. R. 1997. A general template-based method for the preparation of nanomaterials. *J. Mater. Chem.* 7(7): 1075–1087. Reproduced by permission of The Royal Society of Chemistry.)

decrease in the metal volume fraction of the composites. There is shift in the absorption maximum from 518 (for aspect ratio = 7.7) to 738 nm (for aspect ratio = 0.38) [18,19].

Similarly, the optical properties of the Au NW with diameters between 30 and 130 nm embodied in templates were reported and found that the extinction peaks of Au NW arrays possessed a redshift with increasing wires diameter and a blueshift with decreasing angle between incident light and NW arrays. This is due to the surface plasmon resonance effect [20].

7.3.3 Electrical Properties

During the measurement, electrical resistivity of the individual NWs (particularly of copper) increased from a few hundred ohms to several mega ohms (MΩ) over many hours. They become semiconducting due to their oxidation during measurement. Linear current–voltage (*I–V*) curves for the NW (e.g., Cu or Ni) arrays indicate that the NWs have metallic behavior. The electrical resistivity of the Bi NWs with diameters well above 100 nm is up to eight times higher than that of the bulk. For a given diameter, the wire resistivity becomes larger with diminishing grain size due to electron scattering at grain boundaries. For Bi NWs with a diameter comparable to the Fermi wavelength (i.e., diameter ≈ 100 nm), a quantum-size effect was observed revealing a shift of the absorption edge to higher energies with decreasing wire diameter [19]. In the case of Au with diameters larger than 100 nm, the resistivity is nearly constant with an average value 1.8 times larger than the bulk Au. The higher resistivity is due to the grain boundary scattering. However, for diameter <100 nm, the resistivity increases with decreasing diameter due to additional scattering at the wire surface [20–21].

7.4 Applications of NWs

In previous section, we have discussed that NWs exhibit unique thermal, electrical, magnetic, and optical properties which make them useful for applications in electronics (diodes, transistors, FETs, logic gates), thermoelectrics, chemical and biological sensors, thermal interface materials (TIMs), and magnetic storage devices.

7.4.1 Electronics

The microelectronics industry is facing challenges of reducing the feature size below 100 nm. Self-assembly of NWs may eliminate the need for the expensive lithographic techniques normally required to produce the devices of the size of few tens of nanometers. Unlike traditional Si processing, mono- and multicomponent NWs of different semiconductors can be made for the applications in p-n junction diodes, field-effect transistors (FETs), logic gates, and light-emitting diodes. For example, the Schottky diodes can be made by contacting a GaN NW with Al electrodes [6]. In the case of flat-panel displays, an electric field directs the field-emitted electrons from the cathode (i.e., NWs) to the anode, where the electrons hit a phosphorus screen and emit light. Due to small diameter and large curvature at the NW tip, a low threshold voltage is required for the electron emission, that is, a very high field emission currents can be achieved from the sharp Si NW tip [21]. Ag NWs have received attention due to their high-electrical and thermal conductivities for the applications in electronics, photonics, photography, conductive inks, etc. The Cu NWs are an important material for the microelectronic industry due to their low resistivity and low vulnerability to electromigration. One of the recent potential uses of Cu NWs is in 32 nm generation CMOS. Zinc NWs, due to their superconducting and luminescence properties, are attractive in optoelectronics applications such as light emitters and lasers [3,22].

7.4.2 Thermoelectric Generator

Thermoelectric materials can convert thermal energy to electrical energy or vice versa. Hence, they can also be used as solid-state coolers. In the presence of a temperature gradient, majority charge carriers diffuse from the hot side to the cold side of the material. The world uses $\sim 10^{18}$ J energy annually for direct heating or power generation. About 60%–70% thermal energy is lost as waste heat to the environment. Thermoelectric devices utilize a portion of this waste heat for electricity production, leading to enhanced overall conversion efficiency. Moreover, these stationary solid-state devices are simple and silent compared to other waste heat recovery systems that require moving parts [23]. The efficiency of a thermoelectric device depends upon the figure of merit; $ZT = S^2\sigma T/k$, where S is the Seebeck coefficient, σ is the electrical conductivity, k is the thermal conductivity, and T is the absolute temperature. Traditional thermoelectric materials are made up of Bi_2Te_3 and Sb_2Te_3 with $ZT = 1$. An efficient thermoelectric device must have a high ZT which is achieved from a high-electrical conductivity, high Seebeck coefficient, and a low-thermal conductivity of a material. It has been very challenging to increase the ZT to values >3 for the solid-state thermoelectric generators (TEGs) in order to compete with conventional heat engines. The TEG based on a Si NWs array is of much more efficient thermoelectric material than that of bulk Si because of its lower thermal conductivity. The σ and S values of Si NWs (50 nm diameter) do not deviate much from those of bulk Si. However, their thermal conductivity values varied from 1.6 to 25 W/mK depending on the surface roughness, which are significantly lower than that of bulk Si (i.e., 150 W/mK), resulting in a best ZT value of 0.6 at room temperature [24,25]. Thin Si NWs (10–20 nm) prepared by superlattice NW pattern transfer have a maximum ZT of 1.0 at 200 K. This is due to an enhanced Seebeck coefficient. Enhancement of the Seebeck coefficient is due to the confinement of phonons within NWs [26]. Bi NWs with diameters of 4–15 nm have shown the enhancement in Seebeck coefficient of about 1000 times relative to bulk material. The enhancement is due to the sharp density of states near the Fermi energy in a NW [27]. Thus, the NWs of Bi and Bi–Sb alloy are attractive for potential application in thermoelectric devices.

7.4.3 Sensors

A sensor device transforms a physical or chemical signal of the environment into an electrical signal. In general, the sensor device mainly consists of two parts: first is active sensing part, which translates the input signal into an intermediate signal and second is the transducer part that translates the intermediate signal into the final electrical signal. NWs exhibit high-chemical reactivity due to their high surface-to-volume ratio and huge number of surface atoms. This is the reason that sensors made of NWs would be potentially smaller, more sensitive, and faster than their macroscopic counterparts. Moreover, they need less power, and can provide better spatial resolution and accurate real-time information regarding the concentration of a specific analyte. This encourages NWs as the sensing probe (or sensors) for chemical and biochemical substances. For example, for biosensing, NWs have been used in the detection of cancer marker proteins, which are molecules occurring in blood or tissue that are associated with cancer. Functionalized Si NWs have been used for the detection of nucleic acids (DNA) and viruses with high selectivity [28]. Si NW field effect transistors (SiNW-FETs) have recently attracted attention toward biosensors because of their excellent sensitivity, selectivity, and label-free and real-time detection capabilities. Thus, these SiNW-FETs may be useful in biomedical diagnosis and cellular research [29]. Traditional biosensors usually require massive sample preparations and/or have low-detection sensitivity. Huge surface-to-volume ratio and diameter-dependent resistance make the NWs attractive for sensor applications. For example, the resistance of Pd NWs is very sensitive to the presence of H_2 gas and hence provides improved sensitivity compared to the sensors made of bulk counterpart [30]. A Schottky barrier field effect transistor (SBFET) made up of parallel NWs and an aluminum oxide pH-sensitive layer can detect small volumes, which is not possible using a planar transducer. The reactive gases such as carbon dioxide, benzene, and ethyl alcohol could also induce significant changes in electrical conductivity of semiconducting films, due to electron donation from the reactive gas to the film or vice versa. The NW sensor showed a clear and rapid response at a very low concentration (i.e., 1 ppm) of reactive gas. These NWs can be integrated in different sensors to detect H_2, NO_2, and CO. Sensors based on ZnO NWs are nontoxic and biocompatible at low concentrations, which makes them ideal candidates for biosensors [31].

7.4.4 Magnetic Information Storage Devices

The magnetic energy in a grain can be increased by increasing either the volume or the anisotropy of the grain. If the volume is increased, the particle size increases, and thus, the resolution decreased. For spherical magnetized grains, the superparamagnetic limit at room temperature is reached at 70 Gbits/in.[2] In NWs, the anisotropy is very large and yet the wire diameters are small, so the magnetostatic switching energy can easily be above the thermal energy while the spatial resolution is large. For magnetic data storage applications, a large aspect ratio is needed for the NWs to maintain a high coercivity, and a sufficient separation between NWs is needed to suppress the interwire magnetic dipolar coupling. Thus, NWs can form stable and highly dense magnetic memory arrays with packing densities in excess of 10^{11} wires/cm^2. The ordered magnetic NW arrays can permit data storage of about 10^{12} bits/in.^{2} [5]. When the magnetic field is applied parallel to the long axis of the magnetic NW, it exhibits a coercive field, which is inversely proportional to the diameter. The squareness of the hysteresis loop increased from 30% up to nearly 100% by decreasing the wire diameter [22]. The small-diameter and single-domain NWs of Ni and Co have been found to be most suitable for the magnetic information storage medium. In addition, the nickel NWs have their potential in microsensor applications.

7.4.5 Solar Cell

A solar cell (or photovoltaic cell) is an electrical device that converts the energy of light into electricity by the photovoltaic effect. The operation of a solar cell requires three important aspects; (i) absorption of light and generating either electron-hole pairs or excitons, (ii) separation of charge carriers of opposite types, and (iii) separate extraction of those carriers to an external circuit. The high-surface area and the high-electrical conductivity along the length of NWs are suitable for inorganic–organic solar cell. The solar cells made of CdSe NWs have high efficiency [32,33]. Recently, Si NWs with diameter of subwavelength have shown much lower losses compared to those of metallic NWs; hence, they can be used for optical waveguides within visible to near-IR range of spectrum. NWs of Au, Ag, Ni, Pd, etc. can be used as barcode tags [34] for different optical readouts.

7.4.6 Nanocomposites

Inefficient heat dissipation is the limiting factor for the performance and reliability of microelectronics. This problem has become much more severe for the modern microelectronic devices, that is, efficient heat transfer from an integrated circuit (IC) to a heat sink is a critical issue. Moreover, microscopic roughness of the IC and heat sink surfaces results in asperities between the two mating surfaces, which prevent the ideal thermal contact. In view of this, thermal interface materials (TIMs) with high-thermal conductivity and good conformability are generally applied between the two mating surfaces to provide a good heat conduction path. TIMs are typically made of low modulus polymer matrix filled with high-thermally conductive particles. However, traditional TIMs exhibit low-thermal conductivity (i.e., ca. 1–5 W/mK) even at high-volume fraction. The use of $\sim$0.9 vol.% single crystalline Cu NWs (diameter: 80 nm) with large aspect ratio (100–1000) increased the thermal conductivity of Cu NW/polyacrylate nanocomposites to $\sim$2.46 W/mK, that is, increased by 1350% compared to pure matrix. Such an excellent performance makes the NWs attractive fillers for high-performance TIMs. In contrast, Ag NWs/polyacrylate nanocomposites showed thermal conductivity value of 1.4 W/mK at 1.1 vol.% Ag NWs (aspect ratio ~167) despite better thermal conductivity of Ag than that of Cu. The high increase in thermal conductivity for Cu NWs/polyacrylate than that of Ag NW/polyacrylate nanocomposites was attributed to the larger aspect ratio of Cu NWs, which makes easier the 3-D network of NWs in the matrix [35].

Owing to an excellent electrical conductivity (6.3×10^7 S/m) and thermal conductivity (429 W/mK), silver NWs have the ability to enhance electrical and thermal properties of the polymer nanocomposites. Therefore, silver NWs are used as conductive filler and thermal interface material in sophisticated nanodevices. Additionally, due to high surface area and aspect ratio, its small loading is required for the preparation of conducting polymer nanocomposites for use in optoelectronic devices such as touch screen, liquid crystal display, and solar cell [36].

References

1. Lin J. S., Chen C. C., Diau E. W. G., Liu T. F. 2008. Fabrication and characterization of eutectic gold-silicon (Au–Si) nanowires. *J. Mater. Process. Tech.* 206: 425–430.
2. Rao C. N. R., Govindaraj A. 2005. *Nanotubes and Nanowires*. RSC Publishing, Norfolk, UK.
3. Meyyappan M., Sunkara M. (Eds.). 2009. *Inorganic Nanowires: Applications, Properties, and Characterization*. CRC Press, Florida.
4. Whitney T. M., Jiang J. S., Searson P. C., Chien C. L. 1993. Fabrication and magnetic properties of arrays of metallic nanowires. *Science* 261: 1316.
5. Hulteen J. C., Martin C. R. 1997. A general template-based method for the preparation of nanomaterials. *J. Mater. Chem.* 7: 1075–1087.
6. Bhushan B. (Ed.). 2004. *Springer Handbook of Nanotechnology*. Springer-Verlag, Berlin, Heidelberg, Germany.
7. Olson E. A., Efremov M. Y., Zhang M., Zhang Z., Allen L. H. 2005. Size-dependent melting of Bi nanoparticles. *J. Appl. Phys.* 97: 034304.
8. Wu Y., Yang P. 2001. Melting and welding semiconductor nanowires in nanotubes. *Adv. Mater.* 13: 520.
9. Sun X. H., Yu B., Ng G., Nguyen T. D., Meyyappan M. 2006. III-VI compound semiconductor indium selenide (In2Se3) nanowires: Synthesis and characterization. *Appl. Phys. Lett.* 89: 233121.
10. Toimil-Molares M. E. 2012. Characterization and properties of micro- and nanowires of controlled size, composition, and geometry fabricated by electrodeposition and ion-track technology. *Beilstein J. Nanotechnol.* 3: 860.
11. Wang Y., Zhao H., Hu Y., Ye C., Zhang L. 2007. Thermal expansion behavior of hexagonal Zn nanowires. *J. Cryst. Growth* 305: 8–11.
12. Birringer R., Gleiter H., Cahn R. W. (Ed.). 1988. *Encyclopedia of Material Science and Engineering Supplement*. Vol. 1. Pergamon, New York.
13. Zhao Y. H., Sheng H. W., Lu K. 2001. Microstructure evolution and thermal properties in nanocrystalline Fe during mechanical attrition. *Acta Metall.* 49: 365–375.
14. Zhang H. Y., Mitchell B. S. 1999. Thermal expansion behavior and microstructure in bulk nanocrystalline selenium by thermomechanical analysis. *Mater. Sci. Eng.* A(270): 237–243.
15. Li D., Wu Y., Kim P., Li S., Yang P., Majumdara A. 2003. Thermal conductivity of individual silicon nanowires. *Appl. Phys. Lett.* 83: 2934.
16. Ma D. D. D., Lee C. S., Au F. C. K., Tong S. Y., Lee S. T. 2003. Small-diameter silicon nanowire surfaces. *Science* 299: 1874–1877.

17. Gudiksen M. S., Wang J., Lieber C. M. 2002. Size-dependent photoluminescence from single indium phosphide nanowires. *J. Phys. Chem. B* 106(16): 4036–4039.
18. Li Z., Chen Y., Li X., Kamins T. I., Nauka K., Williams R. S. 2004. Sequence-specific label-free DNA sensors based on silicon nanowires. *Nano Lett.* 4, 245–247.
19. Hulteen J. C., Martin C. R. 1997. A general template-based method for the preparation of nano-materials. *J. Mater. Chem.* 7(7): 1075–1087.
20. Yao H., Duan J., Mo D., Gunel H. Y., Chen Y., Liu J., Schäpers T. 2011. Optical and electrical properties of gold nanowires synthesized by electrochemical deposition. *J. Appl. Phys.* 110: 094301.
21. Au F. C. K., Wong K. W., Tang Y. H., Zhang Y. F., Bello I., Lee S. T. 1999. Electron field emission from silicon nanowires. *Appl. Phys. Lett.* 75: 1700.
22. Sarkar J., Khan G. G., Basumallik A. 2007. Nanowires: Properties, applications and synthesis via porous anodic aluminium oxide template. *Bull. Mater. Sci.* 30(3): 271.
23. Dasgupta N. P., Sun J., Liu C., Brittman S., Andrews S. C., Lim J., Gao H., Yan R., Yang P. 2014. 25th Anniversary article: Semiconductor nanowires—Synthesis, characterization, and applications. *Adv. Mater.* 26: 2137.
24. Hochbaum A. I., Chen R. K., Delgado R. D., Liang W. J., Garnett E. C., Najarian M., Majumdar A., Yang P. D. 2008. Enhanced thermoelectric performance of rough silicon nanowires. *Nature* 451: 163–167.
25. Li Y., Buddharaju K., Singh N., Lee S. J. 2012. Top-down silicon nanowire-based thermoelectric generator: Design and characterization. *J. Electr. Mater.* 41(6): 989–992.
26. Boukai A. I., Bunimovich Y., Tahir-Kheli J., Yu J. K., Goddard W. A., Heath J. R. 2008. Silicon nanowires as efficient thermoelectric materials. *Nature* 451: 168.
27. Heremans J. P., Thrush C. M., Morelli D. T., Wu M-C. 2002. Thermoelectric power of bismuth nanocomposites. *Phys. Rev. Lett.* 88: 216801.
28. Wanekaya A. K., Chen W., Myung N. V., Mulchandani A. 2006. Nanowire-based electrochemical biosensors. *Electroanalysis* 18(6): 533–550.
29. Chen K. I., Li B. R., Chen Y. T. 2011. Silicon nanowire field-effect transistor-based biosensors for biomedical diagnosis and cellular recording investigation. *Nanotoday* 6: 131–154.
30. Walter E. C., Penner R. M., Liu H., Ng K. H., Zach M. P., Favier F. 2002. Surface and size effects on the electrical properties of Cu nanowires. *J. Appl. Phys.* 104: 023709.
31. Mikolajick T., Weber W. M. 2015. Silicon Nanowires: Fabrication and Applications. In *Anisotropic Nanomaterials: Preparation, Properties, and Applications.* Quan Li (Ed.) Springer International Publishing, Switzerland.
32. Huynh W. U., Dittmer J. J., Alivisatos A. P. 2002. Hybrid nanorod-polymer solar cells. *Science* 295: 2425–2427.
33. Wu Y., Fan R., Yang P. 2002. Block-by-block growth of single-crystalline Si/SiGe superlattice nanowires. *Nano. Lett.* 2: 83–86.
34. Nicewarner Pena S. R. 2004. *Encyclopedia of Nanoscience and Nanotechnology.* Vol 6. American Scientific Publishers, Valencia, CA. p. 215.
35. Wang S., Cheng Y., Wang R., Sun J., Gao L. 2014. Highly thermal conductive copper nanowire composites with ultralow loading: Toward applications as thermal interface materials. *Appl. Mater. Interfaces* 6: 6481–6486.
36. Abbasi N. M., Yu H., Wang L., Abdin Z.-u., Amer W. A., Akram M., Khalid H. et al. 2015. Preparation of silver nanowires and their application in conducting polymer nanocomposites. *Mater. Chem. Phys.* 166:1–15.

8

Graphene

8.1 Introduction

There has been a rapid increase in interest in the study of the synthesis, properties, and applications of graphene after the first report on graphene in 2004 [1]. It has a very high-charge carrier intrinsic mobility, zero effective mass, and large mean free path distances (i.e. several micrometers) at room temperature. It is the basic building block of graphite or carbon nanotubes (CNTs), which consists of a single atomic layer of sp^2-hybridized carbon atoms arranged in a honeycomb structure. A single graphene layer has a thickness of 0.34 nm, carbon–carbon distance of 0.142 nm, intrinsic mobility at room temperature of 2.5×10^5 cm^2/(V.s), Young's modulus of ~1 TPa, strength of 130 GPa, thermal conductivity in the range of 600–5000 W/mK, and an excellent transparency (i.e., >97%). Its theoretical specific surface area is 2630 m^2/g and it is impermeable to gases. It has current density of approximately a million times higher than that of copper. There are two approaches for the synthesis of graphene: top-down and bottom-up. The top-down approach involves breaking of graphite particles down to graphene by mechanical cleavage or liquid-phase exfoliation, while bottom-up approach involves its synthesize by chemical vapor deposition (CVD), epitaxial growth on silicon carbide, molecular beam epitaxy, etc. The simplest way of preparing small samples of single-layer or few layer graphene (FLG) is by mechanical cleavage. It produces relatively defect-free and unoxidized graphene layers, which are typically of few microns in lateral size. In this chapter, commonly used methods such as mechanical exfoliation, chemical exfoliation, reduction of graphene oxide, dry ice reduction method, and CVD are discussed. Moreover, thermal, mechanical, and electrical properties, and applications of graphene have been discussed.

8.2 Synthesis of Graphene

8.2.1 Mechanical Exfoliation

It involves the exfoliation of the individual graphene sheets by using scotch tape onto graphite and then they are transferred by pressing them onto a substrate (Si, SiO$_2$, or Ni). For this, typically highly ordered pyrolytic graphite (HOPG) is chosen in order to get high-quality graphene crystallites. Crystallites larger than 1 mm, which are visible to the naked eye, can be obtained [2]. This method has the advantage of synthesizing graphene at room temperature using low-cost equipment. However, this method has poor scalability.

A sharp single-crystal diamond wedge can also be used to exfoliate graphene layers from the HOPG [3]. The shear mixing of graphite both in the *N*-methyl-2-pyrrolidone (NMP) and in aqueous surfactant solutions (sodium cholate, NaC) results in large-volume suspensions. After centrifugation, these suspensions contain high-quality graphene nanosheets, including some monolayers [4].

8.2.2 Chemical Exfoliation

It involves dispersion of graphite powder in a solvent, which has a surface energy similar to graphene (~0–50 mJ/m) followed by sonication. This method is based on enthalpy and charge transfer between the graphene layers and the solvent molecules. The enthalpy of the mixing is minimal when the two surface energies are close or equivalent. Under such conditions, the exfoliation will take place with mild sonication. The common solvents which match the graphene surface energy are *N,N*-dimethylformamide (DMF), benzyl benzoate, γ-butyrolactone (GBL), 1-methyl-2-pyrrolidinone (NMP), *N*-vinyl-2-pyrrolidone (NVP), and *N,N*-dimethylacetamide (DMA). There is a charge transfer between the solvent and the graphite layers allowing the exfoliation to take place. Therefore, the graphene sheets could be positively or negatively charged with varying donor and acceptor numbers depending on the solvents. However, the liquid-phase exfoliation method produces graphene for films with transparency from 80% to 90% and a sheet resistance from 8 to 5 kΩ/sq. The increased sheet resistance is due to damages caused by the sonication during exfoliation. Electrochemical exfoliation method can produce a better quality graphene with a faster rate. In this method, the graphite or HOPG is usually connected as anode and a platinum wire as cathode electrode and the setup is placed in an acidic solution. The complete exfoliation takes place in 15–30 min with voltages varying from 4 to 10 V. The thin film made of the synthesized graphene sheets shows a sheet resistance ~210 Ω/sq with 96% transparency [5,6].

8.2.3 Reduction of Graphene Oxide

Graphene oxide (GO) can be synthesized by the treatment of graphite in a mixture of strong mineral acids (H_2SO_4 or HNO_3) and oxidizing agents ($KMnO_4$ or $KClO_3$ or $NaClO_3$) followed by heat treatment in an inert atmosphere at high temperature for few seconds. The reactions disrupt the delocalized electronic structure of graphite and impart a variety of oxygen-based chemical functionalities to the surface. Generally, oxidation results in an increase of the d-spacing and intercalation between adjacent graphene layers, and thus weakens the interaction between adjacent sheets, and finally leads to the delamination of GO in an aqueous solution. The intercalated GO is commonly exfoliated using chemical reduction in appropriate media chemical compounds such as NH_2NH_2, KOH, $NaBH_4$, HI, or thermal exfoliation. This yields an expanded graphite which has layer thickness in the range of few tens to hundreds nanometers. This is called Hummers and Offeman method. The Hummers-modified method produces a higher fraction of well-oxidized hydrophilic carbon material with a more regular structure, where the basal plane of the graphite is less disrupted. GO consists of epoxy (C–O–C) trigonally bonded in sp^2/partially sp^3 configurations, hydroxyl groups (C=O) in sp^3 configuration displaced above or below the graphene plane, and of some carboxylic groups (–COOH) at the edges of the graphene plane. GO exhibits excellent mechanical, optical, thermal, and electronic properties that are similar to graphene because of its specific 2-D structure and the presence of various oxygenated functional groups. However, GO has a high-sheet resistance ~10^{12} Ω/sq. This intrinsic insulating nature is due to the presence of C–O–C and C=O groups in GO. The chemical and thermal treatments are required to reduce the GO in order to remove these oxygen groups with a resulting increase in the electrical conductivity. The resultant product is usually called a reduced GO (rGO). The presence of residual functional groups and defects break the conjugate structure, thus decreasing the carrier mobility and concentration of the electrons. This leads to lower electrical conductivity for rGO compared to pure graphene. The GO can be exfoliated by two preferred routes: chemical method and thermal exfoliation techniques.

The chemical method involves dispersion of GO precursor in water assisted by mechanical exfoliation (i.e., ultrasonication and/or stirring) at concentrations up to 3 mg/mL followed by the addition of an aqueous KOH solution (a strong base). This method gives a colloidal suspension of exfoliated GO having a large negative charge through reactions with the reactive hydroxyl, epoxy, and carboxylic acid groups present on the GO sheets. This results in an extensive coating of the sheets with negative charges and K$^+$ ions. KOH-modified GO is reduced by using hydrazine to produce a stable aqueous dispersion of the individual graphene sheets [5]. Before the reduction of GO, the C/O ratio is typically 4:1 to 2:1, and can be reduced to 12:1 or even to 246:1 after the reduction. However, the sonication treatment fragments the GO platelets, reducing their lateral dimensions by over an order of magnitude down to a few hundred nanometers. In contrast to sonication method, mechanical stirring can produce single-layer GO platelets of much larger lateral dimensions and aspect ratios but it exfoliates GO very slowly with low yield. GO can also be exfoliated at lower concentrations (<0.5 mg/mL) via sonication in polar organic solvents such as DMF, polycarbonate (PC), and NMP. The exfoliation and reduction of GO has become a primary low-cost process that can yield a large quantity of rGO with high processability.

In thermal exfoliation method, the GO powder is typically loaded into a quartz tube and subjected to thermal shock in a tubular furnace by heating to high (annealing) temperatures (500–1100°C) at high heating rates (i.e., 2000°C/min). The heating is carried out in the presence of inert or reducing gases or in a high vacuum. The rapid heating is believed to cause various small molecules (of CO, CO_2, and water) to evolve and generate internal pressure. This high pressure forces the sheets apart and yielding a dry, high-surface area material with a lower bulk density. It has been found that higher the heating rate, the greater are the exfoliation and deoxygenation of graphite oxide. Further, high annealing temperature is essential to remove structural defects. As the annealing temperature increases from 500°C to 750°C, the C/O ratio could increase from more than 7 to 13 and this may increase the electrical conductivity from ~50 to 550 S/cm for annealing temperatures of 500°C and 1100°C, respectively. The increase in electrical conductivity is attributed to the loss of oxygen. The resultant rGO has BET surface area in the range from 700 to 1500 m^2/g, which is less than that of the theoretical limit (i.e., ~2600 m^2/g for graphene). This method has the drawback as it consumes high-energy because of the associated high temperature and longer time duration for reduction. The hydrogen arc discharge exfoliation method (>2000°C) can synthesize high-quality graphene nanosheet from GO with excellent electrical conductivity of ~2 × 10^3 S/cm and good thermal stability of ~601°C [6–8].

8.2.4 Chemical Vapor Deposition

In this, single crystal and atomically smooth metals are usually preferred as substrate to grow high-quality graphene. The metal or catalytic substrate must have negligible or no diffusion of carbon atoms, for example, Cu surfaces are considered the best choice as the substrate because the diffusion of C atoms in Cu is very low (0.001 at% at 1000°C). The CVD process on Cu foils can be scaled using a roll-to-roll technique. Under high vacuum and temperature, small amount of methane is fed, which is disassociated into carbon vapors and absorbed onto the film. The growth automatically stops after the formation of single layer [9]. The method involves mainly three steps: adhesion of polymer supports to the graphene on the copper foil, etching of the copper layers, and release of the graphene layers and transfer onto a target substrate. In the adhesion step, the graphene film grown on a copper foil is attached onto a thin polymer film, which is coated with an adhesive layer by passing between two rollers. In the subsequent step, the copper layers are removed by electrochemical reaction with aqueous 0.1 M ammonium persulfate solution $(NH_4)_2S_2O_8$. Finally, the graphene films are transferred from the polymer support onto a target substrate by removing the adhesive force holding the graphene films. The CVD method is commonly used to produce large-area uniform graphene films. The CVD process is very difficult to control, particularly in polycrystalline metals where the grain boundaries act as nucleation sites for multilayer growth. High-quality FLG exceeding 1 cm² in area has been synthesized by using CVD on a thin nickel film. The synthesized graphene covers the entire substrate surface and the thickness ranged from 1 to 12 graphene layers. In contrast to Cu, the solubility of carbon in nickel is ~0.1% at 900°C. Thus, the growth of graphene on Cu occurs by a surface adsorption process, whereas graphene on Ni grows via a carbon segregation process. Compared to Ni, significantly less solubility of carbon in Cu, the growth of graphene on Cu is self-limiting, resulting in single-layer films. The CVD method has several drawbacks as it is an expensive process due to a large amount of energy used and for the difficulty of transferring the graphene to dielectric or other substrates. Moreover, controlling the crystallographic orientation is critical for several electronic applications.

Heating silicon carbide (SiC) to high temperatures (>1100°C) under high vacuum (~10⁻⁶ torr) reduces SiC to epitaxial graphene. The face of SiC (i.e., silicon or carbon-terminated) used for the graphene formation highly influences the thickness, mobility, and carrier density. Graphene grows with a well-defined orientation and its lattice matches well on a reconstructed interfacial layer on Si-face SiC. Samples produced by this process have smooth surfaces. The carbon layer grown on this surface consists of mainly monolayer graphene [10]. The high cost of the SiC wafers and the high temperature are the major drawbacks of this method.

8.2.5 Dry Ice Reduction Method

It involves burning of magnesium (Mg) metal in dry ice. In this method, Mg ribbon is ignited inside a dry ice (i.e., CO_2) bowl. As soon as the combustion of Mg ribbon is completed, the black product is transferred to a beaker containing 1.0 M HCl acid and the product is stirred at room temperature overnight. Both Mg and MgO react with HCl acid to form $MgCl_2$, which is soluble in water. The mixture is filtered and washed with de-ionized water several times to make the powder free from acid molecules. Finally, the isolated black powder is dried under high vacuum overnight at 100°C. The ignition of Mg in a CO_2 can be represented by Equation 8.1 [11]. This method results in FLG with yields of ~92%.

$$2Mg + CO_2 \rightarrow 2MgO + C \text{ (graphene)} \tag{8.1}$$

8.3 Properties of Graphene

8.3.1 Structural Properties

The Raman spectra of graphene include the G peak located at ~1580/cm and 2-D peak at ~2700/cm, caused by the in-plane optical vibration (degenerate zone center E_2g mode) and second-order zone boundary phonons, respectively. The D peak, located at ~1350/cm due to first-order zone boundary phonons, is absent from defect-free graphene, but exists in defected graphene. Hence, the Raman spectra could be used to distinguish the quality of graphene and to determine the number of layers for n-layer graphene (for n up to 5) by the shape, width, and position of the 2-D peak [12,13].

As shown in Figure 8.1a, in case of single-layer graphene, the G′ (or 2-D) Raman band is twice the intensity of the G band, whereas in the two-layer graphene, the G band is stronger than the 2-D band. Further, the 2-D band is shifted to higher wavenumber in the two-layer graphene and has a different shape. As the number of layers is increased, the 2-D band moves to higher wavenumber and becomes broader and more asymmetric in shape for more than around five layers. The absence of D band in Raman spectra indicates that the graphene has a very high degree of perfection. Raman spectra of the epitaxial graphene grown on SiC show a significant phonon hardening (i.e., blueshift of the G and 2-D peaks), mainly due to the compressive strain that is generated when the sample is cooled down after the growth. In contrast, the substrate plays a negligible role in the Raman spectrum of micromechanically cleaved graphene transferred onto them, indicating the weak interactions between the transferred graphene and such substrates [12–15].

Figure 8.1b shows significant structural changes occurring during the chemical processing from pure expandable graphite (EG) to GO, rGO, and FGO. The Raman spectrum of the pure EG displays a prominent G peak at 1583/cm and a 2-D band at 2725/cm. A small D band adsorption at 1357/cm indicates the presence of defects caused by the intercalation of strong acid into the graphite layers during its synthesis process. In case of GO, the G band is broadened and shifts to 1598/cm, the D band shifts downward to 1355/cm and becomes prominent. This indicates an increased D/G intensity ratio of 1.48 (Table 8.1). For graphene, the D/G intensity ratio is further increased to 1.66. The Raman spectra of FGO are similar to that of GO, with a slight increase of vibration frequency of D and G band and a decrease of D/G intensity ratio to 1.35, which was attributed to the reaction of phenyl glycidol chlorophosphate (PGC) with GO [14].

X-ray diffraction (XRD) patterns of EG, GO, organic-phosphate-functionalized graphite oxides (FGO), and hydrazine-reduced graphene (graphene) are shown in Figure 8.2a. The pure EG shows (002) peak at 26.54°, which corresponds to an interlayer distance of 0.34 nm. In case of GO, the (002) peak shifts to 10.31° with the interlayer distance of 0.86 nm. The absence of any sharp diffraction peaks in case of graphene sheets indicates disorder and exfoliation of graphene sheets. However, one can see a broad weak peak at 23.8° showing some amount of reclustering. The FGO shows XRD pattern similar to GO, indicating that the modification process does not change the d-spacing of GO [14]. Figure 8.2b shows the XRD patterns for GO and rGO (graphene) powders, which indicates a larger interlayer spacing for GO than that of rGO. Intercalation of water and acid molecules, as well as the formation of oxygen-containing groups between the layers increase the d-spacing during the synthesis of GO. This process results in a lower angle reflection peak ($\sin \theta \propto 1/d$) $2\theta = 9.32°$ (or d-spacing = 9.52 Å). The reduction of GO

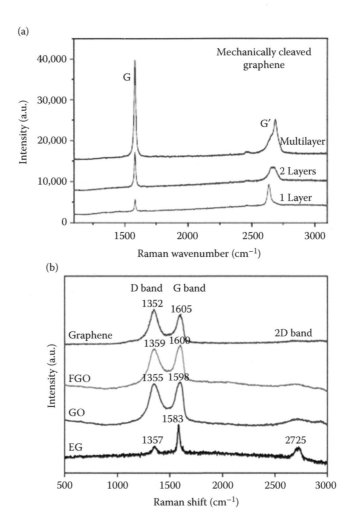

FIGURE 8.1
Raman spectra of (a) monolayer, bilayer, and multilayer graphene. (Reprinted from *Compos. Sci. Technol.*, 72, Young R. J. et al., The mechanics of graphene nanocomposites: A review, 1459–1476, Copyright 2012, with permission from Elsevier.) (b) Comparison of Raman spectra of graphene with EG, GO, and FGO. (Reprinted with permission from Guo Y. et al. In-situ polymerization of graphene, graphite oxide, and functionalized graphite oxide into epoxy resin and comparison study of on-the-flame behavior. *Ind. Eng. Chem. Res.* 50: 7772–7783. Copyright 2011, American Chemical Society.)

to rGO removes the intercalated compound and deoxygenates it, thus decreasing the interlayer spacing in rGO and shifting the angle of reflection peak 2θ angles to 23.56° (or d-spacing = 3.77 Å) [15].

8.3.2 Mechanical Properties

Figure 8.3 shows a stress–strain curve of monolayer graphene suspended over hole on a silicon substrate, using an atomic force microscope through the nanoindentation. It can be seen that the stress–strain curve becomes nonlinear with increasing strain and the fracture occurs at a strain over 25%. The defect-free graphene has a Young's modulus of

TABLE 8.1

Raman Peak Positions and D/G Intensity Ratio of GO, Graphene, and Functionalized GO

Sample	Peak of D Band (cm⁻¹)	Peak of G Band (cm⁻¹)	D/G Intensity Ratio
EG	1357	1583	0.55
GO	1355	1598	1.48
Graphene	1352	1605	1.66
Functionalized-GO	1359	1600	1.35

Source: Reprinted with permission from Guo Y. et al. In-situ polymerization of graphene, graphite oxide, and functionalized graphite oxide into epoxy resin and comparison study of on-the-flame behavior. *Ind. Eng. Chem. Res.* 50: 7772–7783. Copyright 2011, American Chemical Society.

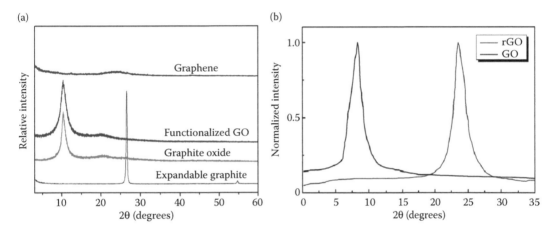

FIGURE 8.2
XRD patterns of (a) EG, GO, FGO, and graphene. (Reprinted with permission from Guo Y. et al. In-situ polymerization of graphene, graphite oxide, and functionalized graphite oxide into epoxy resin and comparison study of on-the-flame behavior. *Ind. Eng. Chem. Res.* 50: 7772–7783. Copyright 2011, American Chemical Society.) (b) GO and rGO. (Reprinted from *J. Colloid Interface Sci.*, 430, Konios D. et al., Dispersion behaviour of graphene oxide and reduced graphene oxide, 108–112, Copyright 2014, with permission from Elsevier.)

1000 ± 100 GPa and a fracture strength of 130 ± 10 GPa. These values are very close to those of theoretical values predicted from the density functional theory [13]. Chemically modified graphene has a mean elastic modulus of 250 GPa $\pm$ 150 TPa, which is much lower than that of defect free graphene. However, this value is higher than that of graphene oxide paper (~32 GPa) [16]. The spring constant of few layer graphene (FLG) with thicknesses between 2 and 8 nm is in the range of 1-5 N/m, which is different from that of bulk graphite [17]. Though, graphene has excellent stiffness and strength, it is relatively brittle with a fracture toughness of ~4.0 MPa√m compared to that of metals which exhibit a value of 15-50 MPa√m. This indicates that imperfect graphene is likely to crack in a brittle manner like ceramic materials. The fracture strength of precrack graphene is ~100 GPa, which is significantly lower than the intrinsic strength of graphene. This is due to the weakening effect of precrack [18]. Table 8.2 gives the properties of graphene and carbon allotropes for comparison [8].

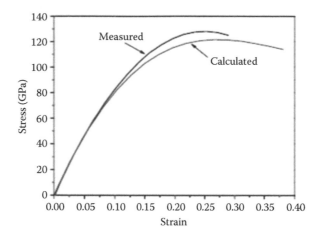

FIGURE 8.3
Stress–strain curve for the deformation of a monolayer graphene. (From Young R. J. et al. 2012. The mechanics of graphene nanocomposites: A review. *Compos. Sci. Technol.* 72: 1459–1476. Reproduced with permission from Elsevier.)

TABLE 8.2

Properties of Graphene and Other Allotropes of Carbon

Carbon Allotropes	Graphite	Diamond	Fullerene	Carbon Nanotubes	Graphene
Hybridized form	sp²	sp³	sp²	sp²	sp²
Crystal structure	Hexagonal	Cubic	FCC	FCC	Hexagonal
Dimensions	Three	Three	Zero	One	Two
Experimental specific surface area (m²/g)	~10–20	20–160	80–90	~1300	~1500
Density (g/cm)	2.09–2.23	3.5–3.53	1.72	>1	>1
Optical properties	Uniaxial	Isotropic	Nonlinear optical response	Structure dependant	97.7% of optical transmittance
Thermal conductivity (W/mK)	1500–2000[a] 5–10[c]	900–2320	0.4	3500	4840–5300
Hardness	High	Ultrahigh	High elastic	High-flexible elastic	Highest (single layer)
Tenacity	Flexible nonelastic	–	Elastic	Flexible elastic	Flexible elastic
Electronic properties	Electrical conductor	Insulator	Insulator	Metallic and semiconducting	Semimetal, zero-gap semiconductor
Electrical conductivity	Anisotropic, 2.3×10^{4a}, 6[b]	–	10^{10}	Structure dependant	2000

Source: Reprinted from Nano Energy, 1, Wua Z. S. et al., Graphene/metal oxide composite electrode materials for energy storage, 107–131, Copyright 2012, with permission from Elsevier.
[a] along a-axis direction.
[b] along b-axis direction.
[c] along c-axis direction

8.3.3 Thermal Properties

8.3.3.1 Thermal Conductivity

Due to relatively low carrier density in pristine graphene, the electronic contribution to thermal conductivity is negligible and hence, the thermal conductivity of graphene is dominated by phonon transport. In other words, there is diffusive conduction at high temperature and ballistic conduction at sufficiently low temperature. The thermal conductivity of suspended monolayer graphene measured by the shift in the Raman G band in confocal micro-Raman spectroscopy is ~5000 W/mK, which is more than double than that of pyrolytic graphite (i.e., ~2000 W/mK) at room temperature. Due to extreme high-thermal conductivity, graphene can outperform CNTs in heat conduction for the electronic applications as thermal management [19]. In contrast, a thermal conductivity of ~2500 W/mK (at 350 K) was obtained from CVD-grown graphene deposited onto a thin silicon nitride membrane having an array of through-holes. The membrane was coated with a thin layer of gold for a better thermal contact. The thermal conductivity of micromechanically exfoliated graphene deposited on a SiO_2 substrate is ~600 W/mK, which is much lower than the values obtained from the suspended graphene synthesized by micromechanically exfoliated or CVD grown. This discrepancy might be due to the phonons leaking across the graphene–support interface and strong interface scattering of flexural modes [16]. The large variation in the reported thermal conductivities can be attributed to large measurement uncertainties (from the various measurement methods) (Table 8.3) [20], temperature of measurement, different processing conditions and methods, aspect ratio, and quality of the graphene.

8.3.3.2 Thermal Expansion

As discussed above, graphene has superior mechanical properties and thermal conductivity. However, its coefficient of thermal expansion (CTE) plays a crucial role in its application in high-density and high-speed integrated electronic devices because of the mismatch between graphene and parts, which are in contact with it. In addition, the strain caused by the CTE mismatch between graphene and the substrate plays a crucial role in determining the physical properties of graphene, and hence its CTE must be accounted for an interpretation of experimental data taken at various temperatures. The CTE of graphene is very sensitive to the substrate. Without substrate, it has a CTE of ca. -6.0×10^{-6}/K at room

TABLE 8.3

Thermal Conductivities of Graphene and Graphene-Oxide-Based Materials

Method	Material	Thermal Conductivity
Confocal micro-Raman spectroscopy	Single layer graphene	4840–5300 W/mK at RT
Confocal micro-Raman spectroscopy	Suspended graphene flake	4100–4800 W/mK at RT
Thermal measurement method	Single layer (suspended)	3000–5000 W/mK at RT (suspended)
Thermal measurement method	Single layer (on SiO_2 support)	600 W/mK at RT (on a silicon dioxide support)
Electrical four-point measurement	Reduced graphene oxide flake	0.14–0.87 W/mK

Source: Reprinted from *Prog. Mater. Sci.*, 56, Singh, V. et al., Graphene based materials: Past, present and future, 1178–1271, Copyright 2011, with permission from Elsevier.

temperature (i.e., 300 K) and a very small positive values at high temperatures. A very weak substrate interaction (about 0.06% of the in-plane interaction) can largely reduce the negative CTE region and increase the resultant CTE. If the substrate interaction is strong enough, the CTE will be positive in whole temperature range and its value reaches to $\sim 2.0 \times 10^{-5}/K$ [21]. The room temperature CTE of single-layer graphene estimated with Raman spectroscopy is $\sim (8.0 \pm 0.7) \times 10^{-6}/K$ [22].

8.3.3.3 Thermal Stability

Thermogravimetric analyzer (TGA) is one of best techniques to assess the level of reduction of GO or thermal stability of graphene. Figure 8.4 shows the weight loss as a function of temperature for dried GO and rGO powders. It can be seen clearly that there is a slight loss in the weight of GO during temperature interval from room temperature to 200°C. It can be attributed to the evaporation of the absorbed water molecules. A further total weight loss of about 40% up to 800°C is due to the decomposition of oxygen-containing functional groups. In contrast, rGO displays total weight loss of about 10% during the same temperature range (up to 800°C), indicating that rGO has higher-thermal stability than GO. The rGO shows higher-thermal stability than GO due to the deoxygenation during the reduction process [15].

8.3.4 Electrical Properties

Graphene is a 2-D hexagonal-packed structure of sp^2 hybridization, and the free electrons of carbon atoms align on the perpendicular to the hexagonal plane, forming the out-of-plane π bond, which is the response of Fermi surface of graphene. The Fermi surface of graphene is in the intersection of completely filled valence band and empty conduction band, as well as in the middle of π band, where electrons are the valence electrons. Graphene is a zero-gap semiconductor with unique electronic properties. The electrical conductivities of graphene produced by different methods are shown in Table 8.4. The conductivity decreases with increasing graphene layer, and finally approaches to the conductivity of graphite.

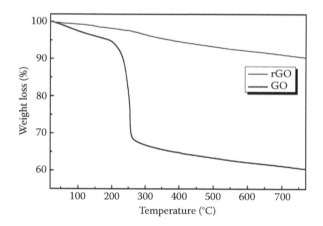

FIGURE 8.4

Decomposition behavior of GO and rGO. (Reprinted from *J. Colloid Interface Sci.*, 430, Konios D. et al., Dispersion behaviour of graphene oxide and reduced graphene oxide, 108–112, Copyright 2014, with permission from Elsevier.)

TABLE 8.4

Electrical Conductivity of Graphene Produced by Different Methods

Material	Method	σ (S/m)	Ref.
Single layer graphene (0.45 nm)	Chemical vapor deposition	$\sim 10^8$	[33]
7–8 monolayer graphene (5–6 nm)	Chemical vapor deposition	$\sim 1.4 \times 10^5$	[33]
10–12 monolayer graphene (10–12 nm)	Chemical vapor deposition	$\sim 1.1 \times 10^5$	[33]
Reduced-graphene oxide paper	Thermally reduced	0.5–1.4×10^5	[34]
Graphene paper	Hydrazine reduced	1.2–3.5×10^{4a}	[35]
Reduced-graphene oxide paper	Thermally reduced	2×10^4	[36]
Graphene oxide paper	Chemical method	0.5–1.5×10^{-2}	[34]

[a] The conductivity increases as the heat treatment temperature increases.

The electrical conductivity of chemically synthesized graphene is low compared to that of defects free-graphene. However, the presence of functional groups and defects enhance the polarization losses which enable them useful for EMI shielding application [23,24].

Graphene has the fastest electron mobility of $\sim 1.5 \times 10^4 \, cm^2/V/cm$, a super high mobility of temperature-independent charge carriers of $2 \times 10^5 \, cm^2/V/s$ (200 times higher than Si), and an effective Fermi velocity of $10^6 \, m/s$ at room temperature. Graphene has an optical transmission of 97.7%, electrical conductivity of up to $10^8 \, S/m$ depending upon the method of synthesis, and good flexibility. Due to its unique mechanical, thermal, electrical, and optical properties, graphene might outperform CNTs, graphite, metals, and semiconductors where it is used as an individual material or as a component in a hybrid or composite material [12].

8.4 Properties of Graphene/Polymer Nanocomposites

After discussing various properties of graphene, this section has discussion on th improvement in mechanical, thermal, electrical and gas barrier properties of graphene filled polymer nanocomposites.

8.4.1 Mechanical Properties of Graphene Composites

Excellent mechanical properties such as high elastic modulus (1.1 TPa) and tensile strength (125 GPa) of graphene sheets have attracted the attention of researchers toward polymer matrix nanocomposites. In addition to intrinsic properties, the mechanical properties of the nanocomposites are dependent on the volume fraction, size, and aspect ratio of the filler (or graphene); distribution and dispersion of filler in the host matrix; and interface bonding. The stress–strain curves for the polystyrene (PS)/graphene nanocomposites with increasing graphene content are shown in Figure 8.5a. It can be seen that there is a significant increase in the tensile strength of the nanocomposite, which can be attributed to effective load transfer between graphene and polymer. However, the strain (or elongation) at break decreases gradually with increasing graphene content. As shown in Figure 8.5b, the Young's modulus and tensile strength of pure PS are 1.45 GPa and 24.44 MPa, respectively. Both the Young's modulus and the tensile strength of the nanocomposite films increase significantly with increasing graphene content in the PS matrix. For the nanocomposite

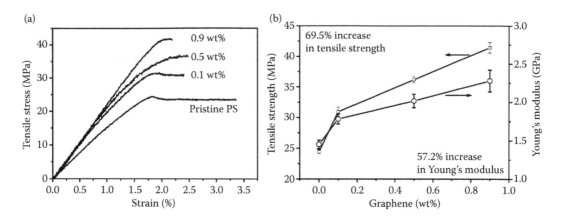

FIGURE 8.5
(a) Representative stress–strain curves of the pristine PS and nanocomposite films with different contents of graphene sheets. (b) Young's modulus and tensile strength changes with increasing graphene content. (Fang M. et al. 2009. Covalent polymer functionalization of graphene nano-sheets and mechanical properties of composites. *J. Mater. Chem.* 19: 7098–7105. Reproduced by permission of The Royal Society of Chemistry.)

film with 0.9 wt% of PS-functionalized graphene sheets (FGS), the Young's modulus and fracture strength increase to 2.28 GPa and 41.42 MPa, corresponding to an increase of 57.2% and 69.5%, respectively, compared to pure PS film [24].

Figure 8.6 shows stress–strain curves for rGO-filled poly(vinyl alcohol) (PVA) at room temperature. It can be seen that there is a significant effect on the stress–strain behaviour with a loading of only 0.3 vol% of rGO but the addition of more loading (i.e., 3.0 vol%) of rGO leads to decrease in elongation. Both Young's modulus and tensile strength of the PVA are found to increase with increasing rGO content but the elongation at break decreases. This is the typical behavior shown by several graphene/polymer nanocomposites in the literature [25,26].

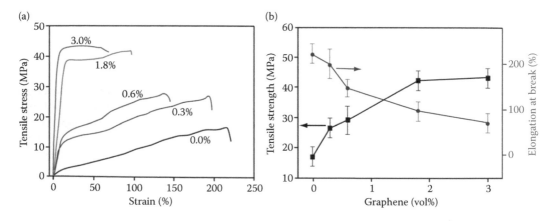

FIGURE 8.6
(a) Stress–strain curves and (b) tensile strength and elongation at break versus graphene loadings for rGO-filled PVA. (Reprinted with permission from Zhao X., Zhang Q., Chen D. Enhanced mechanical properties of graphene-based poly(vinyl alcohol) composites. *Macromolecules* 43: 2357–2363. Copyright 2010, American Chemical Society.)

8.4.2 Thermal Properties of Graphene Composites

The exceptional thermal properties such as thermal conductivity, CTE, and thermal stability of graphene have been exploited to improve thermal conductivity, dimensional stability, and thermal stability of polymers for various applications. Pure graphene is highly thermally conductive, that is, it has room temperature thermal conductivity in the range of 600–5000 W/mK. CNTs show similar intrinsic thermal conductivities, but the sheet-like geometry of graphene-based materials provides lower interfacial thermal resistance and thus produce larger conductivity improvements in polymer composites (Figure 8.7). The geometry of graphite and graphene filler can also impart significant anisotropy to the thermal conductivity of the polymer composite, with the measured in-plane thermal conductivity as much as ten times higher than the out-of-plane thermal conductivity [27–29].

The addition of graphene into composites results in low CTEs, and increasing graphene fraction reduces the CTEs significantly. The addition of 5 wt% GO reduces CTE (measured below the glass transition temperature) from 82×10^{-6}/K for pure epoxy to 56×10^{-6}/K for nanocomposite, that is, there is ~32% decrease in CTE compared to pure epoxy. However, the CTEs of the nanocomposites measured above glass transition temperature are higher than that of pure matrix [28]. There is also significant decrease in CTE of PC nanocomposites with increasing graphite or graphene content (Figure 8.8). However, the FGS is marginally better than micrometer-sized graphite in reducing the CTE of PC, despite its higher-aspect ratio. Probably, the structural distortion and higher flexibility of FGS does not help in reducing its thermal expansion significantly [29]. In addition, the thermal conductivity of the polymer matrix composite containing 5 wt% graphene shows about four-fold increase compared to the pure polymer matrix [28].

The addition of graphene into polymers can also increase the thermal stability, which is characterized by a particular weight loss (5% or 10%) degradation temperature or the maximum weight loss rate temperature in a given atmosphere using TGA. For example, 5% weight loss degradation temperature of 0.5 wt% graphene-filled PC nanocomposites in nitrogen atmosphere is increased ~55°C compared to pure PC. The higher-thermal

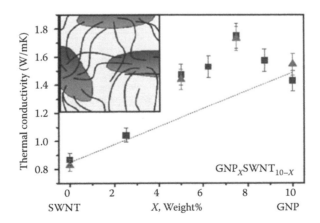

FIGURE 8.7
Synergistically improvement in the thermal conductivity of epoxy composites containing combination of graphite nanoplatelets (GNP) and single-walled carbon nanotubes (SWNT). Inset shows how the presence of SWNTs bridges the gap between dispersed GNP, which may be responsible for the synergistic effect. (Reprinted from *Polymer*, 52, Potts J. et al., Graphene based polymer nano-composites, 5–25, Copyright 2011, with permission from Elsevier.)

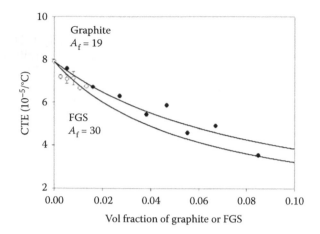

FIGURE 8.8
The CTE of PC composited reinforced with graphite and functionalized graphene sheet (FGS). (Reprinted from *Polymer*, 50, Kim H., Macosko C. W., Processing-property relationships of polycarbonate/graphene composites, 3797–3809, Copyright 2009, with permission from Elsevier.)

stability of graphene-filled polymer nanocomposites compared to pure polymers may be attributed to three main factors: higher-thermal stability of graphene compared to pure PC, restriction of chain mobility of polymers near the graphene surface, and tortuous path for the decomposed low-molecular-weight organic molecules or gases by the platelet-like morphology of graphene. Thus, well-dispersed graphene nanoplatelets delay the escape of volatile degraded products during decomposition [29]. This suggests an application of graphene/polymer nanocomposites for flame retardation. The flame-retardant properties, measured by combustion calorimeter, of both epoxy/graphene and epoxy/FGO nanocomposites are better than epoxy/GO composites. That is there is a maximum reduction of ~24% and ~44% in peak heat release rate (HRR) for 5 wt% FGO and 5 wt% graphene-filled epoxy nanocomposites, respectively (Figure 8.9).

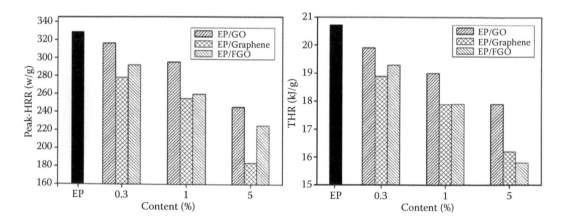

FIGURE 8.9
Peak HRR and total heat release of epoxy/graphene nanocomposites in N_2 atmosphere, where EP represents epoxy. (Reprinted with permission from Guo Y. et al. In-situ polymerization of graphene, graphite oxide, and functionalized graphite oxide into epoxy resin and comparison study of on-the-flame behavior. *Ind. Eng. Chem. Res.* 50: 7772–7783. Copyright 2011, American Chemical Society.)

The dispersion of the GO or functionalized GO (FGO) in the matrix can act as a barrier and effectively slow the HRR and hinder the transfer of combustion gases to the flame zone and energy feedback. In addition, the carbon layer formed during the degradation of the GO or FGO can hinder transfer of heat. Hence, the peak HRR and total heat release (THR) are reduced. The large specific surface area of the GO or FGO may act as a radical trap for the radicals generated during the combustion and degradation process. However, the mechanism of flame retardant by nanosized fillers needs to be studied further [14].

8.4.3 Electrical Properties of Graphene Composites

Owing to excellent electrical conductivity of graphene, it is considered as one of the most promising fillers for polymer nanocomposites for use in electronics applications. The paper made of stacked rGO exhibits electrical conductivities as high as 351 S/cm. Its addition to the polymer matrices increases the electrical conductivity by several orders of magnitude depending upon volume fraction, size, and aspect ratio of graphene, and its degree of dispersion and distribution in the matrix. In order for a nanocomposite with an insulating matrix to be electrically conductive, the volume fraction of the conducting filler (or graphene in present case) must be above the electrical percolation threshold, where a conductive network of filler particles is formed. The electrical percolation threshold is much lower for graphene-filled PC nanocomposites than that of graphite-filled PC composites. Figure 8.10 shows the electrical conductivity of the rGO-filled PS nanocomposites (prepared by *in situ* polymerization method) as a function of rGO content (volume fraction). It can be seen that the nanocomposites have a percolation threshold of 0.1 vol% with an increase in the electrical conductivity more than 10 orders of magnitude. At 1 vol% rGO content, the nanocomposite has an electrical conductivity of $\sim 10^{-3}$ S/cm [30].

Similar to rGO/PS nanocomposites, a remarkably low-percolation threshold of 0.12 vol% has been reported for the rGO/epoxy nanocomposites owing to the uniform dispersion

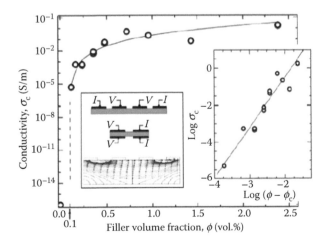

FIGURE 8.10
Electrical conductivity of *in situ* rGO/PS nanocomposites as a function of rGO volume fraction. The *right* and *left* insets, respectively, show the plot of log σ_c against log($\phi - \phi_c$), where ϕ_c is the percolation threshold and the four-probe setup for measurements with the computed distributions of the current density (*contour lines*) with directions and magnitude (*arrows*). (Stankovich S. et al.: Graphene-based composite materials. *Nature*, 2006. 442. 282–286. Copyright Wiley-VCH Verlag GmbH & Co. KGaA. Reproduced with permission.)

of monolayer graphene sheets having extremely high-aspect ratios in the matrix. The self-alignment of rGO into the matrix above a critical volume fraction induces a unique anisotropy in electrical and mechanical properties. Such microstructure results in a high dielectric constant of over 14,000 for a nanocomposite with 3 wt% rGO at 1 kHz. This is due to the charge accumulation at the highly aligned rGO/polymer interface. These nanocomposites exhibit EMI shielding efficiency of ~38 dB [31].

8.4.4 Gas Barrier Properties

The plastics used for soft drinks and pouches of wafers must have excellent barrier against moisture and gases to retain taste and quality of the commodity packaged inside them. The permeation (or permeability) rate of gas molecules such as N_2, O_2, moisture, and CO_2 diffusing through polymeric membranes can be decreased by embedding high-aspect ratio and impermeable particles like clay/graphene that provide tortuous paths and reduce the cross-sectional area available for the permeation. Defect-free graphene sheets are impermeable to all gas molecules. Hence, the addition of graphene sheets into polymer matrices increases their barrier resistance with mechanical integrity significantly over CNTs or CNFs or clay. Kim et al. studied the gas barrier properties of the graphite (BET-specific surface area = 29 m²/g) and thermally treated GO (or FGS) (BET-specific surface area = 800 m²/g)-filled PC composites melt blended by using a twin screw extruder at 250°C under N_2 purge and screw speed of 200 rpm. Permeability of neat PC in He and N_2 at 35°C is about 13.6 and 0.29 Berrers, respectively. The addition of graphite and FGS into PC significantly suppress the N_2 and He permeation indicating that the gas molecules diffusing through the membranes can be decreased by embedding impermeable particles with a high-aspect ratio. As shown in Figure 8.11, for a given volume fraction, the permeability of FGS/PC nanocomposites is much smaller than that of graphite-filled composites. It is due to higher-aspect ratio of FGS than that of graphite, which can provide more tortuous paths and reduce the cross-sectional area available for permeation. Moreover, both graphite and FGS particles appear to be slightly better for blocking diffusion of N_2 than He, probably due to the higher-kinetic diameter of N_2 (3.64 Å) than that of He (2.89 Å) [29].

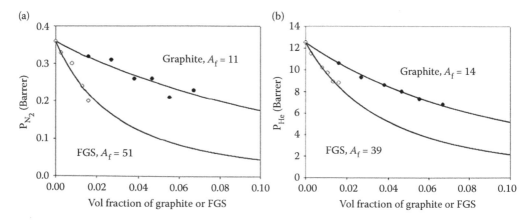

FIGURE 8.11
Permeation constants against (a) N_2 and (b) He of graphite- and FGS-filled PC composites at 35°C, where A_f is the aspect ratio of the particles calculated from the ratio of diameter to thickness. Curves are predictions based on the model of Lape and coworkers. (Reprinted from *Polymer*, 50, Kim H., Macosko C. W., Processing-property relationships of polycarbonate/graphene composites, 3797–3809. Copyright 2009, with permission from Elsevier.)

8.5 Applications of Graphene

Owing to its high-electrical conductivity and thermal conductivity, graphene is recognized as one of the potential nanomaterials for realizing the next-generation electronic devices. The single-layer graphene possesses a theoretical surface area of 2630 m^2/g; hence, it can provide large interfacial volume fraction in the nanocomposites. The addition of graphene and GO/rGO into the polymer matrices has shown significant improvement in electrical conductivity, strength, elastic modulus, thermal conductivity, thermal stability, and reduced permeability toward gas molecules at a very low volume fraction. The significant increase in properties at a lower volume fraction of filler enables a new avenue for developing high-strength-lightweight structural polymer composites for automobile, aerospace, and electronic industry for thermal management. Graphene/polymer nanocomposites can be used in the packaging of medicine, food, electronics, and beverages due to low permeability of gas molecules. The graphene/polymer nanocomposites have been found suitable for the applications in energy storage, electrically conductive polymers, antistatic coatings, and electromagnetic interference (EMI) shielding. Graphene/polyaniline nanocomposites have shown specific capacitance up to 1046 F/g, which is significantly higher than that of other carbon allotropes (i.e., 450–500 F/g) and have good cyclic stability.

The surface resistivity required for electrostatic discharge (ESD) materials ranges between 10^{12} and 10^5 Ω/sq, whereas, for EMI-shielded material, it is lower than 10^5 Ω/sq. There are several materials or electronic devices such as carpeting floor mats, electronics packing to telecommunication antenna, mobile phone parts and frequency shielding coating for aircraft and electronics which need to be protected from EMI. A 15 wt% hydrazine-reduced-GO-filled epoxy nanocomposites showed an EMI shielding efficiency of ~20 dB in the X-band. Graphene-filled polymer nanocomposites usually exhibit a positive temperature coefficient and a negative temperature coefficient depending upon the type of polymer matrix.

Graphene/polymer nanocomposites have promising applications for dye-sensitized solar cells, future touch screens, video displays, and electrochromic devices. However, there are challenges in mass production of the low-cost, high-quality graphene and the control dispersion in the polymer matrix [20].

Graphene/CNT and graphene/polymer nanocomposites have been studied for lithium ion batteries and supercapacitor as energy storage devices. Microbial fuel cells based on polyaniline/graphene are promising green energy source to harvest electricity from various organic matters. Due to the change in conductance of graphene as a function of extent of surface adsorption, large specific area, and low Johnson noise, graphene has shown a promising application to detect a variety of molecules such as gases to biomolecules. Moreover, it can be used for pH, pressure, and temperature sensors. Graphene and graphene-filled nanomaterials have been suggested to be useful for various biomedical applications such as drug and gene delivery, nanomedicine, bioimaging, and potential cancer therapies [32].

References

1. Novoselov K. S., Geim A. K., Morozov S. V., Jiang D., Zhang Y., Dubonos S. V. 2004. Electric field effect in atomically thin carbon films. *Science* 306: 666–669.
2. Geim A. K., MacDonald A. H. 2007. Graphene: Exploring carbon flatland. *Phys. Today* 60(8): 35–41.

3. Buddhika J., Sathyan S. 2011. A novel mechanical cleavage method for synthesizing few-layer graphene. *Nanoscale Res. Lett.* 6(1): 95.

4. Paton K. R. 2014. Scalable production of large quantities of defect-free few-layer graphene by shear exfoliation in liquids. *Nat. Mater.* 13(6): 624–630.

5. Kuilla T., Bhadra S., Yao D., Kim N. H., Bose S., Lee J. H. 2010. Recent advances in graphene based polymer composites. *Prog. Polym. Sci.* 35: 1350–1375.

6. Notarianni M., Liu J., Vernon K., Motta N. 2016. Synthesis and applications of carbon nanomaterials for energy generation and storage. *Beilstein J. Nanotechnol.* 7: 149–196.

7. Potts J. R., Dreyer D. R., Bielawski C. W., Ruoff R. S. 2011. Graphene-based polymer nanocomposites. *Polymer* 52: 5–25.

8. Wua Z. S., Zhou G., Yin L. C., Ren W., Li F., Cheng H. M. 2012. Graphene/metal oxide composite electrode materials for energy storage. *Nano Energy* 1: 107–131.

9. Sukang B., Hyeongkeun K., Youngbin L., Xiangfan X., Jae-Sung P., Yi Z., Jayakumar B., Tian L., Ri K. H. 2010. Roll-to-roll production of 30-inch graphene films for transparent electrodes. *Nat Nanotechnol.* 5(8): 574–578.

10. Sutter P. 2009. Epitaxial graphene: How silicon leaves the scene. *Nat. Mater.* 8(3): 171–172.

11. Chakrabarti A., Lu J., Skrabutenas J. C., Xu T., Xiao Z., Maguire J. A., Hosmane N. S. 2011. Conversion of carbon dioxide to few-layer graphene. *J. Mater. Chem.* 21(26): 9491–9493.

12. Wua Z. S., Zhou G., Yin L. C., Ren W., Li F., Cheng H. M. 2012. Graphene/metal oxide composite electrode materials for energy storage. *Nano Energy* 1: 107–131.

13. Young R. J., Kinloch I. A., Gong L., Novoselov K. S. 2012. The mechanics of graphene nanocomposites: A review. *Compos. Sci. Technol.* 72: 1459–1476.

14. Guo Y., Bao C., Song L., Yuan B., Hu Y. 2011. In-situ polymerization of graphene, graphite oxide, and functionalized graphite oxide into epoxy resin and comparison study of on-the-flame behavior. *Ind. Eng. Chem. Res.* 50: 7772–7783.

15. Konios D., Stylianakis M. M., Stratakis E., Kymakis E. 2014. Dispersion behaviour of graphene oxide and reduced graphene oxide. *J. Colloid Interface Sci.* 430: 108–112.

16. Zhu Y., Murali S., Cai W., Li X., Suk J-W., Potts J. R., Ruoff R. S. 2010. Graphene and graphene oxide: Synthesis, properties, and applications. *Adv. Mater.* 22: 3906–3924.

17. Frank I. W., Tanenbaum D. M., Van Der Zande A. M., McEuen P. L. 2007. Mechanical properties of suspended graphene sheets. *J. Vac. Sci. Technol. B* 25(6): 2558–2561.

18. Peng Z., Lulu M., Feifei F., Zhi Z., Cheng P., Phillip L. E., Zheng L., Yongji G., Jiangnan Z., Xingxiang Z., Pulickel A. M., Ting Z., Jun L. 2014. Fracture toughness of graphene. *Nat. Commun.* 5: 3782.

19. Balandin A. A., Ghosh S., Bao W., Calizo I., Teweldebrhan D., Miao F., Lau C-N. 2008. Superior thermal conductivity of single-layer graphene. *Nano Lett.* 8(3): 902–907.

20. Singh V., Joung D., Zhai L., Das S., Khondaker S., Seal S. 2011. Graphene based materials: Past, present and future. *Prog. Mater. Sci.* 56: 1178–1271.

21. Jiang J., Wang J., Li B. 2009. Thermal expansion in single-walled carbon nanotubes and graphene: Nonequilibrium Green's function approach. *Phys. Rev. B* 80: 205429.

22. Yoon D., Son Y. W., Cheong H. 2011. Negative thermal expansion coefficient of graphene measured by Raman Spectroscopy. *Nano Lett.* 11: 3227–3231.

23. Cao M., Wang X., Cao W., Yuan J. 2015. Ultrathin graphene: Electrical properties and highly efficient electromagnetic interference shielding. *J. Mater. Chem. C* 3: 6589–6599.

24. Fang M., Wang K., Lu H., Yang Y., Nutt S. 2009. Covalent polymer functionalization of graphene nano-sheets and mechanical properties of composites. *J. Mater. Chem.* 19: 7098–7105.

25. Zhao X., Zhang Q., Chen D. 2010. Enhanced mechanical properties of graphene-based poly(vinyl alcohol) composites. *Macromolecules* 43: 2357–2363.

26. Young R. J., Kinloch I. A., Gong L., Novoselov K. S. 2012. The mechanics of graphene nanocomposites: A review. *Compos. Sci. Technol.* 72: 1459–1476.

27. Potts J., Dreyer D., Bielawski C., Ruoff R. 2011. Graphene based polymer nano-composites. *Polymer* 52: 5–25.

28. Wang S., Tambraparni M., Qiu J., Tipton J., Dean D. 2009. Thermal expansion of graphene composites. *Macromolecules* 42: 5251–5255.
29. Kim H., Macosko C. W. 2009. Processing-property relationships of polycarbonate/graphene composites. *Polymer* 50: 3797–3809.
30. Stankovich S., Dikin D. A., Dommett G. H. B., Kohlhaas K. M., Zimney E. J., Stach E. A. 2006. Graphene-based composite materials. *Nature* 442: 282–286.
31. Yousefi N., Sun X., Lin X., Shen X., Jia J., Zhang B., Tang B., Chan M., Kim J. 2014. Highly aligned graphene/polymer nanocomposites with excellent dielectric properties for high-performance electromagnetic interference shielding. *Adv. Mater.* 26: 5480–5487.
32. Das T. K., Prusty S. 2013. Graphene-based polymer composites and their applications. *Polym. Plast. Technol. Eng.* 52: 319–331.
33. Nirmalraj P. N., Lutz T., Kumar S., Duesberg G. S., Boland J. J. 2011. Nanoscale mapping of electrical resistivity and connectivity in graphene strips and networks. *Nano Lett.* 11: 16–22.
34. Lin X., Shen X., Zheng Q., Yousefi N., Ye L., Mai Y. W., Kim J. K. 2012. Fabrication of highly-aligned, conductive, and strong graphene papers using ultralarge graphene oxide sheets. *ACS Nano* 6: 10708–10719.
35. Chen H., Muller M. B., Gilmore K. J., Wallace G. G., Li D. 2008. Mechanically strong, electrically conductive, and biocompatible graphene paper. *Adv. Mater.* 20: 3557–3561.
36. Zhao X., Hayner C. M., Kung M. C., Kung H. H. 2011. Flexible holey graphene paper electrodes with enhanced rate capability for energy storage applications. *ACS Nano* 5: 8739–8749.

9

Magnetic Nanomaterials

9.1 Introduction

This chapter discusses properties, and applications of magnetic nanomaterials for various applications. The magnetic nanomaterials are important for biomedical applications in separation, immunoassay, and drug targeting (due to easy manipulation of their properties by an external magnetic field), and hyperthermia (due to hysteresis and other losses in alternating magnetic fields). The bees and pigeons have magnetic nanomaterials such as biological compasses for navigation. Similarly, magnetotactic bacteria consist of chains of small magnetite nanoparticles (which respond to a magnetic field), which navigate to the surface or bottom of the pools that they live in for using these nanoparticles. The presence of the lipid layer makes these particles biocompatible and they can be readily functionalized for a variety of biomedical applications. The earliest known biomedical use of naturally occurring magnetic materials involves magnetite (Fe_3O_4) or lodestone which was used by the great ancient Indian surgeon Sushruta around 2600 years ago (600 BC). Sushruta wrote in the book *Ayurveda* that magnetite can be used to extract an iron arrow tip [1]. Nowadays, magnetic nanoparticles can be produced with controllable sizes ranging from a few nanometers up to tens of nanometers. Due to such smaller size than or comparable to those of a cell (10–100 μm), a virus (20–450 nm), a protein (5–50 nm), or a gene (2 nm wide and 10–100 nm long), they can get close to a biological entity of interest. Interestingly, they can be coated with biocompatible or biological molecules to make them interact with or bind to a biological entity. Therefore, it provides a controllable means of tagging [2].

9.2 Hysteresis *M–H* loop

In an external magnetic field, the dipoles (S–N poles) of a magnetic material possess a magnetic dipole moment. When an external magnetic field strength (H, A/m) is applied on a material, the material is subjected to induced magnetic induction (B, tesla) and the magnetization (M, A/m), which can be represented by the following equation:

$$
\begin{aligned}
B &= \mu_0(H + M) \\
&= \mu_0(H + \chi H) \quad (\text{since}, M = \chi H) \\
&= \mu_0 H(1 + \chi)
\end{aligned}
\tag{9.1}
$$

where $\mu_0 (= 1.257 \times 10^{-6}$ H/m) is the permeability of free space and χ the volumetric magnetic susceptibility. M is the magnetic moment per unit volume of the material. Its value depends on the type of material, the temperature, and the field H.

A bulk ferromagnetic material displays a distinct nonlinear magnetic response to an applied magnetic field (H), referred to as a hysteresis loop M–H curve or B–H curve (Figure 9.1). In absence of an applied magnetic field, despite spontaneous internal magnetic moment the magnetic material shows zero magnetization (B), that is, the B–H curve starts at the origin. This is due to the random orientation of magnetic domains, which gives net magnetization "zero" to minimize the magnetostatic energy within a bulk ferromagnetic material. When a magnetic field (H) is applied, the field energy can overcome the magnetostatic energy associated with the domain formation, and the magnetization (B) of the material increases until all of the domains are aligned with the field. The domains that are nearly lined up with the field grow at the expense of unaligned domains. In order for the domain (size ~50 µm or less) to grow, the Bloch walls (size ~0.1 µm) must move. The Bloch walls are the boundaries which separate the individual magnetic domains. The Bloch walls may connect magnetic domains with orientation differences of 90° or 180°, and the direction of magnetization within a grain is changed by moving the Bloch walls. Eventually, the unfavorably oriented domains disappear and rotation completes the alignment of the domains within the field, thus producing the greatest amount of magnetization that the material can have. This stage is known as saturation magnetization (M_s). When the field is removed, some of the domains remain oriented near the direction of original field and a residual magnetization, known as the remanence (M_r), is present in the material. A negative magnetic field is needed to reduce the remanence magnetization to zero, which is called coercivity (H_c). When the field is applied in reverse direction, the domains grow with an alignment in the opposite direction. A coercive field (H_c) or coercivity is required to force the domains to be randomly oriented and cancel the one another's effect. Further increase in the strength of the field eventually aligns the domains to saturation in the opposite direction. Finally, a hysteresis B–H loop is obtained, as shown in Figure 9.1. It is to be noted that M–H loop shows saturation, while B–H loop does not. The coercivity (also the susceptibility) of the material is microstructure sensitive property, that is, it depends upon the shape and size of the particle or grain. For example, elongated shape of magnetic particles leads to higher coercivity. In ferromagnetic materials, the cause of hysteresis loop *can be attributed to* the pinning of magnetic domain walls at impurities or grain boundaries within the material, as well as to intrinsic effects such as the magnetic anisotropy of the

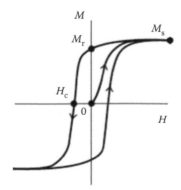

FIGURE 9.1
A typical magnetization versus magnetic field (M–H) curves for ferromagnetic nanoparticles.

crystalline lattice. The shape of these loops is determined in part by particle size. However, the saturation magnetization is structure insensitive. A hysteresis measurement provides information about coercivity, remanent magnetization, and saturation magnetization of a given material [3,4].

The B–H (or M–H) loop enables us to classify the magnetic materials into two broad categories; hard magnetic and soft magnetic materials. The hard magnetic material exhibits larger area within the B–H loop than that of soft magnetic material. Thus, soft magnets such as commercial iron ingot, oriented silicon–iron, and Ferroxcube A (48% $MnFe_2O_4$–52% $ZnFe_2O_4$) can be easily magnetized. In contrast, hard magnets are difficult to demagnetize, hence they are referred to as permanent magnets. The energy product, $(BH)_{max}$, is one of the useful parameters to classify the hard magnets. In general, the value of the energy product of conventional hard magnets is between 2 and 80 kJ/m^3, while that of high energy magnets is >80 kJ/m^3.

9.3 Effect of Particle Size on M–H Loop

The particle size plays an important role in changing the value of coercivity of magnetic materials, as shown in Figure 9.2. To minimize the energy, macroscopic ferromagnetic materials are divided up into domains of parallel magnetic moments. Within a domain, the magnetic moments orient in one direction, while the alignment of spins in neighboring domains is usually antiparallel. The oppositely aligned magnetic domains are separated from each other by the domain wall (the Bloch wall). As the particle size decreases from several hundred micrometers to few micrometers, the coercivity increases to a highest value. For larger particles with size larger than about 1 μm, there are many magnetic domains (called multidomains) with size of about 50 μm or less within each grain which leads to a narrow hysteresis loop (Figure 9.2a) since it takes relatively little field energy to make the domain walls move. Such particles are useful in immunomagnetic separation of pathogenic microorganisms in microbiology. When size of the particles is smaller than about 1 μm, the formation of domain walls becomes energetically unfavorable and

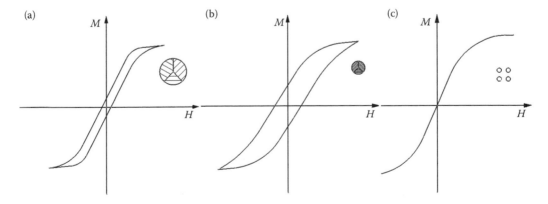

FIGURE 9.2
Effect of particle size on M–H hysteresis loop area for (a) larger size, (b) sub-micron size, and (c) superparamagnetic size (i.e., << single domain size).

the ferromagnetic particle can have single domains. The response of such small particles to a magnetic field is qualitatively different, resulting in a broader hysteresis loop (Figure 9.2b). If the particle size is reduced further to about order of tens of nanometers depending upon the material's composition, the material becomes superparamagnetic. For example, it may vary from about 14 nm for Fe and up to about 170 nm for γ-Fe_2O_3. The magnetization of a particle smaller than the intrinsic domain size tends to follow the applied field freely, thereby enhancing the local field enormously. Such particles are called superparamagnetic because their behavior is that of a paramagnet, the magnetization rising and falling with the applied field. The magnetic susceptibility of these superparamagnetic particles is many orders of magnitude larger than bulk paramagnetic materials.

In a single domain particle, all the magnetic moments within the particles are pointing toward the same direction. A colloidal solution of such particles has no net magnetization above the Bloch temperature in absence of external magnetic field due to thermal agitation. For these very small magnetic nanoparticles (i.e., <20 nm), the energy barrier (KV) becomes small (due to small volume, V, of particle) and comparable to thermal energy (k_BT), where K the anisotropy energy density, k_B Boltzmann's constant, and T the absolute temperature. So for the magnetic particles smaller than critical size (D_c), the energy barrier can no longer pin the direction of magnetization to the timescale of observation and hence, the rotation of the direction of magnetization occurs due to thermal fluctuations. In other words, thermal energy leads to flipping of the magnetic moment of small magnetic particles. Such particles are said to be superparamagnetic. The coercivity of a superparamagnetic particle is zero because thermal fluctuations prevent the existence of a stable magnetization. This leads to the anhysteretic but still sigmoidal M–H curve (Figure 9.2c).

As discussed, thermal fluctuations are strong enough to spontaneously demagnetize a previously saturated assembly. Therefore, these particles have zero coercivity and no hysteresis. Nanoparticles become magnetic in the presence of an external magnet but revert to a nonmagnetic state when the external magnet is removed. This avoids an active behavior of the particles when there is no applied field. However, cooling of a superparamagnetic particle reduces the energy of thermal fluctuation, and, at a certain temperature, the free movement of magnetization becomes blocked by anisotropy. The temperature of the transition from the superparamagnetic to the ferromagnetic state is called the blocking temperature (T_B). Superparamagnetic particles are useful as magnetic biomaterials for MRI, cell separation, and cell labeling applications. When these particles are introduced in the living systems, they are magnetic only in the presence of an external field, which gives them unique advantage in working in biological environments. Importantly, the remanence of superparamagnetic particles is zero and hence, B–H curve shows no hysteresis. This property is important for reducing the tendency of the particles to agglomerate. Conveniently, magnetite and maghemite are superparamagnetic at room temperature below crystal sizes of $\sim$30 nm. The measured saturation magnetization of nanocrystals is typically smaller than that of their corresponding bulk materials, due to surface spin canting, under coordination, or crystal defects [1–6]. Figure 9.3 summarizes the particle (or grain) size dependency on coercivity, which indicates that coercivity increases as the particle size decreases from several hundred micrometers (i.e., multidomain) to sub-micrometer size (i.e., single domain), reaching a peak value at critical size (D_c). For coarse-grained materials, the coercivity increases with inversely proportional to the grain size (D^{-1}). On further reducing the size to less than single domain size, the coercivity drastically falls of with D^6.

Figure 9.4a shows the typical result for cobalt (Co)-magnetic nanoparticles with size of about 7 nm (i.e., <critical size). It can be seen that the broad transition from superparamagnetism to ferromagnetism occurs at 92 K, which can be considered as T_B for Co

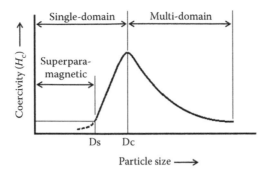

FIGURE 9.3
Effect of particle size on coercivity of magnetic materials.

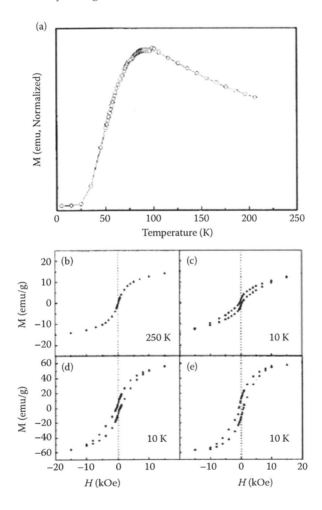

FIGURE 9.4
(a) ZFC magnetization versus temperature of cobalt NPs, (b) hysteresis loop of cobalt NPs compacted into a capsule obtained at 250 K, and (c) 10 K. Hysteresis loop of Co NPs obtained at 10 K for (d) diluted cobalt NPs with wax and (e) cobalt NPs deposited on HOPG and dried under N_2 to prevent oxidation. (Reprinted with permission from Yang H. T. et al., Self-assembly and magnetic properties of cobalt nanoparticles. *Appl. Phys. Lett.,* 82(26): 4729–4731. Copyright 2003, American Institute of Physics.)

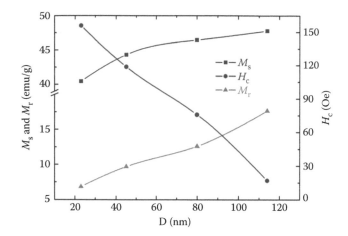

FIGURE 9.5

Size-dependent magnetic parameters of spherical Ni nanoparticle, where saturation magnetization, remanent magnetization, and coercivity are represented by M_s, M_r, and H_c, respectively. (From He X. et al. 2013. *Nanoscale Res. Lett.* 8: 446. Open Access.)

nanoparticles. Figure 9.4b and c shows the hysteresis loop of the Co-nanoparticles powder compacted at 250 and 10 K, respectively. Co nanoparticles show no hysteresis in their magnetization data at 250 K, indicating a superparamagnetism behavior at room temperature. However, at 10 K ($<T_B$), the Co nanoparticles are ferromagnetic, which shows hysteresis loop with the remanent magnetization of about 1.5 emu/g and the coercive field of 163 Oe. The value of magnetization at saturation is about 14.0 emu/g. Figure 9.4d shows the hysteresis loop of Co nanoparticles mixed with wax. There is a clear change in the shape of the hysteresis loop. The remanent magnetization reaches 7.3 emu/g, the magnetization at saturation reaches to 59.6 emu/g, and coercive field increases from 163 to 600 Oe in comparison with the isolated cobalt nanoparticles powder. Figure 9.4e shows the hysteresis loop of ordered arrays of cobalt nanoparticles on HOPG substrate. The magnetization at saturation reaches 12.6 emu/g and coercive field increases to 790 Oe [7].

Figure 9.5 shows the effect of particle size on M_s, M_r, and H_c of spherical Ni nanoparticles prepared by high-temperature reductive decomposition of nickel (II) acetylacetonate with oleic acid and oleylamine both as surfactant and solvent. It can be seen clearly that both the M_s and M_r decrease and the H_c increases monotonously with decreasing particle size (d) of Ni nanoparticle from 114 to 23 nm. As discussed above, the H_c increases ($H_c \propto 1/d$) with decreasing particle size from larger to smaller but above the single domain size (i.e., 21.3 nm) due to decrease in domain walls. The decrease of saturation magnetization and remanent magnetization with decreasing Ni particle size has been attributed to the presence of oxide layer on Ni core. It is known that as the particle size decreases, the surface-to-volume ratio of the nanoparticles increases. This leads to increasing the ratio of the thickness of oxide layer to the diameter of magnetic core. The oxide does not contribute to magnetization [8].

Zhang et al. [9] reported magnetic properties of Ni–Zn ferrite ($Ni_{0.5}Zn_{0.5}Fe_2O_4$) particles synthesized by sol-gel method. The samples are of pure cubic spinel structure with size ranging from 9 to 96 nm. Figure 9.5 shows the effect of particle size on magnetic saturation and coercivity of Ni–Zn ferrite. H_c measured at 300 K, increases rapidly as particle

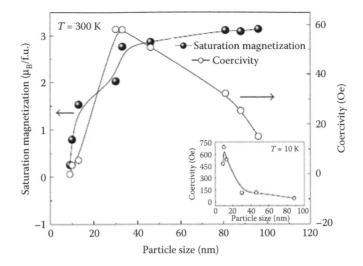

FIGURE 9.6
Size-dependent saturation magnetization and coercivity at 300 K of the nanosized Ni–Zn ferrites. The inset shows particle size dependence of the coercivity at 10 K. (From Zhang M. et al. 2013. *Adv. Mater. Sci. Eng.*, Article ID 609819. With permission from Hindawi Publishing Corporation, under the Creative Commons Attribution License.)

size increases from 9 to 30 nm with a maximum value of 58.2 Oe for particles with size of 30 nm and then decreases with further increasing the particle size from 30 to 96 nm.

The inset of Figure 9.6 indicates that H_c measured at 10 K increases as the particle size increases, reaches a maximum value of about 700 Oe for 10 nm sized particles, and then decreases on further increasing the particle size. It indicates that the critical particle size decreases as the temperature of measurement decreases from 300 to 10 K. In addition, the peak value of H_c shifts to the lower particle size when the measurement temperature decreases [9].

9.4 Surface Modification of Magnetic Nanoparticles

The surface modification of nanoparticles is required to improve their colloidal stability by overcoming van der Waals and magnetic dipole–dipole attractive forces. The magnetic nanoparticles are made biocompatible by applying a coating of biocompatible polymers such as polysaccharides (e.g., Dextran), poly(vinylpyrrolidone) (PVP) or poly(ethylene glycol)s (PEG), or poly(ethylene oxide), etc. onto nanoparticles. Other materials such as gold, activated carbon, or silica can also be used as coating. The coating also facilitates the stabilization of nanoparticles in an environment with a slightly alkaline pH or a significant salt concentration. For example, the silica-coated nanoparticles are negatively charged at the pH of blood, thus inducing electrostatic repulsion that helps in avoiding the nanoparticles-aggregate formation. The coating also acts to shield the magnetic particle from the surrounding environment. Moreover, the external surface of silica coatings can be functionalized by attaching carboxyl groups, biotin, avidin, carbodiimide, and other molecules to allow the binding of biomolecules. As shown in Figure 9.7, these functionalized

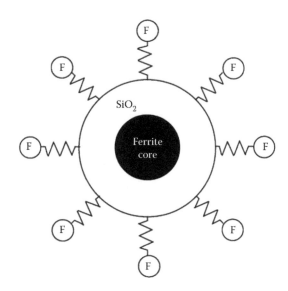

FIGURE 9.7
A schematic diagram of magnetic nanoparticle with biocompatible coating.

molecules then act as attachment points for the coupling of cytotoxic drugs or target anti-bodies to the carrier complex. The presence of hydroxyl surface groups provides intrinsic hydrophilicity and allows surface attachment by covalent linkage of specific biomolecules. The microporous and mesoporous (with pore sizes <4 nm) coatings made up of silica, carbon, and zeolites on magnetic cores are heat resistant with high-specific surface areas and mechanical strength. For nonbiodegradable cores, the coating is needed to avoid the leaching of the magnetic core and to facilitate the intact excretion through the kidneys.

9.5 Applications of Magnetic Nanoparticles

The magnetic properties of magnetic nanoparticles enable them to be used for three main applications: (i) magnetic contrast agents in magnetic resonance imaging (MRI), (ii) hyper-thermia agents, where the magnetic nanoparticles are heated selectively by application of high-frequency magnetic field (e.g., in hyperthermia of tumors), and (iii) magnetic vectors that can be directed by means of a magnetic field gradient toward a certain location such as in the case of the targeted drug delivery. The scientific community is seeking to exploit the intrinsic properties of magnetic nanoparticles in diagnosis, drug delivery, and treat-ment of cancer. In the case of drug delivery, the particle sizes should be small enough (i.e., optimum size = 5–100 nm) to be injected into the bloodstream and then allow to pass through the required capillary systems. For *in vitro* applications, the larger particle sizes can be used, and biocompatibility and toxicity issues are less important than that of *in vivo*.

9.5.1 Hyperthermia

It involves dispersing magnetic nanoparticles throughout the target tissue, and then applying an AC magnetic field of sufficient strength and frequency to heat the particles.

This heat, as explained earlier, is generated by hysteresis losses from the ferromagnetic materials or by the rotation of the particles themselves or the atomic magnetic moments of the superparamagnetic nanoparticles. This heat conducts into the immediate surrounding of the diseased tissue (or malignant cells) where the temperature can be maintained above a therapeutic threshold of 42°C for 30 min or more, for destroying cancerous cells. In contrast to conventional method, this method does not affect healthy cells. Most commonly used magnetic particles for hyperthermia are iron oxides [1–2].

9.5.2 Medical Diagnostics

Usually paramagnetic gadolinium (Gd) and superparamagnetic iron oxide nanoparticles coated with dextran have been used to enhance the contrast in nuclear MRI. MRI contrast relies on the differential uptake of different tissues. NMR imaging functions by measuring the concentration of protons that, in living structures, are quite low. However, the introduction of superparamagnetic particles into the region of interest changes the magnetic field locally and this leads to local variations in the conditions for magnetic resonance; the result is a major improvement in imaging contrast. Magnetic nanoparticles are selectively taken up by the reticuloendothelial system. Thus, if a region is tagged using the magnetic particles, the relaxation time will be lower compared to untagged regions; thus, "contrast" is provided and the particle acts as a contrast agent. The tumor cells do not have the effective reticuloendothelial system of healthy cells so that their relaxation times are not altered by the contrast agents [2].

9.5.3 Magnetic Cell Separation

The separation of entities from their surroundings is based on the difference in the susceptibility between a magnetically labeled entity and the surrounding medium, i.e., shorting of magnetic cell for cellular therapy and immunoassay. Entities that can be labeled include cells, bacteria, and some types of vesicles. It involves labeling of the entities with the nanoparticles followed by the separation of the labeled entities by magnetic separation. This method has been used to select tumor cells from blood as well as to isolate the enzymes, DNA, and RNA from various sources including the body fluids. The nanoparticles may be attached with protein cells forming a type of fur of nanoparticles or beads that consist of many superparamagnetic particles. When an external magnetic field is applied, one type of cell attached to the nanoparticles is removed from the suspension [3].

9.5.4 Drug Delivery

The conventional chemotherapy process is relatively nonspecific, which results in wastage of drugs by being distributed to areas where they are not required. This can lead to undesirable side effects as the drug attacks normal and healthy cells in addition to the targeted tumor cells. The magnetic nanoparticles coated with biocompatible polymer and/or drugs can be introduced in the form of ferrofluid into the bloodstream of the patient and then, a magnetic field gradient is applied close to the targeted region, that is, cancerous tumors within the body. As the drug/carrier complex is concentrated at the target, the drug can be released either via enzymatic activity or changes in physiological conditions such as pH, osmolality, or temperature, and can be taken up by the tumor cells. In this way, magnetically targeted therapy does not affect healthy cells. Besides, it reduces the amount of systemic distribution of the cytotoxic drug, thus reducing the associated side effects as well as the dosage required by more efficient, and localized targeting of the drug [2].

9.5.5 Magnetic Refrigeration

Superparamagnetic nanoparticles can replace ozone-depleting refrigerants and energy-consuming compressors. The power consumption of such a magnetocaloric fridge is <60% of a conventional one [3].

9.5.6 Removal of Heavy Metals

The industry-related paints, pigments, mining and extraction, glass production, plating, and manufacturing of battery release various residuals and/or waste into the water bodies. Hence, heavy metal ions such as Cd^{2+}, Cr^{3+}, Cu^{2+}, Pb^{2+}, Zn^{2+}, As^{3+}, or Hg^{2+} contaminates the water causing hazardous water pollution. The chitosan-coated magnetite nanoparticles can remove Pb^{2+} ions. Fe_3O_4/cyclodextrin polymer nanocomposites have been demonstrated for selective removal of heavy metals from industrial wastewater [10].

9.5.7 Oil Removal

Wastewater containing oils in the form of fats, lubricants, petroleum products, and cutting oils is one of the environmental concerns nowadays. Sometimes, due to accidental oil spills in the water bodies is threatening aquatic life. The use of floating barriers is the most commonly used technique to control the spread of oil, whereas sorbents are used to remove the final traces of oil or in areas that cannot be reached by skimmers. The conventional techniques are expensive and difficult to separate the products. Magnetic materials having affinity for oil are an excellent alternative. The affinity of Fe_3O_4 for oils can be increased by functionalization or modification by polymers which can facilitate the adsorption of oil. For example, the Fe_3O_4 nanoparticles coated with polystyrene have higher affinity for oil and at the same time water repellent action. This nanocomposite has been found to adsorb lubricating oil three times its own weight. Due to hydrophobic polymer in the magnetic nanocomposites, the oil gets easily adsorbed and removed magnetically [10].

References

1. Narayan R. (Ed.). 2009. *Biomedical Materials*. Springer, New York.
2. Pankhurst Q. A., Connolly J., Jones S. K., Dobson J. 2003. Applications of magnetic nanoparticles in biomedicine. *J. Phys. D: Appl. Phys.* 36: 167–181.
3. Vollath D. 2013. *Nanomaterials: An Introduction to Synthesis, Properties and Applications*. 2nd ed. Wiley-VCH Verlag GmbH & Co. KGaA, Germany.
4. Arruebo M., Fernández-Pacheco R., Ibarra M. R., Santamaría J. 2007. Magnetic nanoparticles for drug delivery. *Nanotoday* 2(3): 22–32.
5. Hu A., Apblett A. (Eds.). 2014. *Nanotechnology for Water Treatment and Purification*. Springer International Publishing, Switzerland.
6. Akbarzadeh A., Samiei M., Davaran S. 2012. Magnetic nanoparticles: Preparation, physical properties, and applications in biomedicine. *Nanoscale Res. Lett.* 7: 144.

7. Yang H. T., Shen C. M., Su Y. K., Yang T. Z., Gao H. J., Wang Y. G. 2003. Self-assembly and magnetic properties of cobalt nanoparticles. *Appl. Phys. Lett.* 82(26): 4729–4731.

8. He X., Zhong W., Au C. T., Du Y. 2013. Size dependence of the magnetic properties of Ni nanoparticles prepared by thermal decomposition method. *Nanoscale Res. Lett.* 8: 446.

9. Zhang M., Zi Z., Liu Q., Zhang P., Tang X., Yang J., Zhu X., Sun Y., Dai J. 2013. Size effects on magnetic properties of $Ni_{0.5}Zn_{0.5}Fe_2O_4$ prepared by sol-gel method. *Adv. Mater. Sci. Eng.* Article ID 609819.

10. Kalia S., Kango S., Kumar A., Haldorai Y., Kumari B., Kumar R. 2014. Magnetic polymer nanocomposites for environmental and biomedical applications. *Colloid Polym. Sci.* 292: 2025–2052.

10

Processing, Properties, and Applications
of Polymer Nanocomposites

10.1 Introduction

Polymers are preferred over metals and ceramics because of their low density, high-specific stiffness, high-specific strength (strength-to-density ratio), and ease of fabrication of complex parts on a large scale using traditional injection molding. However, they have very low-electrical conductivity, mechanical properties (strength and Young's modulus) and poor thermal conductivity, and relatively high coefficient of thermal expansion (CTE). These properties can be improved by adding appropriate volume fraction (or loading) of fillers (i.e., fibers or particulates) into the polymer matrix, called as polymer matrix composites or nanocomposites depending upon the size of constituents or matrix. Particulates of $CaCO_3$, silica, clay, and carbon black and fibers of glass and carbon have been used for several years to toughen elastomers and plastics. Tyres are filled with 10–100 nm diameter particles of carbon black. Epoxy resins filled with 10–20 μm diameter glass fiber are molded into seating, boat hulls, and shower stalls. Composites filled with 5–20 μm carbon fibers meet the performance demands for the applications as airplane bodies and sporting equipments. In composites reinforced with micrometer-sized particles or fibers, the volume fraction of polymer–filler interface is small compared to that of nanocomposites. Consequently nano- and microscale fillers have much different effects on the mechanical properties of a polymer. For example, for the same amount of improvement in property, the addition of only 1–5 vol% nanoparticles is required as against 15–40 vol% microsized particles or fibers. However, the dispersion of nanoparticles or nanofibers requires surface modification with chemical groups that are compatible with the polymer to avoid (or minimize) their agglomerations, otherwise the nanoparticles tend to clusters and finally become responsible for cracks formation in the polymer nanocomposites.

Polymer nanocomposites are mixture of two or more constituents or phases, which exhibit properties different from their individual constituents. It is needed that the dispersed constituents in the polymer matrix have at least one dimension in the range of 1–100 nm. Besides, these constituents or phases must be insoluble in each other. The nanoscale reinforcing phase may be nanoparticles (0-D) or nanowhiskers, nanofibers/nanowires/nanotubes (1-D), or nanoplates/nanosheets (2-D). In case of nanoparticles, the particle size and distribution are of great importance. Depending on the type of nanoparticles/nanofibers added, the mechanical, electrical, optical, and thermal properties of polymer nanocomposites can be tuned. Polymer nanocomposites reinforced with nanoparticles exhibit excellent thermal, mechanical, water-resistant, and barrier properties at a low-filler loading, that is,

<10 wt%. The high-filler content decreases the toughness and optical transparency, and increases melt viscosity.

In addition, compared to conventional flame retardants, nanofillers can significantly improve fire retardancy at a typically <5 wt%, that is, the peak heat release rate (HRR) can be reduced by >50% in the presence of only 5 wt% nanoclay compared to the pure polymers, while there is essentially no reduction for a microcomposite or immiscible system. In this chapter, the effect of addition of modified nanoclay, carbon nanotubes (CNTs), hexagonal boron nitride (h-BN), and nanobarium titanate ($BaTiO_3$) on the properties of polymers have been discussed. There are three most commonly used methods such as *in situ* polymerization, solution method, and melt-mixing method for fabricating the polymer matrix micro- and nanocomposites.

10.2 Polymer–Clay Nanocomposites

10.2.1 Structure of Montmorillonite

Montmorillonite (MMT) clay is a hydrous alumina silicate mineral whose lamellae are made from an octahedral alumina sheet sandwiched between two tetrahedral silicate sheets (2:1 layer silicate), as shown in Figure 10.1. However, when some of the aluminum atoms are replaced with magnesium, the difference in valences of Al and Mg creates negative charges (called as *charge exchange capacity*) distributed within the plane of the layers that are balanced by positive counterions (Na^+, Li^+, Ca^{2+}), typically sodium ions, located between the galleries. In a natural state, the clay platelets are ~1 nm in thickness and other dimensions (i.e., length and width) can be of the order of 150 nm, up to as large as 1 or 2 μm. Hydration of the sodium ions causes the galleries (or d-spacing) to expand and the clay to swell. These

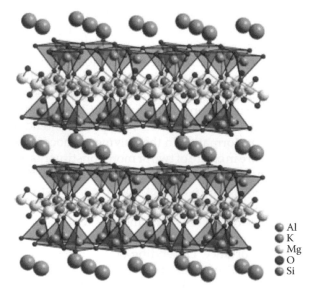

FIGURE 10.1

Structure of sodium montmorillonite. (Kickelbick G. (Ed.): *Hybrid Materials: Synthesis, Characterization, and Applications*, 2007. Copyright Wiley-VCH Verlag GmbH & Co. KGaA. Reproduced with permission.)

hydrated platelets can be fully dispersed in water. The MMT can absorb large amounts of water (i.e., >20 times its volume) and polar liquids, which separate the silicate layers. The sodium ions can be exchanged with organic cations like ammonium salt to form an organo-clay (o-clay). The ammonium cation has hydrocarbon tails and other groups attached and is referred to as a "surfactant" owing to its amphiphilic nature. The resultant o-clay becomes hydrophobic. The longer alkyl tails (of surfactant) give a more hydrophobic o-clay. There are several commercially available o-clays (Cloisite™ Na⁺, 10A, 15A, 20A, 25A, 93A, and 30B), which exhibit thermal decomposition pattern from a single main degradation stage to several degradation stages with a more complex behavior. An onset of the decomposition for Cloisite 10A is the lowest (i.e., 160°C) and for Cloisite 93A, it is the highest (i.e., 212°C) [1,2]. The extent of the negative charge of the clay is characterized by the cation exchange capacity (CEC). In its pristine state, MMT is only miscible with hydrophilic polymers, such as poly(ethylene oxide), polyamide, polyurathane, and poly(vinyl alcohol). To make MMT miscible with other polymers such as polypropylene (PP) or polyethylene (PE), one must modify these polymers with a small amount of a polyolefin that has been lightly grafted with maleic anhydride (MA). An amount of ~1% MA by weight is typical, which acts as a very effective compatibilizer for dispersing the o-clay in the parent polyolefin.

10.2.2 Fabrication of Polymer/Clay Nanocomposites

The most commonly used methods for the synthesis of polymer/clay nanocomposites are *in situ* polymerization, solution method, and melt-mixing method. The steps used in the synthesis of nanocomposites are shown in Figure 10.2.

10.2.2.1 In Situ *Polymerization Method*

It involves the dispersion of clay in the monomer or monomer solution, and the resulting mixture is polymerized in the presence of catalyst or initiator by standard polymerization method. This has the advantage of grafting the polymer onto the clay or filler surface. Several types of nanocomposites including silica/Nylon 6, silica/poly 2-hydroxyethylmeth-acrylate, alumina/polymethylmethacrylate (PMMA), titania/PMMA, and $CaCO_3$/PMMA have been made by *in situ* polymerization. This method provides molecular-level mixing of the modified filler in the monomer or resultant nanocomposites due to easier dispersion of filler in a liquid than in a viscous melt (in melt-mixing method). Due to its hydrophilic nature, the clay can mix readily in nylon, while it does not mix with hydrophobic polymers (or monomers). For better mixing, the o-clay platelets are added into hydrophobic polymers. When the clay platelets are completely separated and dispersed in the matrix, the nano-composite is called "exfoliated nanocomposite." However, exfoliation requires high shear and high temperatures, which is a great challenge for the materials scientists/engineers. If the polymer enters between the o-clay platelets and forces them apart without disturbing the ordering of the platelets, such nanocomposites are called "intercalated nanocomposite."

10.2.2.2 Solution Method

The solution processing is the easiest approach if both the polymer and nanofillers are dispersible in a common solvent. The final nanocomposite is obtained by filtration or precipitation from the solvent or by film casting the solution. This is slow process and requires large amounts of solvent. The most common polymers which can be dissolved in common organic solvents are polycarbonate, polystyrene, PMMA, polyvinyl alcohol (PVA), etc.

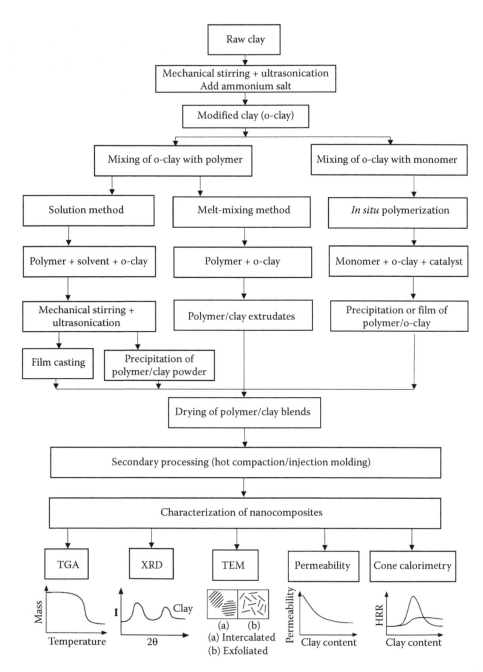

FIGURE 10.2
Fabrication of polymer/clay nanocomposites.

10.2.2.3 Melt-Mixing Method

Melt-mixing method is the most appropriate method for insoluble or immiscible polymers and nanofillers, and for high-production rate. Typically, high-shear mixing (i.e., a twin screw extruder or two-roll mill) and a careful choice of polymer resin and compatibilizer are required to achieve adequate dispersion of the nanofiller in the matrix. The compatibilizer or surfactant is required to control the melt viscosity because the addition of nanofiller increases the viscosity of the polymer in the melt significantly compared to the pure polymer. However, samples with >20 wt% filler content are difficult to process. For example, the processing of nanofiller/PP nanocomposites by twin screw extruder with better dispersion of filler is successful only after the modification of the filler interface. The modification makes the fillers compatible with the matrix. Similarly, a Brabender high-shear mixer has been successfully used to mix nanofillers with several polymers like PET and LDPE. Nonpolar polymers are very difficult to intercalate into clays because the clays are strongly polar. This challenge has been met by first intercalating stearylamine into clay and then melt-mixing the o-clay with MA-modified PP oligomers (PP-MA) [2]. A typical procedure to prepare a polymer/clay nanocomposite is shown in Figure 10.2. The prepared nanocomposites can have four types of structures (Figure 10.3). First, the structure of an intercalated nanocomposite is a tactoid with expanded interlayer spacing ($d_1 > d_0$). Second, exfoliated nanocomposites are formed when the individual clay layers break off the tactoid and are either randomly dispersed in the polymer (a disordered nanocomposite) (Figure 10.3d) or left in an ordered array (Figure 10.3c). However, conventional polymer/clay composites consist of tactoids with no change in interlayer spacing (or d-spacing).

As the layer spacing increases, the process can be monitored by x-ray diffraction (XRD). Intense peaks between 3° and 9° indicate an intercalated composite, but if the peaks are extremely broad or disappear completely, this indicates complete exfoliation.

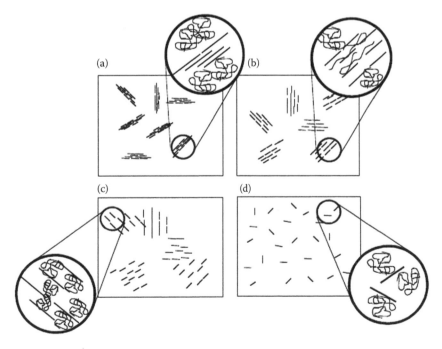

FIGURE 10.3
Schematic of the microstructures of the polymer/clay nanocomposites: (a) a conventional composite with tactoids, (b) an intercalated nanocomposite, (c) an ordered exfoliated nanocomposites, and (d) a disordered exfoliated nanocomposite. (Ajayan P. M., Schadler L. S., Braun P. V. (Eds.): *Nanocomposite Science and Technology,* 2003. Copyright Wiley-VCH Verlag GmbH & Co. KGaA. Reproduced with permission.)

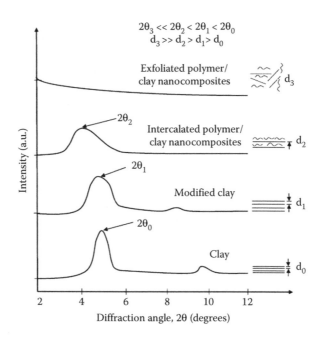

FIGURE 10.4

A typical XRD pattern of raw clay, o-clay, an intercalated polymer/clay, and exfoliated polymer/MMT nanocomposites.

Figure 10.4 shows a typical XRD pattern of raw clay, organically modified MMT (o-clay), an intercalated polymer/clay, and exfoliated polymer/MMT nanocomposites.

10.2.3 Characterization of Polymer/Clay Nanocomposites

XRD is carried out on each sample prepared to determine the change in gallery spacing, which may be seen as an increase in the d-spacing between the clay platelets compared to the unmodified clay. The d-spacing must be confirmed by using TEM. Thermogravimetric analysis (TGA) shows the actual weight % of filler loaded into the polymer matrix, by heating a small amount (ca. 6–10 mg) of the nanocomposite piece in air atmosphere and also used to confirm the improvement in thermal stability of the nanocomposites compared to pure polymer. The differential scanning calorimetry (DSC) of the samples is carried out to find whether the prepared samples consist of amorphous or semicrystalline polymer fraction in the nanocomposites [4]. There are three commonly used indirect methods such as cone calorimetry (ASTM E 1354), radiative gasification, and limiting oxygen index (LOI) (ASTM D2863, ISO 4589) to evaluate the fire properties of polymeric materials, depending on the type of information desired. The UL-94 (ISO 9772 and 9773, ASTM D635) test is a direct burning method used for commercial products to evaluate and qualify a material. Both cone calorimetry and radiative gasification provide information on the heat release rate (HRR), while the oxygen index measures the ease of extinction and the UL protocol evaluates the ease of ignitability. Under cone calorimetry, the heat flux is maintained at 35 kW/m^2, and the spark is continued until the sample is ignited. HRR, total heat released, effective heat of combustion (EHC), mass loss rate (MLR), CO yield, and CO_2 yield data can be recorded. This provides HRR versus time curves for the materials under investigation [5,6].

10.2.4 Properties of Polymer/Clay Nanocomposites

10.2.4.1 Thermal Properties

Heat deflection temperature: The addition of MMT in the PP matrix enhances the heat deflection temperature (HDT) significantly. The HDT of PP-functionalized MMT (f-MMT) is markedly increased from 109°C for the pure PP to 152°C for 6 wt% nanocomposite. However, above 6 wt% of the f-MMT, the HDT of the hybrid levels off. When the same PP is filled with o-MMT, the HDT is also increased but to a smaller extent compared to that of f-MMT. It is due to the lower exfoliation level of the MMT. The increase of HDT originates from the better mechanical stability of the nanocomposite, compared to the pure PP.

Flammability resistance: The resultant "char" provides protection to the interior of the specimen by preventing continual surface regeneration of available fuel to continue the combustion process. The addition of clay reduces the maximum HRR significantly, while the total heat release remains constant with increasing clay content. The relevance of reducing the maximum HRR is to minimize the flame propagation to adjacent areas in the range of the ignited material. In specific cases, the nanoparticle addition can result in reduced flammability rating due to the melt viscosity increase preventing dripping as a mechanism of flame extinguishment (e.g., change UL94 rating from V-2 to HB). The addition of o-clay into PE, nylon, and any other polymers reduces the peak HRR but there is no change in the total heat release with increasing clay content. Figure 10.5a and b shows the effect of clay content on the HRR of PE and nylon, respectively. It is to be noted that the curve position and shape may vary for different polymer matrix and filler content, the generalized behavior of decreased peak HRR with basically no change in the overall heat release (area under the curve) is very typical [7,8].

The improvement in flammability for polymer/clay nanocomposites is not caused by a change of the combustion process in the condensed phase, but it is due to a physical barrier provided by the clay sheets. For example, pure polymer does not form char during

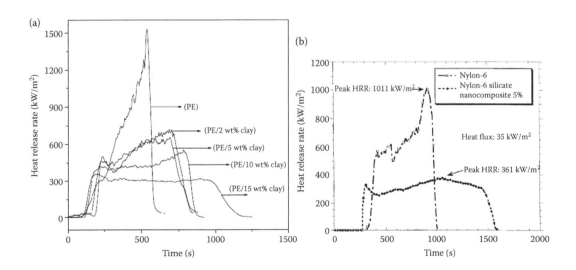

FIGURE 10.5
Effect of clay content on heat release rate of (a) PE/clay and (b) nylon/clay nanocomposites. (Reprinted from *Mater. Sci. Eng. R Rep.*, 28, Alexandre M., Dubois P., Polymer-layered silicate nanocomposites: Preparation, properties and uses of a new class of materials, 1–63, Copyright 2000, with permission from Elsevier; Reprinted from *Polym. Degrad. Stabil.*, 87, Zhao C. et al., Polymer mechanical, thermal and flammability properties of polyethylene/clay nanocomposites, 183–189, Copyright 2005, with permission from Elsevier.)

the combustion. However, polymer/clay nanocomposites leave a black char formed rapidly before ignition even in nanocomposites containing a small fraction of clay (<2 wt%). At the end of combustion, pure polymer leaves no residue, but the nanocomposites leave a solid char-like residue. One could see that the reduction of the peak HRR was ~54% on adding 2 phr sodium-MMT modified by (N-γ-trimethoxylsilanepropyl) octadecyldimethylammonium chloride (JS), but a further increase of modified clay resulted only in a marginal improvement on the peak HRR, that is, as long as the nanocomposite is formed, the protective char can be rapidly formed during the combustion and the peak HRR reduces dramatically, indicating that the formation of the protective char is induced by the presence of the clay [7]. Similarly, for a functionalized-PP nanocomposite with 4 wt% o-clay, there is a 75% reduction in flammability compared to the pure PP. This flame-retardant character is attributed to the formation of a carbonaceous char layer, which is developed on the outer surface during combustion. This surface char has a high concentration of clay layers and becomes an excellent insulator and a mass transport barrier, thus slows down the oxygen supply as well as the escape of the combustion products generated during decomposition. Since a few percent of clay is needed in the PP/MMT nanocomposites, the resulting hybrids are lightweight [9]. However, the incomplete surface coverage, aggregates of fillers, and poor bonding between filler and polymer matrices may lead to poorer flammability resistance [1].

The addition of 3–5 wt% clay into PMMA matrix nanocomposites reduces the peak HRR in the range of 20%–28% compared to pure PMMA. PMMA nanocomposites show a slightly longer time to ignition compared to neat PMMA [9]. Although polymer/clay nanocomposites exhibit improved flame retardancy, they may not show a rating in the UL 94 V protocol, which is required for industrial applications. To achieve an acceptable UL-94 V rating, the polymer/clay nanocomposites must be combined with conventional flame retardants. Ethylene-vinyl acetate (EVA) copolymer nanocomposites containing alumina trihydrate (ATH) with o-clay exhibited higher fire performance than either of them alone as a flame retardant, where the clay can strengthen the char formation [5].

Above mentioned examples indicate that the flammability of polymer/clay nanocomposites is significantly reduced as compared to the pure polymer. Similarly, when a specimen of the nylon-6/5 wt% silicate nanocomposite was heated at 35 kW/m, the maximal HRR was found almost one-thirds than that of the unfilled nylon-6 and the maximal flame temperature was reduced from 820 to <750 K. In addition, the burning time was extended from 1000 to ~1600 s. The reason for this improvement has been attributed to the formation of a ceramic (i.e., silicate) insulating layer at the surface that reduces the heat input to the residual material. This reduced flammability indicates a safety feature, which is extremely important in the automotive and aerospace industry [8].

Dimensional stability: High CTE of pure polymers causes dimensional changes during molding or solidification. The dimensional changes with varying temperatures are either undesirable or unacceptable for the applications where polymers or plastics are integrated with metals or ceramics. The mismatch in the CTEs between the silicon chip and the plastic substrates results in residual stresses, causing the premature thermal fatigue failure of the devices. In view of this, fillers are added to polymers to reduce the CTE. For conventional micrometer-sized or low-aspect-ratio fillers, the reduction in CTE follows, more or less, a simple additive rule of mixtures (RoM). However, when nanosized filler or high-aspect-ratio fillers (i.e., fibers or platelets) are added, the actual CTEs of nanocomposites would be much lower than those of conventional composites or values predicted from RoM. MMT platelets exhibit aspect ratio of >20 and are effective for reducing CTE of polymers as shown in Figure 10.6a. All polymers including nylon-6 have a higher CTE above their T_g than below T_g because of their lower modulus above T_g. Figure 10.6a shows that the addition of 7 wt% clay

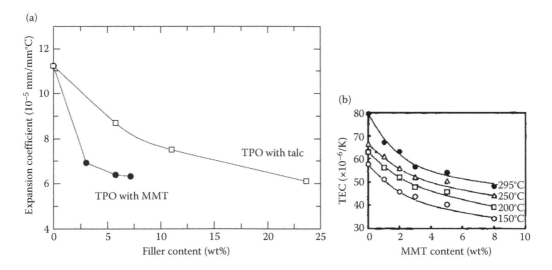

FIGURE 10.6
Linear coefficients of thermal expansion of (a) thermoplastic polyolefin composites as a function of filler (clay or talc) content and (b) polyimide/clay hybrid as a function of clay content and temperatures. (Yano K. et al.: Synthesis and properties of polyimide-clay hybrid. *J. Polym. Sci., Part A: Polym. Chem.*, 1993. 31. 2493–2498. Copyright Wiley-VCH Verlag GmbH & Co. KGaA. Reproduced with permission; Reprinted from *Polymer*, 47, Lee H. S. et al., 3528–3539, Copyright 2006, with permission from Elsevier.)

(or MMT) into a thermoplastic polyolefin (TPO) reduces the linear CTE from $> 110 \times 10^{-6}/°C$ for pure TPO to $< 60 \times 10^{-6}/°C$. The same extent of reduction in CTE was obtained when 20-25 wt% talc powder was added into the TPO. This indicates that a very small amount of clay (i.e., 5–7 wt%) decreases the CTE of the TPO significantly. This is due to the molecular level dispersion and high aspect ratio of the clay platelets which results in a large interfacial area between the clay platelets and the polymer matrix. Inside the interface, the polymer molecules are highly constrained than those of away from the interface. Thus, the CTE of the polymer/clay nanocomposites reduces significantly compared to that of pure polymer. For traditional composites, at low volume fractions of filler, the change in CTE with increasing filler content is almost found linear. In contrast to traditional composites, the CTE dependence on nanofiller is nonlinear due to a huge interfacial volume fraction between the nanofillers such as clay and the matrix. Figure 10.6b shows effect of the addition of MMT content on linear CTE of polyimide-clay hybrid at various temperatures. It can be seen that clay is effective in reducing the CTE of the polyimide measured at all temperatures [1,11,12].

10.2.4.2 Permeability

The absorption of water into polymer matrix composites or nanocomposites decreases the glass transition temperature and mechanical properties. In addition, absorption of water increases the dielectric constant of the composites which in turn decreases the speed of microwave signal in printed circuit boards (PCBs). Therefore, the reduced gas and liquid permeability of the nanocomposites makes them attractive for different application including membranes. There is a large decrease in permeability at ~2 wt% exfoliated clay-filled nanocomposites. Thereafter, the rate of decrease in permeability is reduced due to the formation of aggregates of clay at higher loading. The barrier, however, is sensitive to the degree of dispersion and the alignment of the plates. Significant decrease in permeability

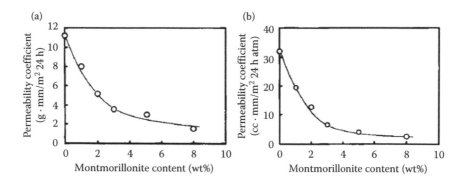

FIGURE 10.7
Effect of MMT content on permeability coefficient of water vapor (a) and oxygen (b) in polyimide–clay hybrid. (Yano K. et al.: Synthesis and properties of polyimide-clay hybrid. *J. Polym. Sci., Part A: Polym. Chem.* 1993. 31. 2493–2498. Copyright Wiley-VCH Verlag GmbH & Co. KGaA. Reproduced with permission.)

has been reported for exfoliated clay-filled PET, nylon, and polycaprolactone nanocomposites. The decreasing trend in permeability with increasing clay content may be liner or nonlinear, depending upon the aspect ratio and the alignment of the clay in the matrix, and the interaction between the clay and the matrix. For example, the higher the aspect ratio of the filler, the larger is the decrease in permeability of the nanocomposites.

The polyimide–clay hybrid has been found to possess superior gas barrier property compared to pure polyimide. Permeability coefficients of water vapor and oxygen gas are shown as a function of MMT content in Figure 10.7a and b, respectively. As the content of MMT in the polyimide/clay hybrid increases, the permeability coefficient of both water vapor as well as oxygen decreases remarkably. An MMT clay content of ~2 wt% decreases the permeability coefficient of water vapor and oxygen gas to a value of about one-thirds of that of pure polyimide. The dispersion of exfoliated clay sheets increases the total path of diffusing gas due to tortuous path. If clay sheets of length L and thickness (W) (L/W is also called aspect ratio) are dispersed parallel in a polymer matrix, the tortuosity factor can be determined by the following equation [12].

$$\tau = 1 + \left(\frac{L}{2W}\right) \cdot V_{\text{f}} \qquad (10.1)$$

where V_{f} is the volume fraction of clay content in the matrix.

Figure 10.8 shows the relative permeability (i.e., ratio of the permeability of the composite to the permeability of the pure polymer, P_c/P_0) of water as a function of silicate content. The permeability in the nanocomposite is significantly reduced compared to the pure polymer. It is interesting to see that the decrease in permeability is much more pronounced in the nanocomposites compared to the conventional composites with much higher-filler (i.e., mica or talc) content. The significant reduction of permeability in the nanocomposites is due to the presence of dispersed large-aspect-ratio silicate layers in the polymer matrix, which are impermeable to water molecules. This forces solutes traversing the film to follow a tortuous path through the polymer matrix surrounding the silicate particles, thereby increasing the effective path length for diffusion [13].

Since nonpermeable clay layers act as a barrier to water molecules or gas diffusion by increasing the tortuosity of the diffusion pathway, the dependence of permeability on loading can be estimated from the following equation [3]:

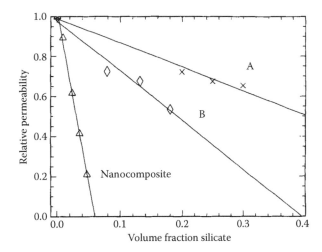

FIGURE 10.8
Relative permeability (P_c/P_0) as a function of filler content for polycaprolactone nanocomposites and conventionally filled silicate composites (A and B). (Giannels E. P.: Polymer layered silicate nanocomposites. *Adv. Mater.*, 1996. 8. 29–35. Copyright Wiley-VCH Verlag GmbH & Co. KGaA. Reproduced with permission.)

$$\frac{P_c}{P_0} = \frac{1}{1 + (L/2W)V_f} \tag{10.2}$$

where P_c and P_p are the permeabilities of the nanocomposite and the pure polymer, respectively. V_m and V_f are the volume fractions of polymer and filler. As indicated by the above equation, a large-aspect ratio will dramatically decrease permeability, provided the filler particles can be oriented parallel to the surface of the sample. For example, the relative permeability of the polyimide–clay nanocomposites with 7.4 vol% o-clay having aspect ratio of 192 is ~0.15 [14]. Furthermore, polymer/clay nanocomposites exhibit self-extinguishing characteristics when exposed to an open flame for ~30 s. The nanocomposite ceases burning after the flame is removed, and thus retains its integrity. In contrast, the pure polymer continues burning, leading to specimen destruction. In the nanocomposites, the silicate layers act most likely as barriers inhibiting gaseous products from diffusing to the flame and shielding the polymer from the heat flux. As more heat, produced from further oxidation of volatiles in the gas phase, is fed back into the polymer, further breakdown of the polymer takes place. In the nanocomposites, this self-sustaining cycle is suppressed, leading to self-extinguishing characteristics [13].

10.2.4.3 Mechanical Properties

The addition of o-MMT into PP leads to an increase in relaxation time and melt viscosity of the polymer. Figure 10.9 shows plots of the storage modulus and loss modulus as a function of frequency with increasing clay content. It can be seen that the storage modulus and the loss modulus increase with increasing clay content. An increase in the storage modulus of the nanocomposites as compared to the pure polymer is due to the possible increase in relaxation time as compared to the polymer. The PP/MMT nanocomposites show the presence of nonterminal behavior, which could be attributed to the presence of an ordered domain structure in the polymer system under flow due to the presence of interacting particles [5].

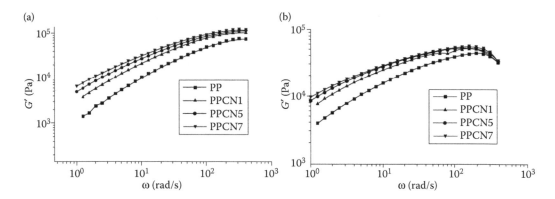

FIGURE 10.9
The variation in (a) storage modulus and (b) loss modulus with frequency for a PP/MMT nanocomposite containing 1%, 5%, and 7% clay, which are represented by PPCN1, PPCN5, and PPCN7, respectively. (From Gupta R. K., Kennel E., Kim K. J. (Eds.). 2010. *Polymer Nanocomposites Handbook.* CRC Press, Boca Raton, FL, USA. Reproduced with permission from Taylor & Francis Group.)

Figure 10.10 shows relative modulus (ratio of modulus of nanocomposite to modulus of matrix) as a function of filler (i.e., o-clay, functionalized MWNT, and modified silica nanoparticles) content in the semicrystalline PA6. It can be seen that the addition of o-clay has intermediate effect, compared to MWNT and modified silica, in enhancing the modulus of the PA6. In contrast, the addition of very small amounts of MWNT, due to its very high aspect ratio, can produce the largest modulus enhancement for PA6. Although o-clay stiffens the PA6 matrix, its addition drastically reduces the elongation at break [15].

A remarkable enhancement in the toughness of Polyvinylidene difluoride (PVDF) filled with o-clay (Cloisite® 30B) has been attributed to the structural and morphological changes induced by the o-clay (MMT) nanoparticles. The addition of o-MMT nanoparticles induces the phase formation of ordered α-phase crystallites to disordered fiber-like

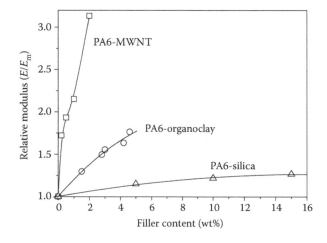

FIGURE 10.10
Relative modulus (E/E_m) versus filler (o-clay, functionalized MWNT, and modified silica) content for PA6 reinforced nanocomposites. E and E_m are Young's modulus of the nanocomposite and matrix, respectively. (Reprinted from *Mater. Sci. Eng. R*, 53, Tjong S. C., Structural and mechanical properties of polymer nanocomposites, 73–197, Copyright 2006, with permission from Elsevier.)

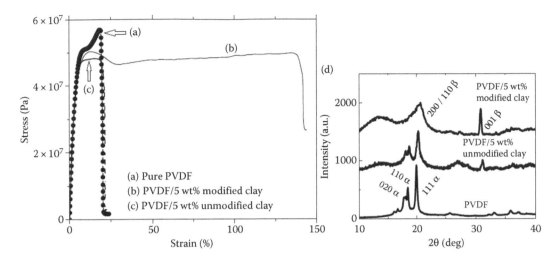

FIGURE 10.11
Stress–strain curves for (a) pure PVDF, (b) PVDF/5 wt% modified clay, (c) PVDF/5 wt% unmodified clay composites, and XRD patterns for (d) PVDF and its composites. (Shah D. et al.: *Adv. Mater.*, 2004. 16. 1173. Copyright Wiley-VCH Verlag GmbH & Co. KGaA. Reproduced with permission.)

β-phase crystallites (average size ~0.6 μm) in PVDF. This transition enhances the stiffness and toughness of PVDF significantly (Figure 10.11). Compared to α-phase crystallite, the fiber-like β-phase is more conducive to plastic flow under applied stress and hence, this could give rise to a more efficient energy dissipation in the nanocomposites, thereby delaying crack formation. As shown in Figure 10.11, Young's modulus of PVDF increases from 1.3 to 1.8 GPa, while the elongation at break increases from 20% (Figure 10.11a) to 140% (Figure 10.11b), giving the o-clay-filled nanocomposite a toughness of ~700% higher than that of the pure PVDF [16]. Moreover, the gallery spacing of the o-clay increased from 1.8 to 2.9 nm in o-clay/PVDF nanocomposites, whereas it remains unchanged for the unmodified clay-filled PVDF composites. This is probably the reason of no increase in elongation or slight decrease in toughness for the 5 wt% unmodified clay-filled PVDF composites (Figure 10.11c).

The XRD pattern (Figure 10.11d) clearly indicates that the (020), (110), and (111) peaks corresponding to the α-phase in pure PVDF disappear and are replaced by the (200/110) and (001) peaks corresponding to the β-phase in the nanocomposites. In contrast, unmodified clay leads to an immiscible composite with PVDF due to a lack of any favorable interaction between the polymer and the clay layers; hence, unmodified clay-filled PVDF composite comprised mostly α-phase with a weak β-phase at $2\theta = 31°$ [16].

In addition, the tensile modulus increased from 925 MPa for pure PMMA to 1225 MPa for a melt-blended nanocomposite of PMMA containing 3.8 wt% o-clay. However, the values of both tensile strength and tensile strain at break decreased, respectively, by 50% and 30% for the same loading of the clay. The decrease in tensile strength was attributed to the presence of agglomerates of clay (or tactoids) particles, which were distributed evenly in the cross section of sample. It is suggested that a strong cohesive force among PMMA chains and attractive forces between the adjacent clay particles in tactoids made PMMA/clay interfaces very weak and allowed for crack propagation during fracture through the interfaces [17]. In contrast, the tensile modulus, tensile strength, and Charpy-notched impact strength of the PMMA/MMT nanocomposites prepared by *in situ* polymerization method were increased from 730 to 1013 MPa, 66 to 88 MPa, and 8.42 to 12.9 kJ/m², respectively, at 0.6 wt% clay content. The significant increase in properties was attributed to the formation of PMMA/ MMT-intercalated nanocomposite with no phase separation between the MMT layers and the PMMA matrix. However, it was also found that MMA monomer did not diffuse into all of the MMT galleries during the polymerization process [18]. In brief, the mechanical properties of polymer matrix nanocomposites are controlled by several parameters such as; properties of the matrix and fillers, types and distribution of the fillers in the matrix, and interfacial bonding between filler and the matrix. In addition, the properties are also affected by the aspect ratio of filler and the processing methods. The interfaces may affect the effectiveness of load transfer from the polymer matrix to fillers. Thus, surface modification of the fillers is generally required to promote better dispersion of fillers in the matrix and to increase the interfacial adhesion between the matrix and fillers.

10.2.4.4 Abrasion and Wear Resistance

The abrasion resistance of filled polymers depends on size, shape, and aspect ratio of the added particles. The addition of nanoparticles (or fillers) has led to some interesting results. For micrometer-sized fillers, filler particles larger than the abrasive particles are stable and increase the abrasion resistance of the composite. As the filler size is decreased to a size similar to that of the abrading particles, filler particles are removed, and the abrasion resistance might decrease. However, in case of nanometer-sized fillers, the abrasion resistance is increased significantly with decreasing filler size. For example, the addition of 3 wt% $CaCO_3$ nanoparticles to PMMA increases the abrasion resistance approximately by a factor of 2 but decreases the coefficient of friction. In contrast, for micrometer-sized fillers, the increased wear resistance is accompanied by an increased coefficient of friction [3]. In most of the cases of nanocomposites, the coefficient of friction measured by dry sliding using pin-on-disk tester against steel decreases monotonically with increasing nanofiller content due to the formation of the tenacious transfer film on the counterface.

10.3 Polymer/CNT Nanocomposites

As discussed in Section 6.5, the CNTs exhibit extremely high-tensile modulus ($\sim$1 TPa) and strength ($\sim$150 GPa), intrinsic electrical conductivities of ca. 10^5–10^6 S/m for metallic CNTs and 10 S/m for semiconducting CNTs, and high-thermal conductivities (3000–6000 W/mK). In addition, the CNTs exhibit high flexibility, low density (1.3–1.4 g/cm^3), and large-aspect ratios (up to 1000 or more). Due to the unique combination of thermal, mechanical, and electrical properties, the addition of CNTs can improve thermal, mechanical, and electrical properties of the polymer matrix nanocomposites. However, the extent of improvement depends upon the degree of dispersion and alignment of the CNTs in the polymeric matrix.

10.3.1 Electrical Properties

The electrical conductivity of SWNT bundles has been found to vary between 1×10^4 S/m and 3×10^6 S/m at room temperature. In contrast, individual multiwalled carbon nanotubes (MWNTs) have shown the conductivity between 20 S/m and 2×10^7 S/m, depending on the helicities of the outermost shells and the presence of defects such as the Stone–Wales defect, vacancies, and impurities [19]. Polymers are known for their very low-electrical conductivity, that is, their electrical conductivity is $<10^{-14}$ S/m. In order to obtain electrical conductivity in the polymer matrix composite or nanocomposites, conductive particles or fibers are introduced into the polymer matrices. The introduction of conductive fillers including nanotubes into these matrices can result in significant improvement in conductivities. The conductive nanocomposites have been used in variety of applications, including housings for cell phones and computers, lightning strike protection for aircraft, chemical sensors, and transparent conductive coatings. Moreover, electromagnetic interference (EMI) shielding of portable electronic devices such as laptop computers, cell phones, and pagers is important to prevent the interference with and from other electronic devices. A basic understanding of the fundamental mechanisms associated with improved electrical conductivities in nanocomposites is important for realization of devices for these applications.

As shown in Figure 10.12, the electrical conductivity of conductive-filler-filled polymer nanocomposites is a strong function of volume fraction of conductive filler. At low-filler loadings, the conductivity of the composite is close to that of the polymer matrix. At or above some critical volume fraction (i.e., V_c), the conductivity increases by several orders of magnitude with slight increase in the volume fraction of filler. Just above the percolation threshold, the filler particles begin to form 3-D (i.e., continuous) conductive networks in the composite. This 3-D network of conductive fillers helps in electrons flow from one end to another end of the sample.

In the matrix, the conductive particles either must touch or at least they must be close to each other so that electrons can hop from one filler particle to another nearest particles, which is possible only when conductive particles form a 3-D conductive network in the matrix. The probability of forming a 3-D conductive network increases with increasing volume fraction of the conductive phase. The critical volume fraction (V_c) at which such a continuous 3-D network formed is called the percolation threshold. It is the volume fraction of the conductive particles required for the onset of electrical conductivity. Above the

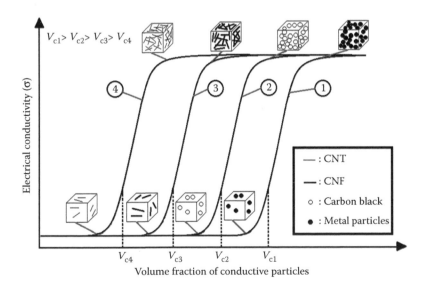

FIGURE 10.12
Schematic diagram for the electrical conductivity as a function of volume fraction of various fillers in a typical polymer matrix composite.

percolation threshold, the electrical conductivity of the nanocomposites can be described by the following equation:

$$\sigma = \sigma_0(V - V_c)^t$$
$$\text{or} \quad \log(\sigma) = t \times \log(V - V_c) + \log(\sigma_0) \tag{10.3}$$

where σ_0 is the conductivity of the conductive phase and V is the volume fraction of the conductive phase. The exponent t is the dimensionality of the network, which varies in the range 1.3 to 3. In some cases, this value may be more or less depending upon the degree of dispersion. One would get a straight line if graph is plotted between $\log(\sigma)$ versus $\log(V - V_c)$. The percolation is extremely sensitive to the shape, size, and aspect ratio of the particles or reinforcement. For example, the concentration of particles necessary for the onset of percolation for fibers is many orders of magnitude less compared to spherical particles. That is, the percolation threshold is inversely proportional to the aspect ratio of the nanofibers or nanotubes. However, high-aspect ratio nanotubes are generally more difficult to disperse in the matrix due to their strong tendency to entangle and to agglomerates. In order to obtain an ultralow-percolation threshold of the order of 10^{-3} vol%, a very high-aspect ratio and sufficient disentanglement of CNTs are required. In addition, when the aspect ratio is reduced to 20, the percolation threshold may jump to 1–10 vol%, regardless of the dispersion states of the CNTs. Hence, long fiber with a given diameter reduces the percolation threshold, as they have a huge aspect ratio. Above the percolation threshold, further addition of the conductive fillers levels off the electrical conductivity, but does not approach to a value close to that of the conductive filler. It is due to the insulating polymer coating onto the conductive fibers or particles and/or high-interfacial contact resistance between the fillers and the matrix [20–23].

CNTs-filled polymer nanocomposites show percolation threshold as low as 0.002 wt% and as high as 15 wt% or higher, depending upon the fabrication process, type, and purity of CNTs, aspect ratio of CNTs, degree of dispersion of CNTs in the matrix, etc. Above percolation, the electrical conductivity increases sharply by several orders of magnitude. The significant difference in both percolation threshold and the highest obtained electrical conductivity for a given nanocomposite system has been found due to the decrease in aspect ratio by the breakage of fibers or formation of fibers bundles (or agglomerates) during processing. The electrical conductivity decreases if the nanofibers or nanotubes are poorly dispersed, while it might be enhanced if dispersion is good. The type of mixing processes such as sonication, mechanically stirring, and calendaring approaches and/or their combination may affect percolation thresholds. The calendaring technique separates the CNTs and sheaths them with an insulating polymer layer more effectively than the dissolver disk. The polymer layer on CNTs may be too thick to allow the tunneling of electrons from neighboring CNTs. Thus, calendaring technique seems to be detrimental for the low-percolation thresholds, but it is a suitable method for improving the mechanical properties of polymeric composites [19].

The percolation threshold in nanocomposites is heavily dependent on the chemical, physical, and geometrical parameters such as the type of CNT, synthesis method, surface treatment, aspect ratio, polymer type (amorphous or semicrystalline), and dispersion method. These factors are not always independent, and it is often difficult to separate their effects on the percolation threshold and the corresponding conductivities. The spatial distribution or dispersion of the nanofillers is also one of the most significant factors that may influence the resulting percolation threshold in nanocomposites. A conductive network or path is more easily achieved when there are a number of conductive particles available throughout the matrix. The strong van der Waals interactions between nanotubes coupled with their large-aspect ratio cause them to bundle together and form agglomerates. As the CNTs bundle together into larger and larger agglomerates, the number of discrete conductive points in the composite is reduced, thereby increasing the percolation threshold. In contrast, a very uniform dispersion can result in the sheathing of the nanotubes by the polymer causing an increased resistance to electron tunneling, resulting again in increased percolation thresholds. Therefore, a uniform spatial distribution of the nanotubes might not provide the optimal conditions from an electrical conductivity performance point of view [19].

The chemical functionalization can significantly affect the dispersability of the CNTs, and the functionalization of CNTs can increase the mechanical properties of the nanocomposite by forming stronger bonds between the matrix and the nanotube. However, the chemical functionalization introduces defects in the CNTs structure, which tend to scatter electrons and generally cause a reduction in their intrinsic electrical conductivity. Moreover, the chemical functionalization tends to reduce the aspect ratio of the nanotubes, which may increase the percolation threshold of the nanocomposites. For example, the lowest percolation threshold for epoxy/CNTs nanocomposites reinforced with nonfunctionalized CNTs is <0.1 wt%, whereas the amino-functionalized CNTs showed a percolation threshold in the range of 0.1–0.5 wt%. In such nanocomposites, the nanotubes' surface groups form an electrically insulating epoxy layer, which increases the distance between individual tubes, thus making the tunneling of electrons from tube to tube difficult [23].

Poor mixing techniques can result in decrease of electrical conductivity by several orders of magnitude than the expected values. The polymer–filler interfaces inhibit electron transport and thus, the interfacial resistance is the dominant effect on nanocomposite electrical performance, and lowers the electrical conductivity by several orders of magnitude. Consequently, polymer/SWNT nanocomposites have not achieved the same levels of conductivity as the polymer/MWNT nanocomposites and polymer/CNF nanocomposites. A very small percolation threshold is possible with an aspect ratio of >10,000. For a given process (i.e., stirring and sonication) for carbon fibers (aspect ratio = ~100)-filled epoxy polymer composites, the electrical conductivity increases in the low-viscosity polymer matrix by more than 10 orders of magnitude over a carbon fiber loading in the range 0 to 5 wt%, whereas the electrical conductivity increases by less than 8 orders of magnitude in the high-viscosity polymer matrix at a given temperature [19]. Sain et al. reported the electrical conductivity of SWNTs- and MWNTs-filled flexible PC/CNTs nanocomposites and found that the percolation threshold was ~0.5 vol% for SWNTs and 4 vol% for MWNTs-filled PC. This is due to the higher-aspect ratio and high purity of SWNTs compared to that of MWNTs [24]. The semiconductive and conductive polymer matrix nanocomposites are widely used for electrostatic discharge (ESD) and EMI-shielding applications. The ESD is a rapid transfer of electrostatic charge between two objects, usually resulting when two objects at different potentials come in contact with each other. Dielectric materials such as polymers can acquire large static potentials and accumulate static charge near vapors of gasoline or other potentially explosive vapors or gases. Hence, polymer nanocomposites containing conductive nanoparticles are less susceptible to ESD. In addition, these electrically conductive polymer nanocomposites help in shielding the electronic devices from unwanted electromagnetic waves, that is, EMI [22].

10.3.2 Mechanical Properties

The addition of 1 wt% MWNTs with aspect ratio >1000 to the epoxy matrix increases Young's modulus and the yield strength of the epoxy/MWNTs nanocomposites by

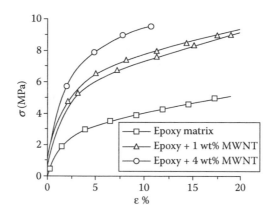

FIGURE 10.13
Tensile stress–strain curves of the pure epoxy and epoxy/MWNTs nanocomposites. (Reprinted from *Compos. Sci. Technol.*, 62, Allaoui A. et al., Mechanical and electrical properties of a MWNT/epoxy composite, 1993–1998. Copyright 2002, with permission from Elsevier.)

twofold and threefold, respectively (Figure 10.13). This is due to the reinforcing effect of high-aspect-ratio MWNTs. However, further addition of MWNT to the epoxy decreases the rate of increase in modulus and strength [25]. This is due to the formation of CNT aggregates, which results in pores in the nanocomposites. These pores are the sites of nucleation of cracks; hence, a drop in rate of increase in strength occurs. The challenges for CNT–polymers are difficulty in uniform dispersion, achieving better nanotube–matrix adhesion, providing stress transfer matrix to CNTs and reducing the intrabundle sliding in SWNTs.

Despite excellent mechanical properties of CNTs, the physical properties of the CNT/polymer nanocomposites are not improved significantly because of the weak interfacial bonding between the CNTs and the polymer matrix. The treatment of CNTs by an acidic solution and subsequent modification by amine treatment or plasma oxidation improves interfacial bonding and dispersion of CNTs in the polymer matrix. Tensile strength and elongation at break of surface-modified CNT/epoxy composites are higher than those of untreated CNT/epoxy composites because functional groups which are formed on the CNT surface improve the interfacial bonding between the CNTs and the surrounding matrix. The interfacial bonding enables an effective stress transfer between the polymer and the CNTs. It is interesting to see that the tensile strength of the plasma-treated CNT/epoxy nanocomposite is increased by 124% compared to that of pure epoxy. Young's modulus is changed slightly because the CNT content is very small but Young's modulus of plasma-treated CNT/epoxy composite is exceptionally high (Table 10.1) [26].

The stress–strain curve of a fiber containing 60 wt% SWNT bounded with polyvinyl alcohol (PVA) matrix is shown in Figure 10.14. The diameter of the tensile specimen was 50 μm. It can be seen that the SWNT/PVA nanocomposite has an ultimate strength, Young's modulus, and yield strength of ~1.8, 80, and 0.7 GPa, respectively. Interestingly, the large plastic deformation can be seen after the yield stress is reached. Such high deformation has been attributed to the absence of necking in SWNT/PVA fibers and/or slippage between the individual SWNTs within the fiber. The ultimate stress of the SWNT/PVA nanocomposite fiber is approximately two- or threefold than that of an average steel and comparable with that of steel wires [20,21]. Similarly, the addition of 3 wt% graphene oxide to polyimide increases the strength to about 0.8 GPa, which is about 10-fold of the pure

TABLE 10.1

Mechanical Properties of CNT/Epoxy Nanocomposites

Fillers	Tensile Strength (MPa)	Young's Modulus (GPa)	Elongation at Break (%)
Pure epoxy	26	1.21	2.33
Epoxy/untreated CNTs	42	1.38	3.83
Epoxy/acid-treated CNTs	44	1.22	4.94
Epoxy/amine-treated CNTs	47	1.23	4.72
Epoxy/plasma-treated CNTs	58	1.61	5.22

Source: Kim A. J. et al. R. 2006. *Carbon* 44: 1898–1905.

polyimide. However, in contrast to the SWNT/PVA nanocomposite, the maximal strain of polyimide/graphene oxide is reduced significantly as compared to pure polyimide [22].

An improvement in modulus and strength for particles-filled polymer nanocomposites depends strongly on the degree of interaction between the particles and the polymer. For example, in case of poly(methyl methacrylate) (PMMA) polymer nanocomposites filled with alumina, the modulus was found to decrease with increasing alumina content, due to the weak interaction between the alumina and the PMMA. In contrast, polystyrene nanocomposites reinforced with silica nanoparticles showed increased modulus with increasing silica content due to a strong physical bonding between the matrix and the silica nanoparticles. Since the nanoparticles have sizes smaller than the critical crack length that typically initiates failure in composites, the addition of the nanoparticles provides improved toughness and strength. However, agglomeration of nanoparticles should be avoided otherwise the presence of agglomeration can decrease the strain to failure. The mechanical properties of the nanocomposites are also affected by the alignment of the fibers including MWNTs. For example, the modulus of the aligned MWNT/polystyrene nanocomposites increases as much as 49%, compared to a 10% increase for randomly oriented nanocomposites. In addition, aligned MWNTs/polystyrene nanocomposites exhibited an increase in yield strength and ultimate yield strength compared to a pure polystyrene polymer [20].

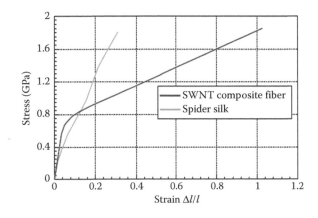

FIGURE 10.14

Stress–strain diagram of an SWNT fiber of 50 μm in size. (Vollath D. *Nanomaterials: An Introduction to Synthesis, Properties and Applications.* 2nd ed. 2013. Copyright Wiley-VCH Verlag GmbH & Co. KGaA. Reproduced with permission.)

10.3.3 Thermal Conductivity

Typically, the thermal conductivity of thermoplastic polymers is ~0.2 W/mK, whereas the thermal conductivity of most metals is more than 100 times than those of polymers. Theoretical thermal conductivity for the CNTs is ~6000 W/mK, which is highest among materials. An experimental value for MWNT increases with increasing temperature up to a maximum value of 3000 W/mK at about 300 K. In contrast, bulk samples of MWNTs exhibit a thermal conductivity of only 25 W/mK at room temperature. The room temperature thermal conductivities for the highly entangled SWNT bundles range from 2.3 to 35 W/mK. The thermal conductivity of aligned SWNT thin films increased with temperature up to 400 K, reaching a maximum value of ~200 W/mK, which is approximately an order of magnitude higher than the highest value obtained for the random entangled samples [19]. The high-thermal conductivities of both SWNTs and MWNTs show tremendous potential for CNTs to be used in thermal management applications such as heat sinks in electrical circuitry.

The thermal conductivity of epoxy resin filled with 1 wt% SWNTs (Figure 10.15) is doubled compared to that of pure epoxy, while the same wt% carbon fibers increase only ~40%. This significant increase in thermal conductivity of the CNTs-filled polymer nanocomposites may be useful for the thermal management applications, including heat sink of silicon processors and in housing for electric motors [27].

Moreover, the addition of a very low amount of MWNTs or SWNTs into a silicone rubber matrix substantially improves the thermal conductivity as shown in Figure 10.16. For the same amount of thermal conductivity enhancement, only 2 vol% CNTs is required against 30 vol% silicon carbide or aluminum nitride. This is due to the fact that the CNTs have a higher-thermal conductivity (as high as 6000 W/mK) as compared to the silicon carbide or aluminum nitride and partly because the CNTs can form a conducting network within the polymer matrix. Figure 10.16 also shows that the acid treatment of CNTs helps to further increase the thermal conductivity, possibly by promoting adhesion between the CNTs and the matrix, and by promoting contacts among the CNTs [5].

The nitric acid treatment can introduce hydroxyl, carboxylic, or carbonyl groups on the CNT surface, which facilitates covalent bond formation between the polymer matrix and the surface of nanotubes. Hence, the acid-treated CNT resulted in greater improvement of

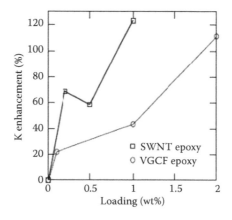

FIGURE 10.15
Thermal conductivity (%) of epoxy resin as a function of SWNTs and carbon fibers content. (Reproduced from Biercuk M. J. et al. 2002. *Appl. Phys. Lett.*, 80: 2767–2769. With permission.)

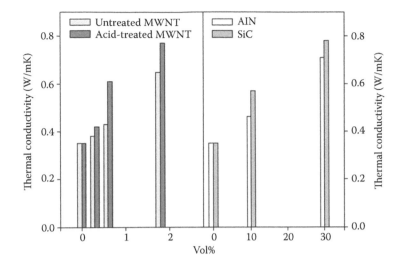

FIGURE 10.16
The enhancement in thermal conductivity of silicone resin by the addition of silicon carbide (SiC), aluminum nitride, and multiwalled carbon nanotubes (MWNTs). (From Gupta R. K., Kennel E., Kim K. J. (Eds.). 2010. *Polymer Nanocomposites Handbook*. CRC Press, Boca Raton, FL, USA. Reproduced with permission from Taylor & Francis.)

the thermal conductivity as shown in Figure 10.17a. The surface treatment helps in better dispersion of the fillers (or CNT in present case), which in turn results in better thermal conductivity improvement for epoxy/MWNT nanocomposite (Figure 10.17b). Figure 10.17c shows the thermal conductivity of the polyimide/MWNT along the longitudinal and transverse direction of the heat flow. It can be seen that the thermal conductivity along the longitudinal direction is much higher than the transverse direction. The difference in their thermal conductivities increases very rapidly with increasing MWNT loading, that is, the thermal conductivity along longitudinally aligned samples (with higher CNT content) is significantly higher than that of transverse direction. For a given filler loading along the longitudinal direction, an almost 40-fold increase in the thermal conductivity was found for expanded graphite-filled polymer compared to only a twofold and 12-fold increase along the transverse direction and the isotropic distribution, respectively. In brief, an increase in trend of thermal conductivity for the nanocomposites with respect to fiber orientation can be arranged as follows: transverse < isotropic < longitudinal. Thus, thermally conductive plastics are therefore suitable for use in microelectronic, heat sinks in electronic devices, tubing for heat exchangers, enclosures for electrical appliances, casings for small motors, and heat exchangers, automotive light reflectors, laser-diode encapsulation, fluorescent ballasts, heat sensors, and switches [5].

10.4 Polymer/BN Nanocomposites

The hexagonal boron nitride (h-BN) is an attractive reinforcement for the polymers due to its low density (2.1 g/cm^3), high-mechanical strength (Young's modulus: 0.7–0.9 TPa, yield strength: ~35 GPa), superior thermal (stable up to 1000°C under atmospheric conditions) and chemical stabilities, and intrinsic electrical insulating property. The h-BN is found in

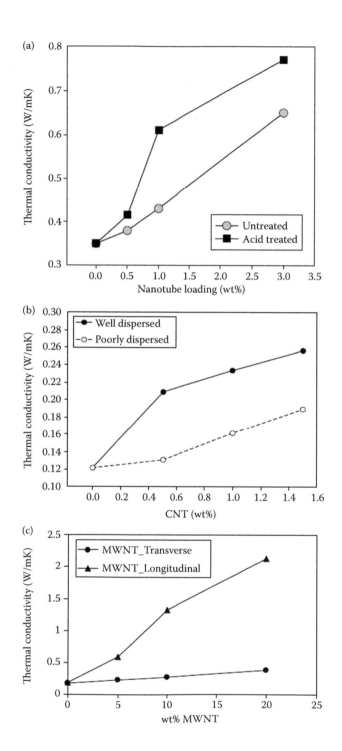

FIGURE 10.17
Variation in thermal conductivity of polymer/CNT nanocomposites as a function of CNT content and the factors affecting the thermal conductivity of the nanocomposites: (a) acid-treated MWNT in silicone resin, (b) dispersion of MWNT in epoxy, and (c) alignment of MWNT in polyimide matrix. (From Gupta R. K., Kennel E., Kim K. J. (Eds.). 2010. *Polymer Nanocomposites Handbook*. CRC Press, Boca Raton, FL, USA. Reproduced with permission from Taylor & Francis.)

various forms such as nanoplatelets, nanotubes, and nanosheets. Owing to these excellent properties, h-BN nanosheets have been used as reinforcement in the polymer matrix nanocomposites to improve the thermal, mechanical, and tribological (wear resistance and coefficient of friction) properties. Due to its layered structure and high-surface area, it has been used for the improvement of wear resistance and friction coefficient [28].

10.4.1 Mechanical Properties

The addition of noncovalently functionalized h-BN nanoflakes (BNNFs) into epoxy resin yielded an elastic modulus of 3.34 GPa, and ultimate tensile strength of 71.9 MPa at 0.3 wt%, corresponding to the increases of 21% and 54%, respectively. The toughness, measured from the area under stress–strain curve, was increased from 0.758 for pure epoxy to 1.573 MJ/m^3 for 0.3 wt% 1-pyrenebutyric acid (PBA) molecules modified-BNNF nanocomposites, that is, the toughness was increased as high as 107% compared to the value of pure epoxy. This implies that the applied load might effectively be transferred to PBA–BNNFs through the interfacial interactions. PBA molecules interact with BNNFs via π–π interactions and inhibit the restacking and aggregation of BNNFs. This results in a relatively large surface area of BNNFs in the nanocomposites, thus increasing the interfacial area between the BNNFs and the epoxy matrix. Additionally, it improves the dispersion of the BNNFs in the epoxy matrix. Furthermore, the creep strain and the creep compliance of the noncovalently functionalized epoxy/BNNFs nanocomposite was significantly less than the neat epoxy and the nonfunctionalized BNNF nanocomposite [29].

10.4.2 Thermal Properties

In the last few decades, due to the fast development and miniaturization of electronic devices, the thermal management has become one of the most critical challenges, particularly in electronic packaging in order to dissipate the large heat flux from high-density/high-power ICs. This ensures high performance and long lifetime for the electronics. In polymeric packaging, the ICs are mostly encapsulated by polymeric materials, which provide an important pathway for heat dissipation. However, the thermal conductivity of the typical polymeric encapsulant is far below the typical needed for efficient heat dissipation. Therefore, thermally conducting ceramic particles are added into the polymer matrix. The thermal conductivity of pure epoxy is reported to be ~0.15 W/mK and it increases with increasing h-BN due to its higher-thermal conductivity. In order to achieve the high-thermal conductivity for the composite, a high loading (i.e., >50 wt%) is required for conventional micrometer-sized spherical or flake-like fillers. Such high loading usually results in poor processability/flowability and high cost, etc., causing the limited practical applications for IC encapsulation. In contrast, due to high-aspect ratio of nanosheets, less loading is required for the same or better improvement. The addition of 5 wt% exfoliated h-BN nanosheet to the epoxy matrix enhanced the thermal conductivity of the epoxy/h-BN nanocomposites by ~113%, but increased to ~28% for micrometer-sized h-BN (h-BN control)-filled epoxy composites compared to pure epoxy (Figure 10.18a). Moreover, the addition of 30 wt% h-BN nanosheet and microsized h-BN particles into epoxy matrix increases the thermal conductivity to about 316% and 247%, respectively. It was also found that the effect of using h-BN nanosheets becomes less obvious at high h-BN loading probably due to the large-thermal boundary resistance [30]. The polyimide/BN composites consisting of 30 wt% of micro- and nanosized BN at the weight ratio of 7:3 showed the highest-thermal conductivity 1.2 W/mk, which is about five times higher than that of

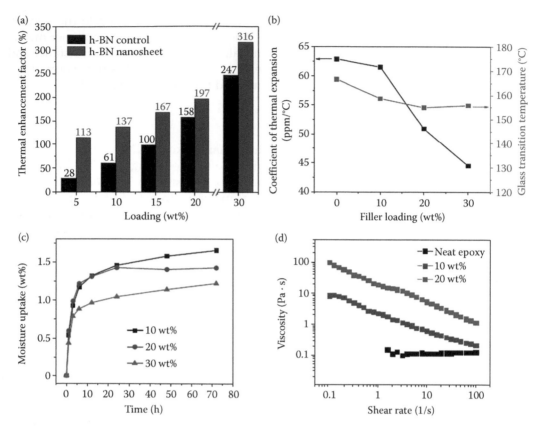

FIGURE 10.18

(a) Thermal enhancement factors of h-BN-nanosheet-based composites and h-BN-control-based nanocomposites; (b) CTE and T_g, (c) moisture adsorption, and (d) the viscosity of h-BN nanosheet/epoxy composites. (Reprinted from *Compos. Sci. Technol.*, 90, Lin Z. et al., Exfoliated hexagonal boron nitride-based polymer nanocomposite with enhanced thermal conductivity for electronic encapsulation, 123–128. Copyright 2014, with permission from Elsevier.)

the polyimide matrix. The high-thermal conductivity at an appropriate ratio of micro- to nanosized BN particles in the polyimide film has been attributed to the formation of a random conductive network. In such cases, the microsized BN particles formed the main thermally conductive path in composites, and the nanosized BN particles probably played the connecting role between microsized BN particles to enhance more contact to obtain high-thermal conductivity [31].

As shown in Figure 10.18b, the CTE of h-BN/epoxy nanocomposites decreases with increasing h-BN nanosheet contents reaching 45×10^{-6}/K at 30 wt% loading. The decreases in CTE may be attributed to the low CTE of h-BN nanosheets compared to pure epoxy ($\sim 62.5 \times 10^{-6}$/K). The T_g of nanocomposites slightly decreases from 167°C for pure epoxy to 156°C for 30 wt% loadings due to the large free volume at the filler–epoxy interfaces. The addition of h-BN also reduces the moisture adsorption of the h-BN nanosheet/epoxy nanocomposite (Figure 10.18c). The nanocomposites with higher h-BN loadings have lower-moisture adsorption due to impermeable h-BN nanosheet for the moisture. The decrease in moisture adsorption for the nanocomposites indicates that the h-BN/epoxy may be suitable for electronic-packaging applications. The addition of h-BN increases the thermal conductivity and strength significantly, and decreases CTE and moisture adsorption;

however, the viscosity of the composites or nanocomposites dramatically increases due to the high-surface area of h-BN (Figure 10.18d) [30].

10.4.3 Wear Resistance

The addition of 0–5 wt% (i.e., 1.33 vol%) h-BN into the poly(aryletherketone) (PAEK)/h-BN nanocomposites prepared by planetary ball mill followed by hot pressing enhanced the microhardness and wear resistance compared to pure matrix. Figure 10.19a shows the microhardness of the PAEK/h-BN nanocomposites. The microhardness of the nanocomposite containing up to 1 wt% h-BN increased sharply, thereafter it increased steadily with increasing h-BN. At higher loading, the nanoparticles form aggregates which results in less effective surface area than the actual, for the interaction with the matrix. Therefore, the rate of increase in microhardness is decreased at higher loadings. The microhardness of the 5 wt% (i.e., 1.33 vol%) h-BN nanocomposite increased to 34.2 kg/mm^2, indicating ~31% increase over the pure matrix. The increase in microhardness was attributed to comparatively higher hardness of h-BN and its better dispersion in the matrix, enhancing

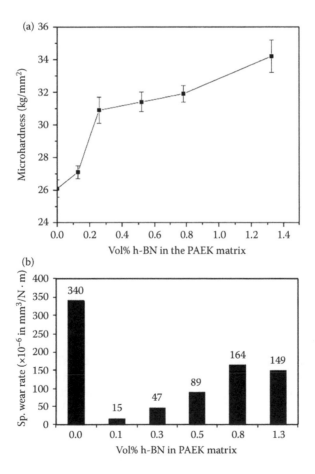

FIGURE 10.19
(a) Microhardness, and (b) specific wear rate of the PAEK/h-BN nanocomposites. (Joshi M. D. et al.: Tribological and thermal properties of hexagonal boron nitride filled high-performance polymer nanocomposites. *J. Appl. Polym. Sci.*, 2016, 133: 44409. Copyright Wiley-VCH Verlag GmbH & Co. KGaA. Reproduced with permission.)

resistance to deformation of the nanocomposites. The specific wear rate (i.e., inverse of wear resistance) of the PAEK/h-BN nanocomposites determined by using pin-on-disk tribometer in dry condition is shown in Figure 10.19b. The specific wear rate reached to a minimum level of 15.2×10^{-6} mm³/N m at 0.5 wt% (0.12 vol%), that is, the specific wear rate of the nanocomposites decreased approximately 22 times as compared to pure matrix. In addition, the coefficient of friction of the PAEK/h-BN nanocomposites was found to increase slightly but it became stable compared to that of pure matrix [28].

10.5 Polymer/BaTiO₃ Nanocomposites

Embedded capacitor technologies play an important role for miniaturization of various electronic/electrical devices because they have major benefits over traditional discrete capacitors. The material for embedded capacitors should be processable at low-processing temperature. They should be tough and flexible in order to be compatible with the PCB manufacturing processes. The polymers tend to have low-dielectric constant; hence, they do not meet the requirements of the capacitors. On the other hand, ferroelectric ceramics have high-dielectric constant but require high-processing temperatures. Polymer–ceramic nanocomposites are one of the best candidates for embedded capacitor because of their unique combination of good processability, flexibility, high-dielectric constant, low-processing temperature, and low cost. Ferroelectric ceramics including BaTiO₃ are widely used in capacitors because of their high-dielectric constant [32].

10.5.1 Introduction

The dielectric constant of BaTiO3 ceramic can be as high as 5000, depending on its grain size, purity, crystallographic direction, measurement temperature range, and method of preparation. It has low-direct current leakage, low-loss tangent (0.009), low-coefficient of thermal expansion, and high-thermal and chemical stability. However, it is brittle and requires high-processing temperatures to compact it. These drawbacks can be overcome by incorporating the suitable amount of BaTiO₃ powders in the polymer matrices. The addition of BaTiO₃ powders in the polymer matrices improves the electrical, thermal, and mechanical properties [33,34]. Out of three most commonly used methods, the solution method is described in Section 10.5.2.

10.5.2 Preparation of PC/BaTiO₃ Nanocomposites

In this section, fabrication of cubic- and tetragonal-phased-BaTiO₃-nanoparticle-filled polycarbonate nanocomposites is discussed. As shown in a schematic flow chart (Figure 10.20), an appropriate amount of BaTiO₃ powder is dispersed in dichloromethane (DCM) solvent by sonicating it in an ultrasonic bath for 1 h in order to break down the agglomerates and to obtain a highly dispersed BaTiO₃ suspension. Thereafter, the PC granules are added slowly to the DCM/BaTiO₃ suspension with concurrent ultrasonication. The solution can be further sonicated for 1 h to produce a homogeneous PC/BaTiO₃ suspension. The suspension is then stirred and heated continuously on a mechanical stirrer cum heater till a viscous solution was obtained. This viscous solution is poured on a glass plate to get its film. The films are further dried in a vacuum oven at a temperature above the glass transition temperature to eliminate completely the residue of solvent and moisture if any. From

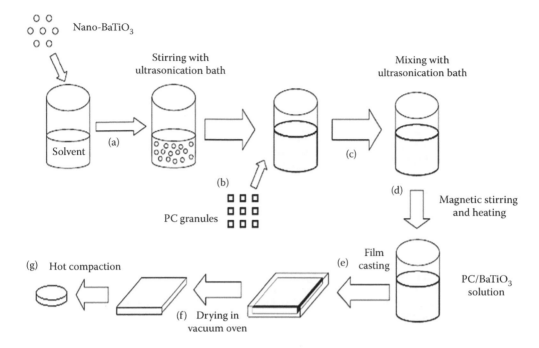

FIGURE 10.20
Schematic diagram of fabrication process of nanocomposites: (a) BaTiO$_3$ nanoparticles were dispersed in DCM solvent using ultrasonication bath; (b) PC granules were dissolved in solvent containing BaTiO$_3$; (c) mixing of PC/BaTiO$_3$/DCM using ultrasonic bath; (d) magnetic stirring and heating for evaporation of solvent; (e) film casting onto a clean glass plate; (f) drying of films in vacuum oven at 200°C; and (g) hot compaction of films for characterization. (Reprinted from *Mater. Chem. Phys.*, 183, Thanki A. A., Goyal R. K., Study on effect of cubic-and tetragonal phased BaTiO$_3$ on the electrical and thermal properties of polymeric nanocomposites, 447–456. Copyright 2016, with permission from Elsevier.)

the nanocomposite film, a substrate of desired shape and size with a suitable strength can be made by hot pressing. For this, weighted amount of dried film pieces is filled in a tool steel die and the samples inside the die are heated at a suitable heating rate. Then, hot pressing is carried out at a temperature above the glass transition under a sufficient pressure to obtain pores free samples [32].

10.5.3 Electrical Properties

Dielectric constant of the cubic BaTiO$_3$ (c-BaTiO$_3$)/PC and tetragonal BaTiO$_3$ (t-BaTiO$_3$)/PC nanocomposites over a frequency range of 1 kHz to 15 MHz at room temperature is shown in Figure 10.21a and b, respectively. It can be seen from Figure 10.21 that the dielectric constant of both series of nanocomposites remains almost stable with increasing frequencies. A weak dielectric constant frequency dependence of the nanocomposites is probably due to nonpolar PC matrix. For a given frequency, the dielectric constant of the nanocomposites increases with increasing BaTiO$_3$ content in the PC due to the increasing number of mini capacitors and higher-dielectric constant of BaTiO$_3$ than that of pure PC. Since the dielectric constant of BaTiO$_3$ nanopowder (~150) is higher than that of pure PC, the addition of BaTiO$_3$ into PC matrix increases the dielectric constant of the resultant nanocomposites. For example, the dielectric constant measured at 1 MHz increased from 2.7 for pure PC to 10.5 for 50 wt% c-BaTiO$_3$/PC nanocomposite and to 16.3 for 50 wt% t-BaTiO$_3$/PC

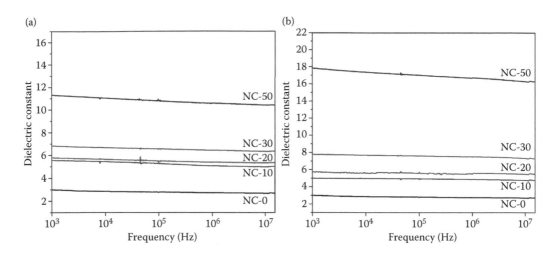

FIGURE 10.21

Dielectric constant of the (a) c-BaTiO$_3$/PC and (b) t-BaTiO$_3$/PC nanocomposites as a function of frequency, where NC-X represents X weight % of BaTiO$_3$ nanopowder in the PC matrix. (Reprinted from *Mater. Chem. Phys.*, 183, Thanki A. A., Goyal R. K., Study on effect of cubic- and tetragonal phased BaTiO$_3$ on the electrical and thermal properties of polymeric nanocomposites, 447–456. Copyright 2016, with permission from Elsevier.)

nanocomposites. For a given volume fraction, the higher-dielectric constant of t-BaTiO$_3$/PC nanocomposites compared to c-BaTiO$_3$/PC nanocomposites has been attributed to the higher-dielectric constant of t-BaTiO$_3$ than that of c-BaTiO$_3$ [32]. Similarly, the dielectric constant of the poly(etheretherketone)(PEEK)/BaTiO$_3$ composites measured at 1 MHz increased approximately 14-fold compared to pure PEEK matrix [33].

Figure 10.22 shows the DC electrical conductivity of the PC/BaTiO$_3$ nanocomposites measured at 100 V and 200 mA at room temperature. It can be seen that the addition of both c-BaTiO$_3$ and t-BaTiO$_3$ nanoparticles into the PC matrix increases the electrical conductivity slightly. However, all nanocomposites have electrical conductivity less than $\sim$9 $\times$ 10^{-12} S/cm, that is, the nanocomposites are insulating. The slight increase in conductivity of the nanocomposites has been attributed to the lower resistivity (i.e., $\sim$10^{10} Ω cm) of pure BaTiO$_3$ than that of PC matrix [32].

The addition of gamma-aminopropyl-triethoxysilane-treated BaTiO$_3$ nanoparticles (average diameter: 100 nm) into the ethylene vinyl-acetate copolymer (EVM) elastomer matrix increases the dielectric constant significantly, but decreases the dielectric breakdown strength. EVM is an ethylene-vinyl acetate copolymer with vinyl acetate content more than 40% by weight and it is primarily used in the wire and cable industry. The dielectric breakdown strength of a material is a crucial design parameter for high voltage device applications such as electric stress control in cable terminations, stress grading for high voltage motor and generator coils. As shown in Figure 10.23, the dielectric strength of the EVM/BaTiO$_3$ tends to sharply decrease from about 28 kV/mm for pure EVM to about 14 kV/mm for 30-vol% BaTiO$_3$-filled EVM nanocomposites. This decrease in dielectric strength has been attributed to the contribution of impurity ions probably existing on the surface of the BaTiO$_3$ nanoparticles. Thereafter, further addition of BaTiO$_3$ nanoparticles does not change the dielectric strength significantly. It is worth noting that the nanocomposites with high (i.e., 30–50 vol%) nanoparticle loadings still have good dielectric strength, which has significant importance for practical applications. The good dielectric strength of these nanocomposites has been attributed to well-dispersed BaTiO$_3$ nanoparticles in the EVM matrix. In other words, BaTiO$_3$ nanoparticles are surrounded by a thin layer of EVM polymer. This may act as the electrical barriers between the neighboring

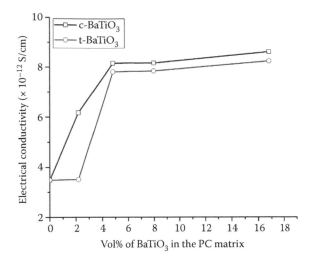

FIGURE 10.22
DC electrical conductivity of the PC/BaTiO$_3$ nanocomposites as a function of BaTiO$_3$ content in the PC matrix. (Reprinted from *Mater. Chem. Phys.*, 183, Thanki A. A., Goyal R. K., Study on effect of cubic- and tetragonal phased BaTiO$_3$ on the electrical and thermal properties of polymeric nanocomposites, 447–456. Copyright 2016, with permission from Elsevier.)

nanoparticles. Moreover, the excellent interfacial adhesion between the nanoparticles and the matrix can lower possibility of formation of voids and pores [35].

10.5.4 Thermal Properties

The addition of gamma-aminopropyl-triethoxysilane-treated BaTiO$_3$ nanoparticles into the EVM–elastomer matrix increases the thermal conductivity significantly. As shown

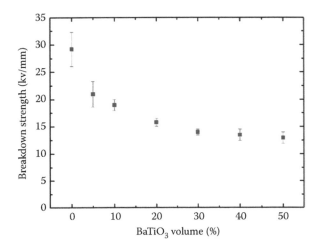

FIGURE 10.23
Breakdown strength (at 50 Hz) of the EVM/BaTiO$_3$ nanocomposites as a function of BaTiO$_3$ content at room temperature. (The source of the material Huang X. et al. Electrical, thermophysical and micromechanical properties of ethylene-vinyl acetate elastomer composites with surface modified BaTiO$_3$ nanoparticles. *J. Phys. D: Appl. Phys.*, 2009 and Institute of Physics is acknowledged.)

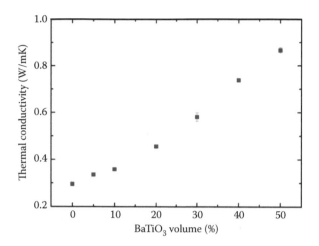

FIGURE 10.24
Thermal conductivity of the EVM/BaTiO$_3$ nanocomposites as a function of BaTiO$_3$ content at room temperature. (The source of the material Huang X. et al. Electrical, thermophysical and micromechanical properties of
ethylene-vinyl acetate elastomer composites with surface modified BaTiO$_3$ nanoparticles. *J. Phys. D: Appl. Phys.*,
2009 and Institute of Physics is acknowledged.)

in Figure 10.24, the thermal conductivity of the EVM/BaTiO$_3$ nanocomposites increases
with increasing BaTiO$_3$ nanoparticles. It can be seen that the thermal conductivity of the
nanocomposites increases from 0.295 W/m/K for pure EVM to 0.87 W/m/K for 50-vol%
BaTiO$_3$-filled EVM nanocomposite. Compared to pure copolymer, a threefold increase in
thermal conductivity of the nanocomposite may be attributed to the higher-thermal conductivity of BaTiO$_3$ compared to pure EVM copolymer and improved interfacial adhesion
(due to treatment of nanoparticles) between the BaTiO$_3$ nanoparticles and the matrix.
The better adhesion minimizes the interfacial phonon scattering and thus decreases the
interface heat resistance [35]. Moreover, compared to c-BaTiO$_3$/PC nanocomposites, a
less drop in T_g was found for the t-BaTiO$_3$/PC nanocomposites, indicating that the PC
chain mobility is more in c-BaTiO$_3$/PC nanocomposites compared to that of t-BaTiO$_3$/PC
nanocomposites. The drop in T_g has been primarily ascribed to the plasticizing effect of
BaTiO$_3$ [32].

10.5.5 Mechanical Properties

Figure 10.25 shows Vickers microhardness of the nanocomposites as a function of BaTiO$_3$
content. A gradual increase in microhardness is observed with BaTiO$_3$ content in the PC
matrix. The microhardness increases from 15.5 for pure PC to 27 kg/mm^2 for 16.8 vol%
(50 wt%) c-BaTiO$_3$, and to 23.4 kg/mm^2 for 16.8 vol% (50 wt%) t-BaTiO$_3$-filled nanocomposites. A uniform dispersion of BaTiO$_3$ nanoparticles in the PC matrix and higher hardness of BaTiO$_3$ (135 kg/mm^2) compared to pure PC (15.5 kg/mm^2) are the reasons for an
increase in microhardness of the nanocomposites. Due to larger size of t-BaTiO$_3$ compared
to c-BaTiO$_3$, for a given vol% of BaTiO$_3$, compared to t-BaTiO$_3$ nanoparticles, more numbers
of c-BaTiO$_3$ nanoparticles are present in a given volume of the nanocomposites. Hence,
compared to t-BaTiO$_3$ nanoparticles, c-BaTiO$_3$ nanoparticles provide more resistance to the
penetration of indentor, which results in higher microhardness for the c-BaTiO$_3$/PC nanocomposites. The experimental microhardness of the nanocomposites is correlated with the
values predicted from the theoretical models such as RoM and modified RoM (M-RoM).

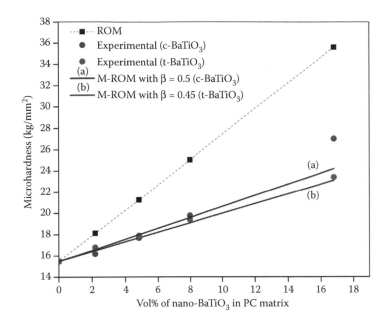

FIGURE 10.25

Correlation of microhardness values of (a) c-BaTiO₃/PC and (b) t-BaTiO₃/PC nanocomposites as a function of BaTiO₃ content with the values predicted from models. (Reprinted from *Mater. Chem. Phys.*, 183, Thanki A. A., Goyal R. K., Study on effect of cubic- and tetragonal phased BaTiO3 on the electrical and thermal properties of polymeric nanocomposites, 447–456. Copyright 2016, with permission from Elsevier.)

As shown in Figure 10.25, there is a large difference between the experimental and the theoretical microhardness values. To bridge the difference, strengthening efficiency factor, $\beta = 0.45$ and $\beta = 0.50$, was introduced in the M-RoM for t-BaTiO₃/PC and c-BaTiO₃/PC nanocomposites, respectively. It was found that $\beta = 0.45$ correlates well with the experimental microhardness for the t-BaTiO₃/PC nanocomposites for the full studied range. In contrast to this, $\beta = 0.50$ correlates well with the experimental microhardness for the c-BaTiO₃/PC nanocomposites containing up to 8 vol% c-BaTiO₃. Thereafter, M-RoM underestimate the microhardness [32].

References

1. Paul D. R., Robeson L. M. 2008. Polymer nanotechnology: Nanocomposites. *Polymer* 49: 3187–3204.
2. Kickelbick G. (Ed.). 2007. *Hybrid Materials: Synthesis, Characterization, and Applications.* Wiley-VCH Verlag GmbH & Co. KGaA, Weinheim, Germany.
3. Ajayan P. M., Schadler L. S., Braun P. V. (Eds.). 2003. *Nanocomposite Science and Technology.* Wiley-VCH Verlag GmbH Co. KGaA, Weinheim, Germany.
4. Armstrong G. 2015. An introduction to polymer nanocomposites. *Eur. J. Phys.* 36: 063001.
5. Gupta R. K., Kennel E., Kim K.J. (Eds.). 2010. *Polymer Nanocomposites Handbook.* CRC Press, Boca Raton, FL, USA.
6. Cervantes-Uc J. M., Cauich-Rodríguez J. V., Vázquez-Torres H., Garfias-Mesías L. F., Paul D. R. 2007. Thermal degradation of commercially available organoclays studied by TGA–FTIR. *Thermochim. Acta* 457: 92–102.

7. Zhao C., Qin H., Gong F., Feng M., Zhang S., Yang M. 2005. Polymer mechanical, thermal and flammability properties of polyethylene/clay nanocomposites. *Polym. Degrad. Stabil.* 87: 183–189.

8. Alexandre M., Dubois P. 2000. Polymer-layered silicate nanocomposites: preparation, properties and uses of a new class of materials. *Mater. Sci. Eng. R Rep.* 28: 1–63.

9. Manias E., Touny A., Wu L., Strawhecker K., Lu B., Chung T. C. 2001. Polypropylene/montmorillonite nanocomposites. Review of the synthetic routes and materials properties. *Chem. Mater.* 13(10): 3516–3523.

10. Zheng X., Jiang D. D., Wilkie C.A. 2005. Methyl methacrylate oligomerically-modified clay and its poly(methyl methacrylate) nanocomposites. *Thermochim. Acta.* 435: 202–208.

11. Lee H. S., Fasulo P. D., Rodgers W. R., Paul D. R. 2006. TPO based nanocomposites. Part 2. Thermal expansion behavior. *Polymer* 47: 3528–3539.

12. Yano K., Usuki A., Okada A., Kuraychi T., Kamigaito O. 1993. Synthesis and properties of polyimide-clay hybrid. *J. Polym. Sci., Part A: Polym. Chem.* 31: 2493–2498.

13. Giannels E. P. 1996. Polymer layered silicate nanocomposites. *Adv. Mater.* 8(1): 29–35.

14. Lan T., Kaviratna P. D., Pinnavaia T. J. 1994. On the nature of polyimide-clay hybrid composites. *Chem. Mater.* 6: 573–575.

15. Tjong S. C. 2006. Structural and mechanical properties of polymer nanocomposites. *Mater. Sci. Eng. R* 53: 73–197.

16. Shah D., Maiti P., Gunn E., Schnidt D. F., Jiang D. D., Batt C. A., Giannelis E. P. 2004. Dramatic Enhancements in Toughness of Polyvinylidene Fluoride Nanocomposites via Nanoclay-Directed Crystal Structure and Morphology. *Adv. Mater.* 16: 1173–1177.

17. Park J. H., Jana S. C. 2003. The relationship between nano- and micro-structures and mechanical properties in PMMA-epoxy-nanoclay composites. *Polymer* 44: 2091–2100.

18. Qu X., Guan T., Liu G., She Q., Zhang L. 2005. Preparation, structural characterization, and properties of poly(methyl methacrylate)/montmorillonite nanocomposites by bulk polymerization. *J. Appl. Polym. Sci.* 97: 348–357.

19. Wernik J. M., Meguid S. A. 2010. Recent developments in multifunctional nanocomposites using carbon nanotubes. *Appl. Mech. Rev.* 63: 050801 (1–40).

20. Vollath D. 2013. *Nanomaterials: An Introduction to Synthesis, Properties and Applications.* 2nd ed. Wiley-VCH Verlag GmbH & Co. KGaA, Germany.

21. Dalton, A. B., Collins, S., Munoz, E., Razal, J. M., Ebron, V. H., Ferraris, J. P., Coleman, J. N., Kim, B. G., Baughman, R. H. 2003. Super-tough carbon-nanotube fibres. *Nature* 423: 703.

22. Wang, J. Y., Yang, S. Y., Huang, Y. L., Tien, H. W., Chin, W. K., and Ma, C. C. 2011. *J. Mater. Chem.* 21: 13569–13575.

23. Gojny F. H., Wichmann M. H. G., Fiedler B., Kinloch I. A., Bauhofer W., Windle A. H., Schulte K. 2006. Evaluation and identification of electrical and thermal conduction mechanisms in carbon nanotube/epoxy composites. *Polymer* 47(6): 2036–2045.

24. Sain P. K., Goyal R. K., Prasad Y. V. S. S., Bhargava A. K. Electrical properties of single-walled/multi-walled carbon-nanotubes filled polycarbonate nanocomposites. *J. Electr. Mater.* 46(1): 458–466.

25. Allaoui A., Bai S., Cheng H. M., Bai J. B. 2002. Mechanical and electrical properties of a MWNT/epoxy composite. *Compos. Sci. Technol.* 62: 1993–1998.

26. Kim A. J., Seong D. G., Kang T. J., Youn J. R. 2006. Effects of surface modification on rheological and mechanical properties of CNT/epoxy composites. *Carbon* 44: 1898–1905.

27. Biercuk M. J., Llaguno M. C., Radosavljevic M., Hyun J. K., Johnson A. T., Fischer J. E. 2002. Carbon nanotube composites for thermal management. *Appl. Phys. Lett.* 80: 2767–2769.

28. Joshi M. D., Goyal A., Patil S. M., Goyal R. K. 2016. Tribological and thermal properties of hexagonal boron nitride filled high-performance polymer nanocomposites. *J. Appl. Polym. Sci.* 133: 44409.

29. Lee D., Song S. H., Hwang J., Jin S. H., Park K. H., Kim B. H., Hong S. H., Jeon S. 2013. Enhanced mechanical properties of epoxy nanocomposites by mixing noncovalently functionalized boron nitride nanoflakes. *Small* 9(15): 2602–2610.

30. Lin Z., Mcnamara A., Liu Y., Moon K. S., Wong C. P. 2014. Exfoliated hexagonal boron nitride-based polymer nanocomposite with enhanced thermal conductivity for electronic encapsulation. *Compos. Sci. Technol.* 90: 123–128.

31. Li T. L., Hsu S. L. C. 2010. Enhanced thermal conductivity of polyimide films via a hybrid of micro- and nano-sized boron nitride. *J. Phys. Chem. B* 114: 6825–6829.

32. Thanki A. A., Goyal R. K. 2016. Study on effect of cubic- and tetragonal phased $BaTiO_3$ on the electrical and thermal properties of polymeric nanocomposites. *Mater. Chem. Phys.* 183: 447–456.

33. Goyal R. K., Madav V. V., Pakankar P. R., Butee S. P. 2011. Fabrication and properties of novel polyetheretherketone/barium titanate composites with low dielectric loss. *J. Electr. Mater.* 40(11): 2240–2247.

34. Goyal R. K., Katkade S. S., Mule D. M. 2013. Dielectric, mechanical and thermal properties of polymer/$BaTiO_3$ composites for embedded capacitor. *Compos. Part B* 44: 128–132.

35. Huang X., Xie L., Jiang P., Wang G., Liu F. 2009. Electrical, thermophysical and micromechanical properties of ethylene-vinyl acetate elastomer composites with surface modified $BaTiO_3$ nanoparticles. *J. Phys. D: Appl. Phys.* 42: 245407.

11

Polymeric Nanofibers

It is well known that the strength of fibers (of polymer, carbon, glass, and ceramic) increases with the decrease of their diameter due to the improvements in microstructure and orientation as well as reduction in the size and defect concentration. Hence, manufacturers of advanced fiber usually prefer the smallest fiber diameter which is technologically and economically feasible. For example, in the last few decades, the diameter of carbon fibers has been reduced from only 7 to 5 μm, leading to significant improvements in strength. Conventional fiber spinning techniques, for example, melt spinning, dry spinning, or wet spinning, rely on mechanical forces to produce fibers by extruding polymer melt or solution through a spinnerette and subsequently drawing filaments as they solidify or coagulate. There is a continuing interest in further diameter reduction. However, conventional techniques are generally not able to synthesis filaments smaller than about 2 μm.

Electrospinning is a simple process that produces polymer fibers with diameter ranging from few tens nanometers to micrometers from melt or polymer solution by jetting polymer solutions in high-electric fields (about 10–20 kV). The resultant fabric has a high-specific surface area (of ∼100 m^2/g or more) and high porosity (∼90%). Polymeric nanofibers often have higher-specific strength and modulus, and low density. These nonwoven fabrics can be used in a variety of applications such as structural, nanogenerators, protective fabric for military, filtration, tissue engineering, fuel cell membranes, catalytic systems, and sensors [1–3].

11.1 Synthesis of Polymeric Nanofibers

The electrospinning process has been known since the 1930s [1]. It involves the introduction of electrostatic charge to a stream of polymer melt or solution in the presence of a strong electric field. When an electric potential is applied to melt or solution of polymer, the charged polymer solution forms a cone shape droplet at the tip of the nozzle. If an electrostatic force is sufficient enough to overcome the surface tension of the solution droplet, the tip of droplet elongates toward a collection plate resulting in a formation of a jet. This charged jet undergoes whipping mode (i.e., instability region) where it splits into multiple fine fibers and travels to the target. The solvent evaporates, while the dry ultrafine fibers are deposited on a collector or drum [1,4]. Figure 11.1a shows a schematic illustration of the basic electrospinning setup, which mainly consists of a syringe filled with polymer solution, a high-voltage source, and a grounded conductive collector screen. In addition, a metering syringe pump can be used to control the flow rate of the polymer solution. The needle of the syringe typically serves as an electrode to electrically charge the polymer solution and the counter electrode is connected to the conductive collector screen. Under the influence of electrostatic field, charges are induced in the solution and the charged polymer is accelerated toward the grounded metal collector. At low-electrostatic

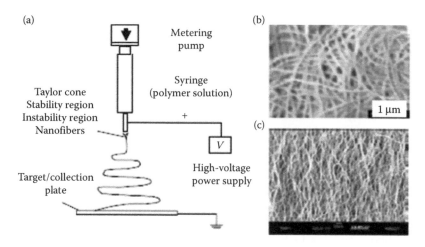

FIGURE 11.1
(a) Schematic drawing of the electrospinning process producing (b) random nanofiber, and (c) aligned nanofibers. (From Gogotsi Y. (Ed.). 2006. *Nanomaterials Handbook*. Taylor & Francis Group, New York. Reproduced with permission from Taylor & Francis Group.)

field strength, the pendant drop emerging from the tip of the pipette is prevented from dripping due to the surface tension of the solution. As the strength of the electric field is increased, the induced charges on the liquid surface repel each other and create shear stresses. These repulsive forces act in a direction opposite to the surface tension, which results in the extension of the pendant drop into a conical shape and serves as an initiating surface. When the critical voltage is reached, a charged jet emanates from the tip of the conical drop. The diameter of the discharged jet decreases with increase in distance between the tip of the jet and the collector [5]. The stretching of the jet is accompanied by rapid evaporation of the solvent molecules. The dry fibers are accumulated on the surface of the collector resulting in a nonwoven random fiber mesh (Figure 11.1b) of fibers with varying diameters. By proper control of process parameters, aligned nanofibers can also be produced (Figure 11.1c).

11.2 Parameter Affecting Diameter of Nanofibers

Diameter of the resultant nanofiber is affected by several factors such as molecular weight, molecular weight distribution, branching or linearity of the polymer, viscosity of polymer solution, electric potential, flow rate, jet diameter, distance between the capillary and collection screen, ambient parameters (temperature, humidity, and air velocity in the chamber), and finally motion of target screen. The polymer solution must have a concentration high enough to cause the polymer entanglements. Fiber diameter increases with increasing solution concentration and molecular weight of the polymer [3,6]. A higher viscosity results in a larger fiber diameter. However, too high polymer viscosity prevents flow of polymer induced by the electric field. In addition, the solution must have a surface tension low enough, a charge density high enough, and a viscosity high enough to prevent the jet from collapsing into droplets (or beads) before the solvent has evaporated. The diameter of the beads become bigger and the average distance between beads in the fibers increases as the viscosity increases. Increasing net charge density and the surface tension coefficient of the solvent favors the formation of small diameter fibers. Increasing the distance or decreasing the electrical field decreases the bead density, regardless of the concentration of the polymer in the solution. Applied fields can influence the morphology in periodic ways, creating a variety of new shapes on the surface. The fiber diameter is proportional to the cube of the polymer concentration. The diameter of nanofibers also depends on the Berry number, which is the product of the efficiency of conversion of electrical power to mechanical work and the inverse of resistance of the system. Below the Berry number of 10, the electrospinning produces droplets or fibers with droplet. However, when the Berry number increases from 9.78 to 23.6, the nanofiber diameter will increase from 65 to 1313 nm. For the synthesis of nanofibers of aqueous poly(ethylene oxide) (PEO), the viscosities in the range of 1–20 poises and surface tension between 35 and 55 dynes/cm are suitable. At viscosities above 20 poises, high cohesiveness of the solution causes instability in flow of solution, while droplets are formed when the viscosity is too low (<1 poise). The increase in electrical potential causes rougher nanofibers. A higher applied voltage ejects more fluid in a jet, resulting in a larger fiber diameter. Under optimized process conditions, mean fiber diameters of the order of 100 nm is obtained, which is well below the diameter of conventional-extruded fibers (10–100 μm) [4,7].

11.3 Properties of Nanofibers

11.3.1 Mechanical Properties

Electrospun nylon-6 nanofiber mats (12 mm width × 30 mm length × 0.011 mm thick) show a tensile strength of 10.45 MPa, strain at break of 250%, and yield strain of 48%. Its Young's modulus is 19.4 MPa, which is about 56 times lower than that of an undrawn single nylon-6 filament, 770 times lower than that of a conventional nylon-6 fiber, and 1540 times lower than that of a single nylon-6 nanofiber at diameter of 85 nm. The lower mechanical properties for nanofiber-mat can be attributed to the (i) poor orientation of molecular chains along the fiber axis inside the electrospun fiber and (ii) weak entanglements inside the nanofiber-mat. As shown in Figure 11.2, the addition of 1 wt% MWNT into nylon-6 increases the tensile strength and yield strength of the nylon-6/MWNTs mat ∼25% (from 10.45 to 13.05 MPa) and 34% (from 6.7 to 9 MPa), respectively. The Young modulus of the nylon-6/MWNTs mat increased by ∼46% (from 19.4 to 28.34 MPa) compared to pure nylon fiber mat. Thus, the composite nanofiber mat has a higher strength and stiffness but lower ductility [8].

The Young modulus and the tensile strength of a single nylon-6 nanofiber (diameter of 800 nm) obtained from stress–strain curve are ∼902 and 304 MPa, respectively. The breaking strain of a single nylon-6 nanofiber is ∼40% indicating that both ultimate tensile strength and Young's modulus of a single nylon-6 nanofiber are improved up to 30 times and 47 times, respectively, than that of a nonwoven nylon-6 nanofiber mat. This remarkable improvement is attributed partly due to the orientation of the internal molecular structure of nylon-6 nanofiber [8]. The Young modulus of a single nylon-6 nanofiber with diameter of 800 nm is still lower by more than 33 times than that of an electrospun single nylon-6 nanofiber with diameter of 85 nm. In latter case, Young's modulus measured by bending test method is about 30 GPa [9].

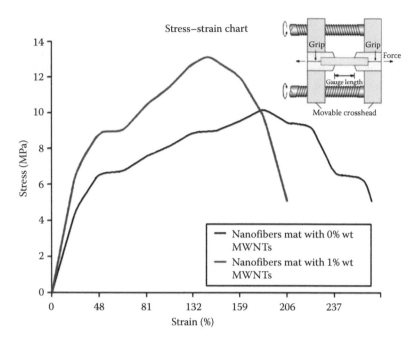

FIGURE 11.2
Typical stress–strain curves of nylon-6/MWNTs nanofiber mats. (Bazbouz M. B., Stylios G. K.: The tensile properties of electrospun nylon 6 single nanofibers. *J. Polym. Sci. Pol. Phys.*, 2010. 48. 1719–1731. Copyright Wiley-VCH Verlag GmbH & Co. KGaA. Reproduced with permission.)

Figure 11.3 shows the stress–strain curves for electrospun fibers of poly(trimethyl hexamethylene terephthalamide) (PA 6[3]T) in uniaxial extension for four representative individual fibers of different diameters. The Young modulus and the yield strength are found to increase with decreasing fiber diameter (<500 nm), that is, smaller diameter fibers tend to have higher Young's moduli and yield strengths but break at smaller strains than the larger diameter fibers. The increase in Young's modulus and yield strength with decreasing fiber diameters is due to the increasing molecular level

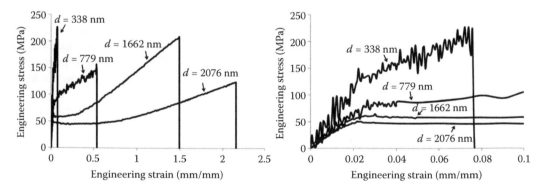

FIGURE 11.3
Stress–strain curves for electrospun fibers of poly(trimethyl hexamethylene terephthalamide) (PA 6[3]T) in uniaxial extension for four individual fibers. (Reprinted from Polymer, 52, Pai C. L., Boyce M. C., Rutledge G. C. Mechanical properties of individual electrospun PA 6(3)T fibers and their variation with fiber diameter, 2295–2301, Copyright 2011, with permission from Elsevier.)

orientation of polymer chains within the fibers with decreasing fiber diameters from 3600 to 170 nm [2,10].

The decrease of fiber diameter from 2.8 μm to <100 nm increases elastic modulus from 0.36 to 48 GPa and fracture strength from 15 to 1750 MPa. In addition, a significant increase in toughness is also reported. The largest increase in strength was found for the nanofibers smaller than 250 nm. The ultrahigh ductility (i.e., average failure strain > 50%) and toughness may be attributed to the reduced crystallinity of nanofibers which resulted from rapid solidification of ultrafine electrospun jets, while the increased chain molecular orientation caused by intense jet stretching is responsible for high strength and modulus. In contrast, the commercial polymer fibers, polyaramid (Kevlar) and ultra high molecular weight polyethylene (Spectra or Dyneema), spun from a solution result in high crystallinity of 75%–95%, which increase strength and modulus, but reduce macromolecular mobility in the crystalline phase and lead to low deformations to failure compared to bulk polymers [11]. The tensile strength of nanofibrous mat is found similar to that of a natural skin [12].

Typical stress–strain curves of electrospun poly(ε-caprolactone) (PCL) mat and bulk PCL (nonspun) systems are shown in Figure 11.4. A significant difference in the stress–strain curve of both samples can be seen, that is, electrospun sample does not display the necking phenomenon whereas the bulk sample shows clear necking. This can be attributed to the oriented and stretched polymer chains in the spun fiber fabrics. Typically, the tensile strength and modulus of the nonwoven fabrics are lower than the mats with uniaxially oriented fibers. This is attributed to the highly porous nature of the nonwoven fabrics. The breaking strength of the nonwoven fabric depends highly on the fiber-to-fiber bonding. The lack of bonding between the fibers facilitates easy orientation and stretching of the fibers when loaded and can give high degree of elongation before failure. The use of highly volatile solvents during electrospinning can produce nonwovens with little or no fiber fusions. Fibers cannot fuse together when the solvent evaporation is high and this also results in weak intermolecular interaction. However, when there is fusion between fibers, the modulus of the nonwovens increases and the elongation to break decreases. The fusion

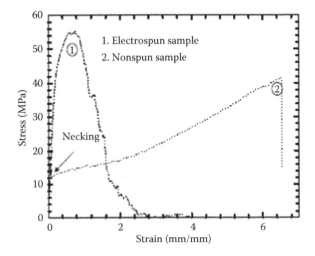

FIGURE 11.4
Stress–strain curves obtained from tensile tests performed on electrospun PCL and nonspun PCL samples. (Reprinted from *Compos. Sci. Technol.*, 70, Baji A. et al., Electrospinning of polymer nanofibers: Effects on oriented morphology, structures and tensile properties, 703–718, Copyright 2010, with permission from Elsevier.)

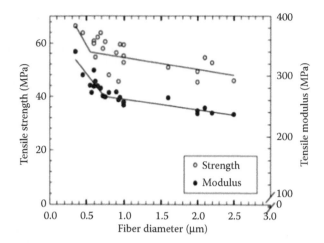

FIGURE 11.5
Effect of PCL fiber diameter on the tensile modulus and strength. (Reprinted from *Compos. Sci. Technol.*, 70, Baji A. et al., Electrospinning of polymer nanofibers: Effects on oriented morphology, structures and tensile properties, 703–718, Copyright 2010, with permission from Elsevier.)

of fibers is obtained if the solvent is not completely evaporated during the fiber-forming process [5].

Figure 11.5 shows the effect of fiber diameter on the tensile strength and modulus of the individual PCL fibers. The fibers with diameters more than 2 μm have the modulus or tensile strength almost similar to those of bulk fibers. The increased strength of finer diameter (<2 μm) fibers is due to the gradual ordering of the molecular chains and modest increase in the crystallinity of the fibers. In addition, reduced diameter leads to densely packed lamellae and fibrillar structures, which align themselves along the fiber axis and thus, play a critical role in enhancing the strength and modulus of the fibers [5].

In brief, the mechanical properties of electrospun nanofibers can be increased by increasing the degree of orientation of polymer chains and crystallites along the fiber axis, and deceasing diameter of nanofiber. The mechanical properties of nanofibers are often well below those of fibers made by conventional processes such as melt- or solution spinning. The main reason for this is the competition between flow-induced chain orientation and chain relaxation before fiber solidification, leading to low degrees of molecular orientation in as-spun fibers. In conventional polymer fiber processing, chain alignment is induced by drawing the as-spun fiber in the solid state below the melting point into a highly oriented structure, as here relaxation times are infinite. In order to achieve similar high levels of chain orientation and chain extension in nanofibers based on flexible chain polymers, it is vital to apply a poststretching step [13,14].

11.3.2 Magnetic Properties

In addition to improvement in mechanical properties (i.e., strength and modulus), the addition of magnetic nanoparticles into the polymer nanofibers provide superparamagnetic nanocomposite and display enhanced magnetic-field-dependent superparamagnetism at ambient temperature as shown in Figure 11.6. Inset of Figure 11.6 shows a TEM micrograph indicating uniform dispersion of magnetic particles in the nanofiber. Moreover, the use of biodegradable polymeric nanofiber as the carrier matrix can be useful in biomedical, magnetic resonance imaging, and drug delivery applications [5].

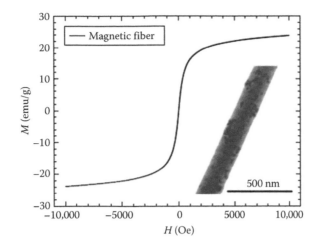

FIGURE 11.6
Magnetic hysteresis loop of magnetic filler-reinforced electrospun fiber. (Reprinted from *Compos. Sci. Technol.*, 70, Baji A. et al., Electrospinning of polymer nanofibers: Effects on oriented morphology, structures and tensile properties, 703–718, Copyright 2010, with permission from Elsevier.)

11.3.3 Thermal Properties

The thermal properties such as glass transition temperature, melting point, thermal stability, coefficient of thermal expansion (CTE or dimensional stability), and thermal conductivity of nanofibers are important properties, which are closely related to fiber's postprocessing and their performance. Due to decrease in the percentage crystallinity and imperfect crystallization of electrospun cellulose diacetate nanofibers, the thermal stability as well as glass transition temperature decreases compared to raw cellulose diacetate [15]. In contrast, Kim and Lee [16] reported an increase of crystallinity and decrease of glass transition temperature (T_g) and crystallization peak temperature (T_c) for poly(ethylene terephthalate) (PET) and poly(ethylene naphthalate) (PEN) electrospun nanofibers. The change in T_g and T_c of electrospun neat PET and PEN are primarily resulted from a decrease of molecular weight after the electrospinning by thermal as well as mechanical degradation. However, the crystalline melting peak temperature (T_m) of PET and PEN remained almost the same before and after electrospinning. Interestingly, the T_g, T_c, and T_m of the electrospun PET/PEN blend nanofibers were found lower than those of the bulk. The decrease in thermal properties of PET/PEN blend nanofibers was attributed to exchange reactions between PET and PEN [17]. Figure 11.7a shows the DSC heating curves of nylon-6,6 electrospun fibers and nylon-6,6 pellets. It can be clearly seen that the melting enthalpy of electrospun nylon-6,6 is higher (i.e., 107 J/g) than that of nylon-6,6 pellets or unspun sample (i.e., 91 J/g) indicating an increase in the degree of crystallinity (Figure 11.7b). In addition, the melting point of electrospun fiber is slightly higher than that of unspun sample.

The thermal conductivity of a single nylon-11 fibers electrospun at voltages of 6–7 kV could be as high as 1.6 W/m · K, which is approximately one order of magnitude higher than that of a bulk value for nylon-11 (i.e., ~0.2 W/m · K) [18]. Similarly, the thermal conductivity of individual polystyrene nanofibers electrospun at voltages of 7–10 kV varies from 6.6 to 14.4 W/m · K, which is significantly higher than that of typical bulk values for polystyrene [19]. Thermal conductivity of individual polyethylene (PE) electrospun nanofibers increases from 1 (electrospun at 10 KeV) to 9.3 W/m · K (electrospun at 45 KeV) indicating an almost

(a)   (b)

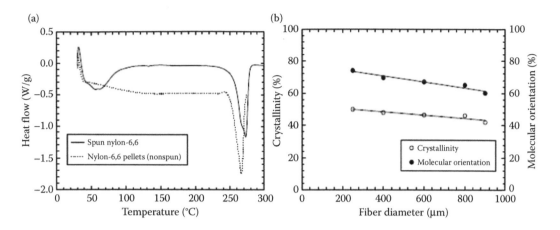

FIGURE 11.7
(a) DSC heating curves of nylon-6,6 electrospun fibers and nylon-6,6 pellets and (b) percent crystallinity or molecular orientation versus diameter of nanofiber. (Reprinted from *Compos. Sci. Technol.*, 70, Baji A. et al., Electrospinning of polymer nanofibers: Effects on oriented morphology, structures and tensile properties, 703–718, Copyright 2010, with permission from Elsevier.)

linear increase in thermal conductivity with the voltage. The highest-thermal conductivity of a single PE nanofiber (i.e., 9.3 W/m · K) is >20 times higher than the typical bulk value [20]. The increased thermal conductivity of electrospun nanofibers can be attributed to the highly oriented polymer chains resulted from the high-strain rates within the electrospinning jets and increased crystallinity in the electrospun nanofibers of semicrystalline polymers.

Figure 11.8 shows the thermal conductivities of the high-quality ultra-drawn (draw ratios from 160 to 410) polyethylene nanofibers with diameters of 50–500 nm and lengths up to tens of millimeters. For comparison, thermal conductivities of micrometer-sized polyethylene fibers with draw ratios from 0 to 350 are also shown. In case of micrometer-sized fibers, the thermal conductivity saturates when the draw ratio is above 100, while the thermal conductivity of nanofibers does not show saturation indicating that there is still room for enhancement in thermal conductivity. The highest-thermal conductivity of the nanofibers

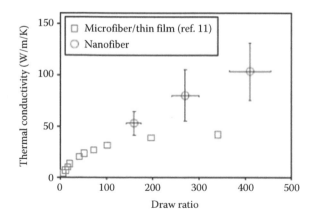

FIGURE 11.8
The thermal conductivities of individual polyethylene nanofibers. (Reprinted by permission from Springer Nature. *Nat. Nanotechnol.*, Shen S. et al. 2010. Polyethylene nanofibres with very high thermal conductivities. 5: 251–255, copyright 2010.)

is found to be as high as ~104 W/m/K, which is larger than the conductivities of about half of the pure metals. Moreover, this highest-thermal conductivity of the nanofibers (i.e., 104 W/m/K) is about three times higher than the reported values for micrometer-sized fibers (i.e., ~30–40 W/m/K at around room temperature) and ~300 times that of bulk polyethylene (~0.35 W/m/K). The high-thermal-conductivity-drawn polyethylene nanofibers have been attributed to the restructuring of the polymer chains by stretching, which improves the fiber quality toward an ideal single crystalline fiber. These thermally conductive polymer nanofibers or nanofilms are potentially useful as heat spreaders and could supplement conventional metallic heat-transfer materials, which are used in applications such as solar hot-water collectors, heat exchangers, and electronic packaging [21].

11.3.4 Electrical Properties

The electrical conductivity of the PVDF/MWNT nanocomposite electrospun nanofiber with diameters ranging from 200 to 1200 nm increases with increasing MWNT as shown in Figure 11.9. The room temperature conductivity of the PVDF/MWNT nanofiber with 0.6 wt% MWNT is 1×10^{-14} S/cm and increases to 1×10^{-6} S/cm at 1.2 wt% MWNT, that is, the conductivity is increased to about eight orders of magnitude compared to that of the 0.6 wt% counterpart. The significant improvement in the conductivity of the polymer nanofibers is due to the presence of highly conductive and high-structured CNT paths to the PVDF, thus resulting a percolation threshold of 1.2 wt% [22].

The room temperature DC electrical conductivity of poly(methyl methacrylate) (PMMA)/multiwalled carbon nanotubes (MWNTs) nanocomposite single fiber (diameter: 6 μ–200 nm) with MWNT (diameter: 10–30 nm) of 0.05% (w/w) is about ten orders of magnitude higher than that of pure PMMA (~10^{-12} S/m) and it further increases slightly with increasing MWNT concentration. The increase in conductivity can be attributed to the alignment of CNTs in the direction of the fiber axis. The *I–V* characteristics of the single PMMA/CNT nanofiber show a linear ohmic behavior [23]. The electrical conductivity of polypyrrole (PPy)/poly(ethylene oxide) (PEO) nanofiber can be improved by varying the content of PPy in PEO matrix (Figure 11.10). The conductivity through the thickness of the

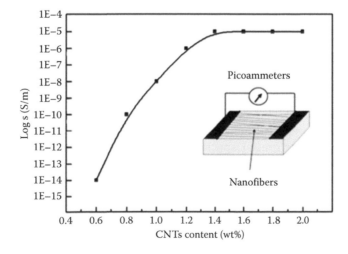

FIGURE 11.9
Electrical conductivity versus MWNT of PVDF/MWNT ultrafine fibers. The inset shows the diagram of electrical measurement. (From Wang S. H. et al. 2014. *Nanoscale Res. Lett.* 9: 522. Open Access.)

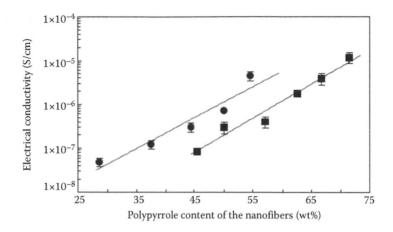

FIGURE 11.10
The electrical conductivity of PPy/PEO nanofiber as a function of PPy content in PEO matrix. (Reprinted from *Polymer*, 47, Chronakis I. S., Grapenson S., Jakob A., Conductive polypyrrole nanofibers via electrospinning: Electrical and morphological properties, 1597–1603, Copyright 2006, with permission from Elsevier.)

electrospun PPy/PEO nanofibers increases from 4.9×10^{-8} for pure PEO to 1.2×10^{-5} S/cm for PPy/PEO. As the content of PPy increases in the PEO matrix, the probability of contacts between conducting polymer regions increases and thus, facilitates electrical conduction. At higher content of PPy, a continuous 3-D network (or pathways) of the conductive PPy molecules is formed which leads to high conductivity due to the high charge-carrier mobility of PPy molecules along the fibers [24]. The DC electrical conductivity of the nanofibers (diameter: 500 nm–5 μm) (a bunch of countable fibers) of polyaniline (PANI) doped with camphor sulfonic acid (CSA) and dispersed in poly(methyl methacrylate) (PMMA)/chloroform solution is found to be ~0.289 S/m, which is about three orders of magnitude less compared to that of pure PANI (CSA) sample (i.e., 200–400 S/m) [25].

11.4 Applications

Owing to high-aspect ratio, pores, and specific surface area (100 m^2/g), electrospun nanofibers can bridge gaps between the nanoscale and the macroscale materials and thus they have been suggested as important candidates for the applications in multifunctional membranes, biomedical field (tissue engineering, wound dressing, drug delivery, artificial organs, vascular grafts), speciality fabrics, filters, nanogenerators, nanocomposites, and electronic machines [26–27]. Figure 11.11 shows various potential applications of electrospun polymer nanofibers [7].

11.4.1 Nanocomposite Fibers for Structural Applications

One of the most important applications of nanofibers is in nanocomposites. For a given improvement in property, compared to conventional fibers, the addition of nanofibers provides superior structural properties such as high-specific modulus and specific

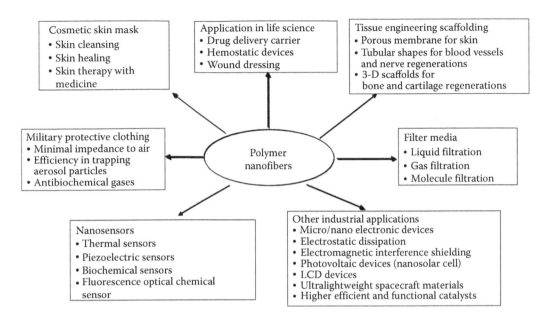

FIGURE 11.11
This figure shows various potential applications of electrospun polymer nanofibers. (Reprinted from *Compos. Sci. Technol.*, 63, Huang Z. M., et al., A review on polymer nanofibers by electrospinning and their applications in nanocomposites, 2223–2253, Copyright 2003, with permission from Elsevier.)

strength. Moreover, if there is a difference in refractive indices between fiber and matrix, the resulting composite becomes opaque or nontransparent due to light scattering. However, this limitation can be circumvented by using nanofibers with diameter smaller than the wavelength of visible light. The epoxy composites reinforced with electrospun nylon-4,6 nanofibers of diameters 30–200 nm are transparent due to the fibers with sizes smaller than the wavelength of visible light and hence, scattering of light is reduced. The addition of electrospun polybenzimidazole (PBI) nanofibers of diameters 300–500 nm in between a unidirectional graphite/epoxy prepregs increases both Mode I critical energy release rate (G_{Ic}) and Mode II critical energy release rate (G_{IIc}) for fracture by ~15% and 130%, respectively [7]. Figure 11.12 shows the stress–strain curves of electrospun pure polyacrylonitrile (PAN) and 1 wt% SWNT-filled PAN nanofibers (diameter 40–300 nm). It can be seen that the random fiber mats containing 1 wt% SWNT shows a more than twofold increase in the tensile strength and elastic modulus compared to that of random PAN fibers without SWNT. Interestingly, aligned 1 wt% SWNT/PAN fibers shows a 33% increase in tensile strength and almost double strain as compared to the random PAN/SWNT composite fibers. The significant enhancement in strength of PAN/SWNT implies that, for the same performance, replacing of commercial carbon fibers with SWNTs may lead to significant reduction in the volume and weight of the structural composites currently used in space applications. Moreover, the addition of SWNT into PAN increases the toughness significantly indicating that the fibril as well as SWNT alignments play an important role in improving the mechanical properties. The post treatment such as drawing and heat treatment can also remarkably improve the properties by inducing molecular orientation. Furthermore, heat treatment enhances interaction between the reinforcement and matrix leading to better load transferring across reinforcement–matrix interface [1]. However, the increase in elastic modulus is not significant.

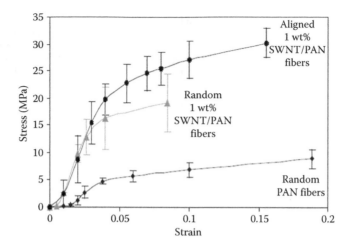

FIGURE 11.12
Stress–strain curves of electrospun PAN/1 wt% SWNT fibers. (From Gogotsi Y. (Ed.). 2006. *Nanomaterials Handbook*. Taylor & Francis Group, New York. Reproduced with permission from Taylor & Francis Group.)

11.4.2 Nanogenerators

A flexible hybrid piezoelectric fiber-based 2-D fabric can be integrated with clothing, which converts the mechanical energy of human body motion into electric energy, as wearable nanogenerators for wearable devices and portable electronic devices. A hybrid woven fabric of piezoelectric $BaTiO_3$ nanowires-filled poly(vinyl chloride) (PVC) exhibits good piezoelectric properties with flexibility. The copper wires (as electrodes) and cotton threads were woven into the fabric to construct the nanogenerator. If such flexible fabric nanogenerator is attached on an elbow pad, the human arms on bending can generate output voltage of 1.9 V and current of about 24 nA which are sufficient to power an LCD [27].

11.4.3 Nanofilters

Nanofibers used for filter media provide high-filtration efficiency and low-air resistance. The channels or pores size of a filter must be matched to the scale of the particles or droplets that are to be removed in the filter. Due to the very high-specific surface area and associated high-surface cohesion, tiny particles of the order of <0.5 mm or oil droplets as small as 0.3 μm can be easily trapped efficiently in the filters made of electrospun nanofibers [7].

11.4.4 Biomedical Application

The human tissues and organs including bone, dentin, collagen, cartilage, and skin are characterized by well-organized hierarchical nanofibrous structures. Due to unique properties such as high specific surface area, small pore size, high porosity, flexibility, strength, etc., the polymeric nanofiber mats find their promising applications in several biomedical areas including medical prostheses, tissues template, wound dressing, and cosmetics.

Medical prostheses: Electrospun polymer nanofibers have been suggested for a number of soft tissue prostheses applications such as blood vessel, vascular, breast, etc. These electrospun biocompatible nanofibers can also be deposited as a thin porous film onto a hard tissue prosthetic device and then implanted into the human body. This porous film works as an interphase between the prosthetic device and the host tissues. It reduces the stiffness mismatch at the tissue/device interphase and hence, prevents the device failure after the implantation [7].

Tissue template: For the treatment of tissues or organs due to malfunction in a human body, one of the challenges to the field of tissue engineering is the design of ideal scaffolds that can mimic the structure and biological functions. The tissue engineering is involved with the creation of reproducible and biocompatible 3-D scaffolds for cell growth, resulting in biomatrix composites for various tissue repair and replacement procedures. The scaffolds made up of synthetic biopolymers and/or biodegradable polymer nanofibers can mimic native structures [7].

Wound dressing: Polymer nanofibers can be used for the treatment of wounds or burns of human skin. Fine fibers of biodegradable polymers can be directly sprayed or spun onto the injured location of skin to form a fibrous mat dressing. This can let wounds heal by encouraging the formation of normal skin and its growth thereby eliminating scar tissue as found in a traditional treatment. Nonwoven nanofibrous membrane mats for wound dressing usually have pore sizes ranging from 500 nm to 1 μm, small enough to protect the wound from bacterial penetration via aerosol particle-capturing mechanisms. High-specific surface area of 5–100 m²/g is extremely efficient for fluid absorption and dermal delivery [7]. Silver nanoparticles (Ag NPs) show antibacterial activity toward germs on contact without release of toxic biocides, and hence, Ag NPs are considered non-toxic and environmentally friendly in biomedical applications. Therefore, the Ag NPs/ polymer nanofibers have been applied for infected burns, purulent wounds, and as a wound–healing matrix [28].

Cosmetics: The cosmetic skin mask made of the electrospun nanofibers can be applied gently and painlessly as well as directly to the 3-D topography of the skin to provide healing or care treatment to the skin.

11.4.5 Protective Fabric for Military

The military needs protective clothing to get survivability, sustainability, and combat effectiveness of the individual soldier system against extreme weather conditions, ballistics, and NBC (nuclear, biological, and chemical) warfare. Conventional protective clothing containing charcoal absorbents has limitations of water permeability and higher weight. A lightweight and breathable fabric, which is permeable to both air and water vapor, insoluble in all solvents, and highly reactive with nerve gases and other deadly chemical agents, is highly desirable. Owing to high-specific surface area, nanofiber fabrics are capable of deactivating the chemical agents without impairing the permeability to the air and water vapor to the clothing. Electrospun nanofibers have high porosity with a very small pore size, which provide good resistance to the penetration of chemically harmful agents in aerosol form compared to conventional textiles. Therefore, nanofiber fabric shows strong promises as ideal protective fabric for military [7].

11.4.6 Functional Sensors

Owing to high specific surface area, electrospun polymeric nanofibers may be useful for functional sensors. Poly(lactic acid co glycolic acid) nanofiber films have been studied as a new sensing interface for chemical and biochemical sensor applications. The sensitivities of nanofiber films to detect ferric and mercury ions and a 2,4-dinitrotulene (DNT) are two to three orders of magnitude higher than those of thin film sensors [7].

References

1. Gogotsi Y. (Ed.). 2006. *Nanomaterials Handbook*. Taylor & Francis Group, New York.
2. Pai C. L., Boyce M. C., Rutledge G. C. 2011. Mechanical properties of individual electrospun PA 6(3)T fibers and their variation with fiber diameter. *Polymer* 52: 2295–2301.
3. Basu S., Gogoi N., Sharma S., Jassal M., Agrawal A. K. 2013. Role of elasticity in control of diameter of electrospun PAN nanofibers. *Fibers Polym.* 14: 950–956.
4. Yang H. 2007. Fabrication and characterization of multifunctional nanofiber nanocomposite structures through co-electrospinning process, PhD Thesis, submitted to the Drexel University, Philadelphia.
5. Baji A., Mai Y. W., Wong S. C., Abtahi M., Chen P. 2010. Electrospinning of polymer nanofibers: Effects on oriented morphology, structures and tensile properties. *Compos. Sci. Technol.* 70: 703–718.
6. Frenot A., Chronakis I. S. 2003. Polymer nanofibers assembled by electrospinning. *Curr. Opin. Colloid Interface Sci.* 8: 64–75.
7. Huang Z. M., Zhang Y. Z., Kotaki M., Ramakrishna S. 2003. A review on polymer nanofibers by electrospinning and their applications in nanocomposites. *Compos. Sci. Technol.* 63: 2223–2253.
8. Bazbouz M. B., Stylios G. K. 2010. The tensile properties of electrospun nylon 6 single nanofibers. *J. Polym. Sci. Pol. Phys.* 48: 1719–1731.
9. Li L., Bellan L. M., Craighead H. G., Frey M. W. 2006. *Polymer* 47: 6208–6217.
10. Macturk K. S., Eby R. K., Adams W. W. 1991. Characterization of compressive properties of high performance polymer fibers with a new microcompressive apparatus. *Polymer* 32(10): 1782–1787.
11. Papkov D., Zou Y., Andalib M. N., Goponenko A., Cheng S. Z. D., Dzenis Y. A. 2013. Simultaneously strong and tough ultrafine continuous nanofibers. *ACS Nano.* 7(4): 3324–3331.
12. Huang Z. M., Zhang Y. Z., Kotaki M., Ramakrishna S. 2003. A review on polymer nanofibers by electrospinning and their applications in nanocomposites. *Compos. Sci. Technol.* 63: 2223–2253.
13. Wong S. C., Baji A., Leng S. W. 2008. Effect of fiber diameter on tensile properties of electrospun poly(ε-caprolactone). *Polymer* 21: 4713–4722.
14. Arinstein A., Zussman E. 2011. Electrospun polymer nanofibers: Mechanical and thermodynamic perspectives. *J. Polym. Sci. Pol. Phys.* 49: 691–707.
15. Qin X., Wang H., Wu S. 2011. Investigation on structure and thermal properties of electrospun cellulose diacetate nanofibers. *J. Ind. Text.* 42(3): 244–255.
16. Kim J. S., Lee D. S. 2000. Thermal properties of electrospun polyesters. *Polym. J.* 32(7): 616–618.
17. Huang Z. M., Zhang Y. Z., Kotaki M., Ramakrishna S. 2003. A review on polymer nanofibers by electrospinning and their applications in nanocomposites. *Compos. Sci. Technol.* 63: 2223–2253.
18. Zhong Z., Wingert M. C., Strzalka J., Wang H. H., Sun T., Wang J., Chen R., Jiang Z. 2014. *Nanoscale* 6: 8283–8291.
19. Canetta C., Guo S., Narayanaswamy A. 2014. Measuring thermal conductivity of polystyrene nanowires using the dual-cantilever technique. *Rev. Sci. Instrum.* 85: 104901.
20. Ma J., Zhang Q., Mayo A., Ni Z., Yi H., Chen Y., Mu R., Bellan L., Li D. 2015. Thermal conductivity of electrospun polyethylene nanofibers. *Nanoscale* 7: 16899–16908.
21. Shen S., Henry A., Tong J., Zheng R., Chen G. 2010. Polyethylene nanofibres with very high thermal conductivities. *Nat. Nanotechnol.* 5: 251–255.
22. Wang S. H., Wan Y., Sun B., Liu L. Z., Xu W. 2014. Mechanical and electrical properties of electrospun PVDF/MWCNT ultrafine fibers using rotating collector. *Nanoscale Res. Lett.* 9: 522.
23. Sundaray B., Subramanian V., Natarajan T. S., Krishnamurthy K. 2006. Electrical conductivity of a single electrospun fiber of poly(methyl methacrylate) and multiwalled carbon nanotube nanocomposite. *Appl. Phys. Lett.* 88: 143114.
24. Chronakis I. S., Grapenson S., Jakob A. 2006. Conductive polypyrrole nanofibers via electrospinning: Electrical and morphological properties. *Polymer* 47: 1597–1603.

25. Veluru J. B., Satheesh K. K., Trivedi D. C., Ramakrishna M. V., Srinivasan N. T. 2007. Electrical properties of electrospun fibers of PANI/PMMA composites. *J. Eng. Fibers Fabr.* 2: 25–32.
26. Frenot A., Chronakis I. S. 2003. Polymer nanofibers assembled by electrospinning. *Curr. Opin. Colloid Interface Sci.* 8: 64–75.
27. Zhang M., Gao T., Wang J., Liao J., Qiu Y., Yang Q., Xue H., Shi Z., Zhao Y., Xiong Z., Chen L. 2015. A hybrid fibers based wearable fabric piezoelectric nanogenerator for energy harvesting application. *Nano Energ.* 13: 298–305.
28. Nguyen T. H., Lee K. H., Lee B. T. 2010. Fabrication of Ag nanoparticles dispersed in PVA nanowire mats by microwave irradiation and electro-spinning. *Mater. Sci. Eng.* 30: 944–950.

12

Characterization of Nanomaterials

There are two basic methods: direct method and indirect method for determining the particle size, particle size distribution, morphology, and composition. Direct method inspects the particles and makes actual measurements of their dimensions, while indirect method utilizes the relationship between particle behavior and its size. The direct method implies an assumption of equivalent spherical size, relating it to a linear dimension. Equivalent spherical diameters are the diameters of spheres that have the same or equivalent length or volume as irregular particles themselves. The main aim of this chapter is to discuss the basic principle and the applications of various characterization techniques such as powder x-ray diffraction (XRD), Raman spectroscopy, UV–visible (UV–vis) spectroscopy, Photoluminescence spectroscopy (PL), Fourier transform infrared spectroscopy (FTIR), surface area analysis (BET method), dynamic light scattering (DLS), scanning electron microscopy (SEM), transmission electron microscopy (TEM), energy-dispersive spectroscopy (EDS), and scanning probe microscopy (SPM). Spectroscopy-based techniques such as UV–vis, XRD, FTIR, and Raman are considered indirect methods for determining the data related to particle size, composition, structure, crystal phase, and properties of the nanoparticles. SEM and TEM are the direct methods for determining the size and morphological features of the nanoparticles. DLS spectroscopy is used to determine the size distribution and quantify the surface charge of nanoparticles suspended in a liquid. The elemental composition of nanoparticles can be determined by EDS and/or wavelength-dispersive spectroscopy (WDS). XRD produces a diffraction pattern which is subsequently compared to data given in a standard crystallographic database. This exercise identifies crystallite size, structure, preferred crystal orientation, and phases present in samples. FTIR spectroscopy is used to investigate the functional groups attached to the surface of the nanoparticles. Raman spectroscopy is the most important technique to distinguish the allotropes of carbon and number of carbon layers in graphene sheets. The topography of the nanomaterials is investigated using SPM, atomic-force microscopy (AFM), or scanning tunneling microscopy (STM). In general, AFM, STM, and TEM have better resolution than other techniques. On the other hand, XRD is a powerful tool for measuring the crystallite size, crystal structure, strain fields, defects, and imperfections.

12.1 X-Ray Diffraction

Powder XRD method has been widely used to determine the crystal structure, lattice parameter, stresses, and crystallite size of the nanoparticles. It is the main method among indirect methods for determining the crystallite size. X-rays are electromagnetic radiation with typical photon energies in the range of 100 eV to 100 keV. For diffraction applications, only short wavelength x-rays in the range of a few angstroms to 0.1 Å (1–120 keV) are used. These x-rays with short wavelengths are comparable to the interplanar distances; hence, they are ideally suited for probing the structural arrangement of atoms and molecules in a wide range of materials. The energetic x-rays can penetrate deep into the materials and provide information about the bulk structure. In XRD, a monochromatic beam of x-rays with a typical wavelength ranging from 0.7 to 2 Å is incident on a specimen. The x-rays are diffracted by the crystalline phases in the specimen, if Bragg's law (Equation 12.1) is satisfied:

$$\lambda = 2d\sin\theta \qquad (12.1)$$

where d is the spacing between atomic planes in the crystalline phase and λ is the x-ray wavelength. The intensity of the diffracted x-rays is measured as a function of the diffraction angle 2θ and the specimen's orientation. This diffraction pattern is used to identify the specimen's crystalline phases and to measure its crystallite size. XRD is a nondestructive technique and does not need elaborate sample preparation. The XRD pattern is like a fingerprint, and the mixtures of different crystalline phases can be easily distinguished by comparing their XRD pattern with reference to data given in electronic databases such as the Inorganic Crystal Structure Database (ICSD).

As shown in Figure 12.1, ideal (free from defects) crystalline materials provide sharp diffraction lines (Figure 12.1a) and the real crystalline materials exhibit diffraction peaks with slight peak broadening (Figure 12.1b) at the base of diffraction lines compared to

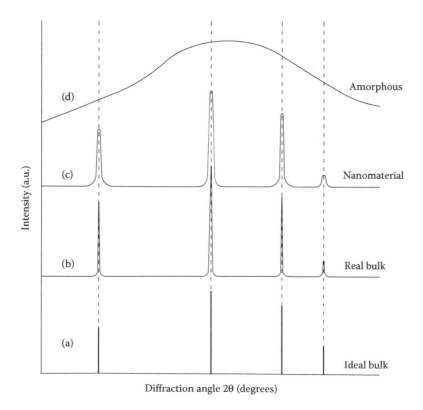

FIGURE 12.1
Typical XRD patterns of (a) ideal bulk, (b) real polycrystalline bulk, (c) polycrystalline nanomaterials, and (d) amorphous metallic materials.

that of ideal materials. For a crystal containing thousands of planes, under some restrictions/conditions, there is destructive interference leading to cancellation of the intensity at all the angles other than Bragg's angle. However, in case of finite sized crystals (size <1000 Å), where the number of diffracting planes is limited, the diffraction peak will broaden significantly (Figure 12.1c and 12.2a) due to the insufficient number of planes causing destructive interference at the diffraction angles other than Bragg's angle. In other words, some diffraction peaks with low intensity also occur at angles smaller and higher than the Bragg's angles, thus resulting in peak broadening. This peak broadening helps in determining the crystallite size of nanomaterials. In contrast to this, amorphous materials show broad hump (Figure 12.1d) in the XRD pattern due to an almost complete lack of periodicity of atoms or ions.

The broadening of diffraction peaks is also caused by microdeformations, inhomogeneity and instrumental error. The broadening, caused by microdeformations and randomly distributed dislocations, depends on the order of diffraction and is proportional to tan θ. The substitutional or interstitial solid solutions (or inhomogeneity) also cause diffraction peak broadening, which is independent of the order of diffraction. The randomly oriented crystals in small particles (size: <100 nm) cause broadening of diffraction peaks due to the absence of complete constructive and destructive interferences of x-rays in a small (finite-sized) particle. One of the reasons of instrumental error is K_α-line which itself has a width of about 0.001 Å, that is, a range of wavelengths extending over 0.001 Å ($K_{\alpha 1} = 1.54056$ Å and $K_{\alpha 2} = 1.54439$ Å) leads to an increase in diffraction line width, for $\lambda = 1.5$Å and $\theta = 45°$, of about 0.08° over the width one would expect if the incident beam were strictly monochromatic. Secondly, actual x-ray beam contains divergent and convergent rays, in addition to parallel rays, which causes diffraction at angles other than the Bragg's angles. The peak broadening due to instrumental error, inhomogeneous lattice strain, and structural faults cannot be ignored. The size calculated from XRD peak broadening is the smallest coherently scattering domains (or single grain or crystallite) of the

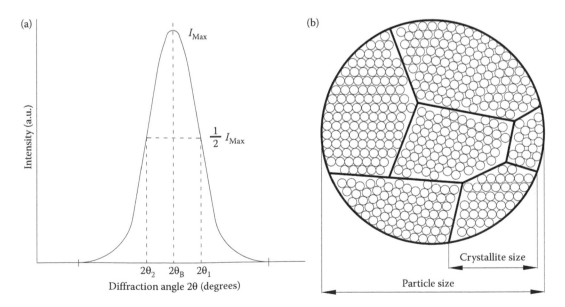

FIGURE 12.2
(a) A typical XRD peak broadening of a nanomaterial powder and (b) crystallites in a particle.

material (Figure 12.2b). The values of the broadening caused by the small grain size, deformations, and inhomogeneity are proportional to sec θ, tan θ, and (sin 2θ)/cos θ, respectively.

Due to their angular dependence, these peaks broadening can be separated. The shape of diffraction peak depends on both the size and shape of the nanoparticles. It is to be noted that the homogeneous (or uniform elastic) strain shifts the diffraction peak positions due to change in lattice constants (or d-spacing of planes). From the shift in peak positions, one can calculate the change in d-spacing. There are three methods such as Williamson and Hall, Warren and Averbach, and Debye–Scherrer used to calculate the crystallite size and strain of the sample. If there is no inhomogeneous strain, the average crystallite size is determined from the full width at half maximum (FWHM) of a diffraction peak broadening using the Debye–Scherrer equation:

$$t = \frac{K\lambda}{\beta_c \cos\theta} \tag{12.2}$$

$$\beta_c = \sqrt{\beta_0^2 - \beta_i^2} \tag{12.3}$$

where t is the crystallite size and β_c is the corrected FWHM. The value of Scherrer's constant (K) for diffractions with different crystallographic Miller indices (hkl) of a cubic crystal lattice varies between 0.9 and 1.15, and it is generally taken 0.9 for a cubic materials for simplicity. Equation 12.2 can be used if the powder or sample under investigation is free from the instrumental error (β_i). The β_0 can be determined from observed FWHM from the nanomaterials by convoluting Gaussian profile and the β_i is measured on an annealed and completely homogeneous powder with a particle size of about 1–10 μm. It can also be measured from the broadening of standard sample such as silicon or quartz. Annealed alpha-quartz crystalline powder with particles size of 25 μm is generally used as a standard sample. The β_i and β_0 are usually measured, in radians, at an intensity equal to half of the maximum intensity. If the sample has strain due to crystal imperfection and/or distortion, the strain can be calculated using the following equation.

$$\varepsilon = \frac{\beta_c}{4\tan\theta} \tag{12.4}$$

Total diffraction peak broadening due to smaller crystallite size and strain can be calculated by combining Equations 12.2 and 12.4:

$$\beta_c = \frac{K\lambda}{t\cos\theta} + 4\varepsilon\tan\theta \tag{12.5}$$

By rearranging the above equation, we get Equation 12.6, which is called as Williamson–Hall equation.

$$\beta_c \cos\theta = \frac{K\lambda}{t} + 4\varepsilon\sin\theta \tag{12.6}$$

Williamson–Hall is a simplified integral breadth method for deconvoluting size and strain contributions to line broadening as a function of 2θ. In this method, the crystallite size is determined from the y-intercept, while strain is calculated from the slope. In this method, the crystallite size and strain contributions to line broadening are assumed to be independent to each other and both have a Cauchy-like profile. In addition, the strain is assumed to be uniform in all crystallographic directions of the crystals. From this broadening, it is possible to determine an average crystallite size by the Debye–Scherrer equation. This equation may give crystallite size with error up to 50%, which may further increase for the relatively smaller crystallites [1,2]. Warren and Averbach's method which takes into account both the peak width and the shape of the peak provides both the crystallite size distribution and lattice microstrain. Figure 12.3 shows XRD patterns of coarse grained nickel and nickel nanopowder, which were compared with diffraction lines of ideal nickel crystal (JCPDS 65-2865).

The peak positions (2θ) of the diffraction peaks of coarse-grained nickel and nanopowder nickel coincides almost completely with those of ideal nickel. In general, the intensity of the diffraction peaks obtained from the nanopowder is decreased compared with those of coarse-grained sample. Even after making the correction using Equation 12.3 or 12.6, the average grain size values determined by XRD might be somewhat larger than the value determined by TEM histogram. This is due to the difference in their basic principles, that is, XRD estimates the average grain sizes, while TEM determines the actual grain sizes. Figure 12.4 shows the powder XRD pattern of a series of indium phosphide (InP) nanoparticles with different sizes. It is clearly seen that the intensity and the sharpness of the diffraction peaks decreases with decreasing crystallite size of InP nanoparticles from 43 to 19 Å [3,4]. The powder XRD method is also used for the characterization of core–shell nanoparticles to find out the diffraction pattern from the prominent lattice planes of the crystalline nanoparticles. It is interesting to see that $Pt@Fe_2O_3$ core–shell nanoparticles exhibit diffraction peaks from the planes of Pt nanoparticles (thickness of ∼10 nm) whereas Fe_2O_3 coating or shell (thickness of ∼3.5 nm) does not show any diffraction peak. This might be due to the strong scattering from Pt nanoparticles..

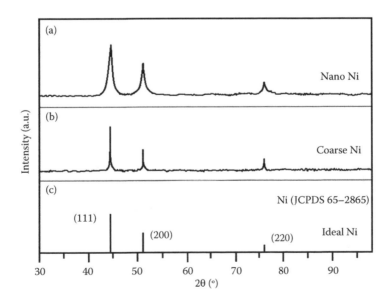

FIGURE 12.3
Powder XRD patterns of (a) ideal Ni, (b) coarse-grained bulk Ni, and (c) Ni nanopowder.

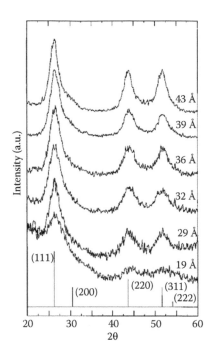

FIGURE 12.4
Powder XRD patterns of InP nanocrystals with different sizes. The stick spectrum shows the bulk diffractions with relative intensities. (Reprinted with permission from Guzelian A. A. et al. 1996. *J. Phys. Chem.*, 100: 7212. Copyright 1996, American Chemical Society.)

Compared to electron diffraction, one of the disadvantages of XRD is the low intensity of diffracted x-rays, particularly for low-atomic number (Z) materials. XRD is more sensitive to high-Z materials. For low-Z materials, neutron diffraction or electron diffraction is suitable. Typical intensities for electron diffraction are about 10^8 times larger than for XRD. Due to small diffraction intensities, XRD generally requires large specimens and the information acquired is an average over a large amount of material. Measurement of grain size by using XRD line broadening is suitable when the average grain size is <100 nm and the grain size distribution is narrow. The instrumental broadening correction becomes critical for grain sizes more than ~30 nm. If the materials have inhomogeneous grain size distribution with some grains >100 nm, TEM should be used [1–5].

12.2 Optical Spectroscopy

Optical spectroscopy has been widely used for the characterization of nanomaterials. This technique can be generally categorized into two groups: absorption and emission spectroscopy and vibrational spectroscopy. The former determines the electronic structures of atoms, ions, molecules, or crystals through exciting electrons from the ground to excited states (absorption) and relaxing from the excited to ground states (emission). UV–vis spectroscopy is an example of absorption and emission spectroscopy, which involves wavelength from 300 to 800 nm to get the optical properties of the metallic or semiconducting nanoparticles with the size from ~2 to 100 nm. Vibrational spectroscopy involves the interactions of photons with species in a sample that results in energy transfer to or from the sample via vibrational excitation or de-excitation. The vibrational frequencies provide the information of chemical bonds in the detecting samples. Infrared spectroscopy and Raman spectroscopy are the examples of vibrational spectroscopy.

12.2.1 Raman Spectroscopy

Raman spectroscopy is based on the phenomenon called Raman scattering, named after the Indian scientist Raman who first discovered it in 1928. Raman spectroscopy measures the inelastic scattering of monochromatic radiation (UV, visible, or near IR) by a sample. Raman scattering has been found to be sensitive to isotope composition, crystal structure, biomolecular interactions, and noncovalent interaction between molecules. In Raman scattering measurement, a monochromatic light (or laser beam) shines on the sample and scattered light is measured off the angle with respect to the incident light to minimize the Rayleigh scattering. The inelastically scattered light with lower (Stokes' scattering) or higher (anti-Stokes' scattering) frequencies can be measured with a photodetector. The energy difference between the scattered and incident light, so-called Raman shift (usually represented by a wave number), equals to the vibrational or phonon frequencies of the sample, as long as selection rules allow. The spectrum is typically presented between the intensity of the Raman scattered light and the Raman shift. Raman shift typically appears in the region of a few hundred to a few thousand per centimeter. The signal detected is usually visible if the incident light is in the visible range. The Stokes–Raman signal is most often detected and appears lower in energy (or frequency) compared to the incident light frequency. Raman peaks for most nanoparticles are comparatively narrower than photoluminescence (PL) peaks. UV Raman has been used to study the phase transitions (anatase and rutile) of TiO_2 nanoparticles [6].

When a sample is illuminated by a monochromatic light, in addition to elastically scattering, a small portion of the light is inelastically scattered. The difference in energy of the emerging photons relative to the incident photons is absorbed or released by collective vibrations of the atoms in the sample, phonons in solids, and normal vibration modes in molecules, liquids, or gases etc. This scattering is known as the Raman effect. The vibration levels of a sample are intrinsically dependent on its atomic structure; hence, the Raman effect can be used as an effective tool for structural and chemical characterization.

The intensity of Raman scattering is about three to eight orders of magnitude weaker than the Rayleigh scattering and about six to eight orders of magnitude weaker than the incident excitation light beam. Rayleigh scattering depends on the polarizability of the material at its equilibrium configuration, while Raman scattering depends on the sensitivity of the polarizability to changes in the atomic configuration along the direction of the normal coordinates of vibration. For example, a particular normal mode will be Raman inactive if the polarizability of the material is zero. For IR absorption, a particular mode of vibration is Raman active if the dipole moment changes with that vibration. Both the Raman and IR techniques are complementary characterization tools. For example, in case of a homonuclear diatomic molecule, its stretching mode will be Raman active but since there is no change of the dipole moment, it will be IR inactive, while a heteronuclear diatomic molecule will be both, Raman and IR active. Raman spectroscopy can be used to differentiate the various phases of a material having same chemistry or regions of different chemical composition of a sample. It can be used to study the incorporation of impurities on a crystalline matrix, to measure the stress distribution in a sample, or to monitor

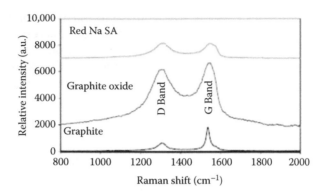

FIGURE 12.5
Raman spectra of graphite, GO, and reduced graphene (Red Na SA). (From Ciszewski M., Mianowski A., Nawrat G. 2013. *J. Mater. Sci.: Mater. Electron.* 24: 3382–3386. Open Access.)

the phase transitions with temperature or pressure [6]. For example, Raman spectroscopy is a useful method to distinguish the various allotropes of carbon and to determine the number of carbon layers in graphene sheet. It can be used to investigate crystallite size, hybridization of carbon atoms, defects, and crystal disorder. Figure 12.5 shows the Raman spectra for graphite, graphene oxide (GO), and reduced graphite (Red Na SA), indicating vibration band "D" around 1355/cm and "G" around 1600/cm. The D band corresponding to disorder is very small in highly ordered graphite structure but increases during oxidation and reduction. The G band that corresponds to sp^2 domains is broadened after oxidation due to defects formation and small crystallite size. The intensity ratio (I_D/I_G) is a convenient measurement method of disordered carbon that simply corresponds to amount of sp^3 and sp^2 carbon atoms. The I_D/I_G for graphite is ~0.36, indicating the ordered and layered structure with high-crystallite size and this ratio rises to 0.9 for GO due to an increase in disorder of graphene layers and diminishing of crystallite size. Upon reduction of GO, the value of I_D/I_G ratio becomes 1.04 due to an increase in sp^2 domains by retrieving the ordered structure and decrease in the average size of crystallites [7].

The Raman spectrum of single crystal silicon has shown a symmetrical profile with a maximum at 521/cm and an FWHM of ~5.0/cm. However, nanostructured silicon has shown significant changes in the spectrum such as a decrease of peak intensity, an increase of FWHM to 14.2/cm, an increase in peak asymmetry, and a peak shift of ~2.5/cm toward lower frequencies [8]. In addition to characterization of the CNTs, Raman spectroscopy has been used extensively to probe the changes in nanoparticle size. For example, the Raman spectrum tends to broaden asymmetric and shifts to lower frequencies as the particle size of CeO_{2-y} nanoparticles decreases (Figure 12.6). The Raman shift is attributed to an increase in lattice constant with decreasing particle size [9].

12.2.2 UV–vis Spectroscopy

UV–vis spectroscopy is widely used to quantitatively characterize the nanomaterials. It is based on the measurement of light absorption due to electronic transitions in a sample. The wavelength (λ) of light required for electronic transitions is typically in the ultraviolet ($\lambda = 200–390$ nm) and visible region ($\lambda = 390–780$ nm) of the electromagnetic radiation spectrum. Hence, electronic absorption spectroscopy is often called UV–vis spectroscopy. According to Beer's law, the absorbance (*A*) can be determined from the

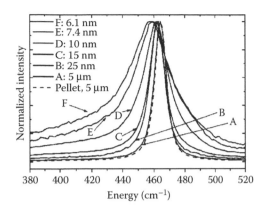

FIGURE 12.6

Normalized Raman spectra from the various sized nanoparticles and the bulk material indicating the peak broadening with decreasing size of particles. (Reprinted with permission from Spanier J. E. et al. 2001. Size-dependent properties of CeO_{2-y} nanoparticles as studied by Raman scattering. *Phys. Rev. B* 64: 245407. Copyright 2001 by the American Physical Society.)

incident light intensity (I_0) and the transmitted light intensity (I) using the following equation.

$$A = \log I_0/I = \varepsilon l c = \alpha c \tag{12.7}$$

where c is the concentration of a solution, l the path length of the sample, α the absorption coefficient, and ε the molar absorptivity. In an experiment, both I_0 and I can be measured and thus A can be determined experimentally. If both l and c are known, the absorption coefficient can be determined by Equation 12.7. The absorption coefficient is wavelength dependent and the spectroscopy provides a spectrum or a plot between α and λ for a given sample solution or solid. The solid sample should be thin to avoid the saturation of absorption.

UV–visible spectroscopy is one of the simplest and most useful optical techniques for studying optical properties of nanomaterials. As shown in Figure 12.7, a beam of monochromatic light (300–800 nm) is split into two beams, one of them is passed through the sample and the other passes a reference. The sample should be of low concentration and free from any kinds of foreign particles or floaters or aggregates. After transmission through the sample and reference, two beams are directed back to the light detectors, photomultiplier tube (PMT) or charge-coupled device (CCD), where they are compared. If the sample absorbs light at some wavelength, the transmitted light will be reduced. Then, the intensity of the transmitted light is plotted as a function of the wavelength of light. Liquid samples are usually contained in a cuvette (or cell) that has flat and fused quartz faces. Quartz is commonly used as it is transparent to both UV and visible lights [10,11].

The UV–vis spectrum of nanomaterials is generally different than that of their bulk counterparts. Figure 12.8 shows the UV–vis spectra of the isolated CdS nanoparticles having different particle sizes. It can be seen that the nanoparticles show broad absorption spectrum due to quantum confinement effects and the absorption peaks shift toward smaller wavelengths (higher energies) as the crystal size of the nanoparticles decreases, while the molar absorption coefficient increases with decreasing size. Peak widths of

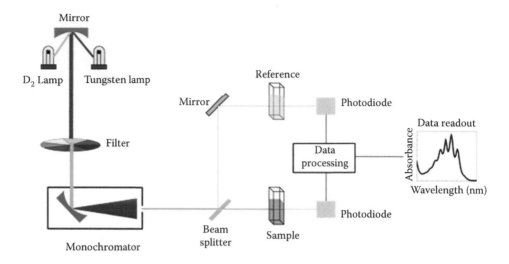

FIGURE 12.7
Schematic diagram of UV–vis spectroscopy for a sample. (From https://en.wikipedia.org/wiki/Ultraviolet%E2%80%93visible_spectroscopy)

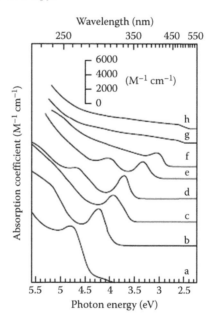

FIGURE 12.8
UV–vis spectra of the isolated CdS nanoparticles with size of (a) 6.4, (b) 7.2, (c) 8.0, (d) 9.3, (e) 11.6, (f) 19.4, (g) 28.0, and (h) 48.0 (Å). (Reprinted with permission from Vossmeyer T. et al. 1994. CdS Nanoclusters: Synthesis, characterization, size dependent oscillator strength, temperature shift of the excitonic transition energy, and reversible absorbance shift. *J. Phys. Chem.* 98: 7665–7673. Copyright 1994, American Chemical Society.)

the nanoparticles also depend on the size distribution of the nanoparticles. The shift of absorption peak from higher to shorter wavelength is called a blueshift [12,13].

The confined electrons have higher energy than those in bulk crystal. The change in energy bandgap (ΔE) for quantum dots (QDs) can be calculated with their sizes using the following equation:

$$\Delta E = \frac{\eta^2 h^2}{8ma^2} \tag{12.8}$$

where η is the principal quantum number, h is the Planck's constant, m is the effective mass of electron, and a is the radius of QDs. Atoms have discrete energy levels, while semiconducting crystals have bands of energy. Due to intermediate structures, semiconductor nanocrystals are called *artificial atoms*. Their optical properties can be tailored to emit in the visible or infrared region by selecting the appropriate particle size. The semiconducting QDs having energy bandgap in the range of from 1.8 to 3.1 eV would absorb and transmit a wavelength lying in this range. If the energy bandgap is <1.8 eV, then black color will be transmitted since all photons are absorbed; and similarly, if the energy bandgap of a semiconductor is >3.1 eV, then it would emit white color. From a given energy bandgap of a semiconductor, the wavelength of the color transmitted is calculated by using the following equation:

$$\Delta E = \frac{hc}{\lambda} \tag{12.9}$$

where c is the velocity of light. The position of absorption maxima for a molecule depends on the difference in the energy of the ground-state level to that of excited state. Larger is the difference between the energy of ground state and excited state, higher is the frequency of absorption and thus smaller will be the wavelength. UV–vis spectroscopy is used to determine the concentration, size, purity, and aggregation state of the constituents, and functional groups in molecules. It also provides size- and composition-dependent spectrum for the nanomaterials. The UV–vis absorbance of Ag and Au nanoparticles exhibits a strong absorption in the visible region of 400–450 nm and 500–550 nm, respectively, due to the surface plasmon resonance (SPR). The SPR band is sensitive to size, shape, aspect ratio, composition, aggregation state, and the refractive index of the medium. The protein binding to Au and Ag NPs shifts the peak of absorption band slightly, that is, ~5 nm and in some cases larger shifts due to subsequent aggregation [12].

12.2.3 Photoluminescence Spectroscopy

Photoluminescence (PL) spectroscopy is a very sensitive tool for the investigation of electronic states in the bandgap of semiconductors. It involves optical excitation of electrons from the valence band to the conduction band of the semiconductor. It uses photons of a higher energy than the bandgap energy. This results in the creation of electron–hole pairs. After some time electrons and holes recombine, thereby releasing energy in the form of phonons, photons, or Auger electrons. Photons emitted from radiating recombination centers can be detected by PL. Generally, PL measurements are carried out at low temperatures (<77 K). The temperature analysis avoids the ionization of color centers and broadening of peaks by phonon processes. It is suitable for the materials that exhibit PL. PL spectroscopy is suitable for the characterization of both organic and inorganic materials of virtually any size, and the samples can be in solid, liquid, or gaseous forms. Both UV and visible wavelengths of light are utilized in PL spectroscopy. The PL emission properties of a sample are characterized by four parameters such as intensity, emission wavelength, bandwidth of the emission peak, and the emission stability. As the size of a semiconductor nanoparticle decreases, the bandgap increases, resulting in shorter wavelengths (or

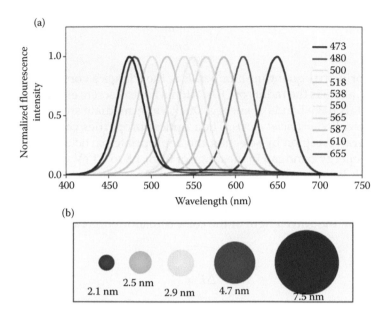

FIGURE 12.9
Size-tunable fluorescence spectra of CdSe quantum dots (a), and illustration of the relative particle sizes (b). (Smith A. M., Nie S. 2004. Chemical analysis and cellular imaging with quantum dots. *Analyst* 129: 672–677. Reproduced by permission of Royal Society of Chemistry.)

blueshift) of light emission. Figure 12.9 shows size-tunable fluorescence spectra of CdSe QDs throughout the visible light spectrum [14].

This shift in wavelength of PL spectra can be utilized to study the bandgap and impurity levels of the nanomaterials. In a typical PL measurement, the sample is irradiated with a specific wavelength of light, and the intensity of the luminescence emission is recorded as a function of wavelength. The light emitted from the sample is collected through lenses, dispersed by another monochromator, and detected by a photodetector. The spectrum is obtained between the intensity of emitted PL light and the wavelength of emitted light. PL usually appears at longer wavelength (or redshifted) with respect to the incident excitation light. The fluorescence of a sample can also be monitored as a function of time, after excitation by a flash of light. This technique is called time-resolved fluorescence spectroscopy [14].

12.2.4 Fourier Transform Infrared Spectroscopy

Fourier transform infrared spectroscopy (FTIR) is one of the most widely used tools for the detection of functional groups in pure compounds and mixtures. It simultaneously collects high spectral resolution data over a wide range of wave number extending from 100–104 cm^{-1}. Infrared spectroscopy is associated with vibrational energy of atoms or group of atoms in a material. The possibility that two compounds have the same infrared spectrum is extremely small. Therefore, the FTIR spectrum is called the fingerprint of a molecule. The FTIR spectrum of nanoparticles differs considerably from that of the bulk sample due to very high surface-to-volume ratio compared to the bulk. Additionally, the large number of atoms constituting the surface of nanoparticles can influence the vibrational spectra and exhibits distinguished features than that of the bulk.

For FTIR analysis, solid sample powder is examined by mixing it with an alkali halide (i.e., KBr). The KBr powder does not absorb infrared radiation in the region 4000–650 cm^{-1}.

When a sample interacts with electromagnetic energy, it absorbs the energy in the infrared region of the spectrum. It causes atoms or group of atoms of the sample to vibrate faster about the bonds, which connect them. However, infrared radiation does not have sufficient energy to cause the excitation of electrons. The sample absorbs energy from a particular region since the vibrations are quantized. The position of a particular absorption band is represented by a particular wave number. Thus, the infrared radiation of successively increasing wavelength is passed through the sample and the percentage of transmittance is measured using an infrared spectrophotometer. The graph (i.e., infrared spectrum) is recorded between the variation of the percentage of transmittance and the wave number. Each dip in a spectrum is called a peak and it represents absorption of infrared radiation at that frequency.

FTIR spectroscopy is mainly used to identify the elements and the phase of the elements. It is an effective tool in detecting the shape of nanometer-sized materials. As particle size increases, the width of the peak decreases and intensity increases. The decrease in the width of the peaks of the spectra indicates change in phase transformation and an increase in crystallinity. It is also used to study the nature of surface adsorbents in nanoparticles. Owing to these adsorbents, the nanoparticles possess additional peaks in comparison with the FTIR pattern of a bare nanoparticle. Moreover, due to high surface-to-volume ratio, the activity of the nanoparticles would be significantly different from that of the bulk. For example, Fe nanoparticles synthesized at oxygen partial pressure of <200 Torr contain only Fe_3O_4 while at higher oxygen partial pressure they contain γ-Fe_2O_3 and its FTIR peak occurs at 450 cm^{-1}.

12.3 Surface Area Analysis (BET Method)

The specific surface area of the particles is the summation of the areas of the exposed surfaces of the particles per unit mass. The gas adsorption method is used to determine the specific surface of the powder. The measured value of the specific surface may be used to estimate the mean particle size. To measure the specific surface, helium or nitrogen is passed through the prepared powder in a special chamber. Helium or nitrogen molecules are adsorbed by the particle surface. Helium-saturated powder is heated to remove the entire amount of adsorbed helium and the amount of adsorbed helium is determined from the variation of the mass. Assuming that helium atoms form a monolayer on the surface of the particles, from the volume of the adsorbed gas it is possible to determine the total surface area of the particles and the specific surface of the powder (in m^2/g). For the spherical particles and narrow size distribution of powder, the specific surface area can be calculated using the following equation:

$$S = \frac{\text{Surface area of particle}}{\text{Mass of powder}} = \frac{4\pi R^2}{(4\pi/3)R^3 \rho} = \frac{6}{\rho \cdot D} \qquad (12.10)$$

where S is the specific surface area D is equivalent particle size, and ρ is the density. It is to be remembered that m^2/g is not an SI unit for the specific surface area, but it is accepted universally. From Equation 12.10, it can be seen that there is an inverse relationship between the average particle size and surface area. Thus, as the particle size decreases, the specific surface increases. The loose primary particles in a powder can provide access to the gas for most of the surface area of the powder and thus, provide an actual surface area or particle size. However, if the primary particles (or crystallites) are in the form of hard agglomerates, the instrument would provide lower surface area than the actual surface area because gas molecules do not have access to most of the surface area of primary particles of which the agglomerate is made up. Thus, the degree of clustering can be determined by the ratio of the expected surface to the measured surface area. Due to clustering or aggregation of nanoparticles, in most of the cases, the experimentally determined specific surface area is significantly smaller than that theoretical values found out from Equation 12.10.

BET method is used to measure the specific surface of powders by gas adsorption at the surface. It is named after the first letters of the names of its three inventors; Stephen Brunauer, Paul Hugh Emmett, and Edward Teller. This method is based on the Langmuir adsorption isotherm with some assumptions of: gas molecules physically adsorb on a solid in layers infinitely, there is no interaction between each adsorption layer, and the Langmuir theory can be applied to each layer. The BET equation can be expressed by the following equation:

$$\frac{P}{v(P_0 - P)} = \frac{1}{v_m c} + \frac{c-1}{v_m c} \frac{P}{P_0}$$

(12.11)

$$c = \exp^{\left[\frac{(E_1 - E_l)}{RT}\right]}$$

where P and P_0 are the equilibrium and the saturation pressure of adsorbates (or gas) at the temperature of adsorption, v is the volume of adsorbed gas, v_m is the gas volume adsorbed at the surface in one monolayer. c is the BET constant, E_1 is the heat of adsorption for the first layer, and E_l is that for the second and higher layers and is equal to the heat of liquefaction. This Langmuir's adsorption isotherm is observed when adsorption (of nitrogen or krypton) to the surface is significantly stronger than the condensation to the liquid. It can be seen from Equation 12.11 that a plot of $P/[v \cdot (P_0 - P)]$ versus P/P_0 gives a straight line, whose intercept is $1/(v_m \cdot c)$ (at y-axis) and the slope is $(c - 1)/(v_m \cdot c)$. This plot is called a BET plot. The linear relationship of BET equation is maintained only in the range of $0.05 < P/P_0 < 0.35$. Thus, from the slope and intercept, the two constants, that is, volume of adsorbed gas (v_m) from a monomolecular absorbed layer and BET constant (c), can be determined. The surface area of the material under investigation can be calculated from the number of gas molecules (N_s) in a monolayer at the surface:

$$N_s = \left(\frac{v_m}{V_M}\right) \cdot N_A$$

(12.12)

$$\text{Specific surface area} (A) = \frac{(N_s \cdot a_s)}{m}$$

where V_M is the volume of 1 mole of gas measured under standard temperature and pressure conditions. N_A is Avogadro's number, a_s is the area covered by one molecule (for nitrogen, its value is 0.158 nm^2) and m is mass of solid sample or adsorbent [15,16].

12.4 Light Scattering Method

Photon correlation spectroscopy (PCS) or dynamic light scattering (DLS) is the most common method for determining the particle size distribution of a powder sample and size of particles, aggregates, and large molecules such as polymers and proteins. It is a relatively simple technique based on elastic Rayleigh scattering from the nano- and/or submicron-sized particles dispersed in a suitable liquid. It measures the speed at which particles move under the Brownian motion by monitoring the intensity of light scattered by the sample at some fixed angle. The Brownian motion causes constructive and destructive interference of the scattered light. It uses the method of autocorrelation to uncover information contained in the light intensity fluctuations. Autocorrelation measures how well a signal matches a time-delayed version of itself, as a function of the amount of time delay. A quicker autocorrelation function decay indicates faster motion of the particles. The autocorrelation function can be related to the hydrodynamic radius (R_H) or hydrodynamic diameter (D_H) ($D_H = 2R_H$) of the particles, which is the information of interest, by the Stokes–Einstein equation. The hydrodynamic diameter is the diameter of a sphere that has the same translational diffusion coefficient as the particle. Larger particles correspond to slower Brownian motion and slower delays of the autocorrelation function. DLS has been used frequently in the study of nanoparticles [1].

The light scattered by the nanoparticles in suspension is fluctuated with time and can be related to the particle diameter. If the scattered light is collected as a function of scattered direction, the method is called static light scattering (SLS). In contrast, if the light scattered intensity is recorded as a function of time from several directions, the method is called DLS method. DLS measures the Brownian motion and correlates that information with the size of the particles. In this dilute suspension of a powder (i.e., 0.0001 to 1% v/v) is prepared in an appropriate solvent (or water) with suitable wetting agent in flow cells or cuvette. The suspension is subsequently sonicated for few tens of minutes to isolate the primary particles from the attractive forces (i.e., van der Waals attraction) between dry particles or break agglomerates. The cuvette is placed in a thermostat for temperature stabilization. The measurements are carried out using a He–Ne laser ($\lambda = 633$ nm, 10–35 mW) at two scattering angles in the center of a cuvette containing a suspension (Figure 12.10a). The interaction of light (or laser beam) with the nanoparticles or molecules leads to a shift in the frequency of the light and angular distribution of the scattered light, both of which are related to the size. The light scattered at a given angle is received by a photoelectric multiplier. A multichannel correlator then measures the autocorrelation function G(τ) of scattered light. The random motion of the particles is correlated with the scattered light intensity fluctuations. From this correlation, the particles' diffusion coefficient is obtained. The equivalent sphere particle size is then calculated from the Stokes–Einstein equation:

$$R_H = \frac{k_B T}{6\pi\eta D} \tag{12.13}$$

where R_H is the van der Waals (or hydrodynamic) radius of the molecule (in m), k_B is the Boltzmann constant (1.38×10^{-23} J/K), T is the absolute temperature (in K), η is the viscosity (in pascals per second or centipoises), and D is the self-diffusion coefficient (in m^2/s) (or translational diffusion coefficient). The D depends on the size and surface structure of particle, and the concentration and type of ions in the medium. The D is obtained from the correlation function by using various algorithms.

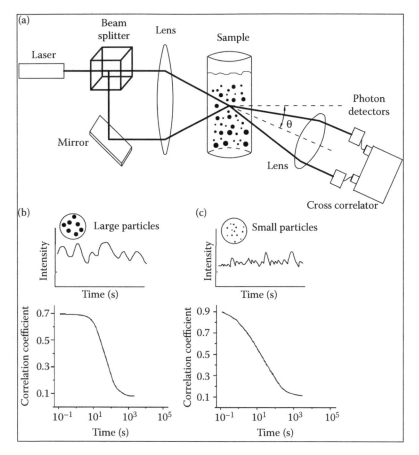

FIGURE 12.10
A schematic set-up for measurement of (a) particle size distribution by light scattering method. The fluctuation in the intensity of scattered light as a function of time for (b) larger particles and (c) smaller particles.

The D is determined from the curve obtained from the fluctuation in the intensity of scattered light as a function of time. From these curves, a typical correlogram is drawn between the correlation function and time. For larger particle, the signal will change slowly (Figure 12.10b) while it changes rapidly and thus correlation reduces more quickly for smaller particles (Figure 12.10c). The correlation function ($G(\tau)$) for monodisperse and polydisperse samples can be represented by equation (12.14) and (12.15), respectively. The D is calculated from the equation (12.16) and wave vector (q) (equation 12.17).

$$G(\tau) = A[1 + B \exp(-2\Gamma\tau)] \tag{12.14}$$

$$G(\tau) = A[1 + B \, g_1(\tau)^2] \tag{12.15}$$

$$D = \frac{\Gamma}{q^2} \tag{12.16}$$

$$q = \left(\frac{4\pi n}{\lambda_0}\right)\sin\left(\frac{\theta}{2}\right) \tag{12.17}$$

where,

A = baseline of the correlation function
B = intercept of the correlation function
n = refractive index of dispersant
λ_o = wavelength of the laser
θ = scattering angle
$g_1(\tau)$ = the sum of all the exponential decays contained in the correlation function

Approximation of the autocorrelation function by the exponential dependence makes it possible to determine the width of the spectral line and then find the diffusion coefficient and the particle size. An increase in radiation power results in an improvement of the accuracy and accelerates the measurements. Industrial analyzers of the particle size, using the PCS method, take the measurements under several angles and make it possible to determine the size of the particles in the range from 3 nm to 3 μm and also the size distribution of the particles. To determine the size of small particles in colloid solutions or suspensions, it is also efficient to use the mass distribution of particles using an ultracentrifuge. The method of separation of small particles by the means of a centrifugal force in an ultracentrifuge is used. If the particle density is known, measuring the mass of the particle, it is easy to find its volume and linear size. Figure 12.11 shows the differential particle size distribution of SiO_2 nanopowders determined using a laser particle size analyzer. It can be seen that the powder has a bimodal particle size distribution for SiO_2, that is, powder has particle size from 30 to 56 nm with mean particle size of 43 nm and second histogram varies between 270 and 590 nm indicating for cluster's size as confirmed by the TEM image (see inset). The inset of Figure 12.11 clearly indicates that the primary particle size of SiO_2 powder is <50 nm with irregular-shaped morphology [17].

In PCS, the scattering of light is proportional to sixth power of the particle size (<100 nm) according to the Rayleigh theory. It means a 50 nm particle will scatter 1 million times more light than a 5 nm particle. In order to obtain equal intensities of scattering from two particles with the size of 5 and 50 nm, the presence of 1 million 5 nm particles for

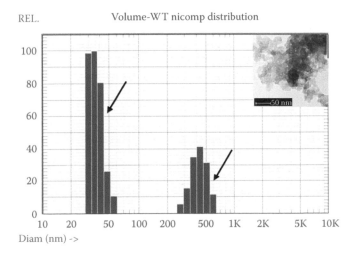

FIGURE 12.11
Differential particle size distribution of silane-treated SiO_2 nanoparticles. (Reprinted from *Compos.* B, 50, Goyal R. K., Kapadia A. S., Study on phenyltrimethoxysilane treated nano-silica filled high performance poly(etheretherketone) nanocomposites. Copyright 2013, with permission from Elsevier.)

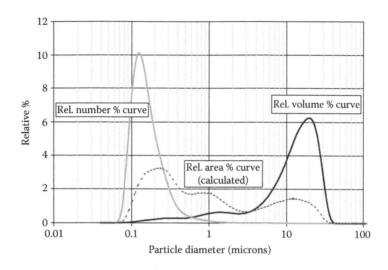

FIGURE 12.12
Laser diffraction size data for nanoscale aluminum powder. (Powers K. W. et al., Research strategies for safety evaluation of nanomaterials. Part VI. Characterization of nanoscale particles for toxicological evaluation. *Toxicol. Sci.*, 2006, by permission of Oxford University Press.)

every 50 nm particle is required. For example, Figure 12.12 shows how different a particle size distribution might appear if represented as a number- or area-based distribution, rather than a volume-based distribution. By viewing only one representation of the data, researchers should not make erroneous conclusions regarding the breadth of the distribution or the state of agglomeration of this system. Multiple-sizing techniques (i.e., microscopy) are also helpful to resolve these ambiguities. This occurs because volume scales as the cube of the particle diameter and calculated area scales as the square. Each curve, if presented by itself, would give an incomplete picture of the particle size distribution/state of agglomeration of the sample. The three curves will overlay only for an ideal spherical, monodisperse, unagglomerated system [18].

Therefore, in DLS, the results are more sensitive to the large particles or agglomerates as they dominate the scattering of light. The PCS method is well suited to the measurement of narrow particle size distributions in the range of 1–500 nm.

12.5 Electron Microscopy

Electron microscopy became one of the most important techniques to characterize the material's morphology on the nanometer to atomic scale. There are two main types of electron microscopes: SEM and TEM. When a high-energy primary electron beam enters a specimen, electrons undergo elastic scattering and inelastic scattering along with transmitted beam (in case of thin foil). The interactions of primary electrons with a specimen's atoms generate several signals such as secondary electrons (SEs), backscattered electrons (BSEs), Auger electrons, Bremsstrahlung (continuous) x-rays, characteristic x-rays, etc. (Figure 12.13). High-energy electrons go forward and form the transmitted beam. Some

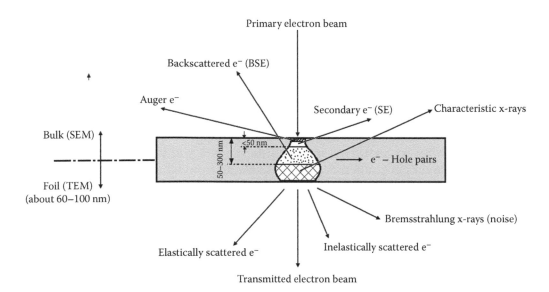

FIGURE 12.13
Typical schematic diagram showing the interaction zone between the primary electron beam and specimen surface.

of the electrons are diffracted or scattered. Both transmitted and diffracted beams are elastically scattered and coherent. When the scattering angle is high, scattered electrons lose coherence.

12.5.1 Scanning Electron Microscopy

12.5.1.1 Basic Principle of SEM

When the electron beam (energy 5–30 keV) impinges on the specimen, many types of signals are generated including SEs and BSEs. Most of the electrons are scattered at large angles (from 0° to 180°) when they interact with the positively charged nucleus. These elastically scattered electrons are usually called BSE. Some electrons are scattered inelastically due to the loss in kinetic energy upon their interaction with orbital shell electrons. Incident electrons may knock off loosely bound conduction electrons out of the sample. These electrons are called SEs. The SEs as well as BSEs are widely used for SEM topographical imaging. Both SEs and BSEs signals are collected when a positive voltage is applied to the collector screen in front of the detector. In contrast, when a negative voltage is applied on the collector screen, only BSEs signal is captured and the low-energy SEs are repelled. The electrons captured by the scintillator/photomultiplier are then amplified and used to form an image in the SEM. As shown in Figure 12.13, the interaction of primary electron beam with a sample generates SEs, BSEs, and characteristic x-rays, which are used to obtain 3-dimensional (3-D) image, elemental contrast and elemental composition or elemental mapping, respectively. The SEs are the products of inelastic scattering, and they have an energy level of several keV. In an interaction zone, SEs can escape only from a volume near the specimen surface and subsurface with a depth of 5–50 nm. In contrast, BSEs are the products of elastic scattering, and they have an energy level close to that of incident electrons. Their high energy enables them to escape from a much deeper level in the interaction zone, that is, from depths of ca. 50–300 nm. The lateral spatial resolution of a SEM image is affected by the size of the volume from where the signal electrons escape. Thus, the images formed by SEs have a better spatial resolution than that formed by BSEs [19]. The backscattering energy is strongly dependent on nuclear interactions; hence, the SEM backscattering mode is quite sensitive to the atomic number.

The spatial resolution of the SEM depends on various parameters: working distance, acceleration voltage (or wavelength of electrons), beam size, probe current, and astigmatism. A typical SEM has an image resolution of 5–10 nm, but a modern state-of-the-art SEM like field emission SEM is capable of providing an image resolution of 1 nm. However, the resolution in the backscattered mode is inferior (~10 nm) to that in secondary mode because of the larger penetration depth (~50–300 nm) from which the electrons are emitted. This resolution is much better than that from an optical microscope. The SEM has advantages of providing a large depth of focus due to the short wavelength of the primary electron beam used, a larger imaging field of view, and 3-D image or topography of the sample than the TEM. SEM is often combined with energy-dispersive x-ray (EDX) spectroscopy detector to get the elemental composition of the nanomaterials. The detection of BSEs or the electron backscatter diffraction (EBSD) gives crystallographic information about the sample. One more electron microscopy, that is, environmental SEM (ESEM), allows sample imaging under low pressure, fairly high humidity, and without the requirement of a conducting overcoat but has limited use for the study of nanomaterials bioconjugates.

12.5.1.2 Specimen Preparation for SEM

A specimen for SEM analysis can be in any form (powder or pieces) and size, which is easily fitted on specimen holder. The size of specimen holder varies (i.e., in tens of centimeters) with manufacturer. However, SEM requires electrically conductive specimen to avoid the surface charging of specimens, and dehydration for biological samples to prevent the destroying of the surface morphology. For topographic examination, the surface preparation involves sizing the specimen if it is not fitting on SEM specimen holder and removing the surface contaminants like hydrocarbons (oil and grease) present on the specimen surfaces. During an interaction of primary electron beam with the specimen, an electron beam decomposes any hydrocarbon and leaves a deposit of carbon on the surface. The deposit generates an artifact in SEM images. Therefore, the surfaces with oil or grease contamination should be cleaned by an organic solvent (methanol or acetone) with an ultrasonic cleaner. The cleaned specimens should not be touched with bare hands otherwise fingerprints containing the volatile hydrocarbon compounds may leave a dark mark on a well-cleaned specimen.

Surface charging is most likely encountered for the electrically insulating specimens such as polymers, ceramics, and biological samples. It occurs when excessive electrons are accumulated on the specimen surface, while primary electron beam impinges on surface. This accumulation of electrons builds up charged regions, which generate distortion and artifacts in SEM images. These charged regions deflect the incident electron probe in an irregular manner during scanning. The charging alters SE emission, and causes the instability of signal electrons. Surface charging is generally not a problem for metallic specimens because their good electrical conduction ensures the removal of excess electrons through electrical conduction between the specimen surface and ground. However, the presence of a metal oxide layer and insulating phases in metallic specimens may also generate surface charging. The charging of the specimen can be prevented by applying a thin (i.e., 10–20 nm) conductive film onto specimen surfaces by either vacuum evaporation or sputtering in a vacuum chamber. The sputtering is more widely used method than vacuum evaporation for conductive coating of the SEM specimens. Gold and gold–palladium are commonly used as the target material. Biological specimens need to be dehydrated before examining it in a SEM. If water is not removed, the high-energy electron beam may become one of the causes of destroying the surface morphology of the specimen. Dehydration is generally accomplished either by critical point drying or freeze-drying. Alternatively, conductive coating is not needed if one uses very low-accelerating voltages (0.5–2.0 kV), depending upon the sample material. In such cases, the electron hitting the sample surface and those being ejected (SEs and BSEs, respectively) balance and thus there is no charge accumulation on the surface. The technique, also called low-voltage scanning electron microscopy (LVSEM), is now being widely used for polymers, but this technique uses a high-brightness source such as LaB_6 or field emission source.

12.5.2 Transmission Electron Microscopy

12.5.2.1 Basic Principle of TEM

TEM utilizes high energy (100 keV to 1 MeV) electron beam to provide morphology, composition, and crystallographic information from a sample. In TEM, images are produced by focusing an electron beam on a very thin specimen (thickness of ca. 60–100 nm) which is partially transmitted and carries information about the specimen. In a TEM, a collimated beam of accelerating electrons with high energy is interacted with a thin sample or fine powder sample uniformly over the illuminated area. As the electrons travel through the sample, they are either scattered or are transmitted unaffected through the sample. The scattering can be elastic or inelastic. The diffracted electrons, deflected away from the optical axis of the microscope, are blocked using an aperture and thus the transmitted electron beam generates a contrast on the fluorescent screen. In case of nanomaterials, the crystalline structures interact with the electron beam mainly by diffraction rather than absorption, though the intensity of the transmitted beam depends largely on the density and thickness of the material through which it passes. The intensity of the diffraction also depends on the orientation of the planes of atom in the crystal relative to the electron beam. The transmitted electron beam strikes the fluorescent screen or a charge-coupled-device (CCD) camera, and generates an image with varying contrast.

In a simple mode of imaging, when all transmitted and diffracted rays leaving the sample are combined to form an image at the viewing screen, the sample shows a little contrast. Contrast is the appearance of a feature in an image. As shown in Figure 12.14, a TEM can be operated in the "diffraction mode" and the "image mode" by changing the optical path. In the diffraction mode (Figure 12.14a), the diffraction pattern formed on the back-focal plane is projected onto a fluorescent screen. In TEM, the diffraction pattern exhibits a plane of the reciprocal lattice of crystal. Such a pattern is obtained by diffraction from a selected area in the sample using selected-area diffraction (SAD). The diffraction mode can reveal whether a material is polycrystalline, single crystal, or amorphous. A single crystal will produce a diffraction pattern consisting of a center spot associated with the transmitted beam and other diffracted spots. Polycrystalline material will show diffraction spots falling on rings of constant radius. In the image mode (Figure 12.14b), the image of the sample is focused and projected onto a fluorescent screen. TEM can generate bright-field image and dark-field image. The contrast in bright- and the dark-field images is usually due to the variation in intensity of diffracted electron beams across the sample. A bright-field image is obtained by allowing only the transmitted beam to pass the objective aperture. In contrast, a dark-field image is obtained by allowing the diffraction beam

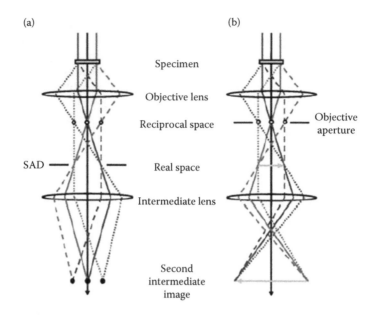

FIGURE 12.14
(a) Diffraction mode and (b) image mode in TEM. (Leng Y.: *Materials Characterization: Introduction to Microscopic and Spectroscopic Methods*. 2008. Copyright Wiley-VCH Verlag GmbH & Co. KGaA. Reproduced with permission.)

to pass the objective aperture by tilting the electron gun or sliding objective aperture. In TEM, image contrast is obtained from mass–density differences within the sample and deflection of primary electrons by the sample.

In TEM image formation, absorption of electrons plays a very minor role while the deflection of electrons from their primary transmission direction when they pass through the sample plays an important role. The amount of electron scattering at any specific point in a sample depends on the mass–density at that point. Thus, difference in thickness and density in a sample will generate variation in electron intensity received by an image screen, when the transmitted beam is allowed to pass the objective aperture. The deflected electron with scattering angle larger than about 0.5° is generally blocked by the objective aperture ring. The brightness of the image depends on the intensity of the electron beam leaving the lower surface of the sample and passing through the objective aperture. For example, the darker areas with higher contrast are those from where fewer electrons have been transmitted due to high density or thickness of the sample. The image area of lower contrast indicates that the samples have less density or thickness, and thus more number of transmitted electrons is present.

For crystalline materials, diffraction contrast is the main mechanism of TEM image formation. Diffraction contrast is the variation in intensity of electron diffraction across the sample. When the Bragg conditions are satisfied at certain angles between electron beams and crystal orientation, constructive diffraction occurs, which results in strong electron deflection. Thus, the intensity of the transmitted beam is reduced when the diffracted beams are blocked by the objective aperture; such an image is called a dark-field image. The diffraction contrast is very sensitive to sample tilting with reference to primary electron beam direction. However, the mass–density contrast is only sensitive to total mass in thickness per surface area. Diffraction contrast is observed by inserting an objective aperture in the beam. Thus, features in the image become much better than that of an aperture-less image. Without objective aperture the image is comparatively gray and featureless. The contrast of aperture-less image increases with the atomic number, and the thickness of the materials. The mass-thickness contrast is particularly useful for biological samples.

In TEM images, three-dimensional defects (i.e., second phase particles) appear as two-dimensional regions having different contrast than the surrounding matrix. The diffraction contrast from two-dimensional interfaces between crystals and domains appears as rows of one-dimensional bands (or fringes). The appearance of different types of fringes changes with the tilt of the beam or the crystal. The one-dimensional crystalline defects (i.e., dislocations) cause severe local distortions of the surrounding crystal. It is the strain (not the core of the dislocation) in the crystal which provides the diffraction contrast of the dislocation. Zero-dimensional point defects (such as vacancies and impurities) are generally not visible in conventional TEM images, but strain effects around nanometer-sized clusters of atoms or vacancies can be imaged and understood semi-quantitatively.

The organic constituents of the sample give low contrast because of their weak interaction with the electrons. This problem can partially be overcome by the use of stains such as heavy metal compounds. The dense electron clouds of the heavy atoms interact strongly with the electron beam. Sometimes, the organic constituents of the sample are decomposed by the electron beam during scanning and hence they are not detected. This problem can be minimized using cryogenic-TEM (cryo-TEM), which keeps the sample at liquid nitrogen or liquid helium temperatures. Similar to SEM, analytical TEM can determine elemental composition and crystal orientation in the specimen. Modern high-resolution TEM (HRTEM) goes down to a resolution better than 0.2 nm (i.e., <0.2 nm) due to the small effective electron wavelengths (λ), which is given by the de Broglie relationship shown in the following equation:

$$\lambda = \frac{h}{\sqrt{2mqV}} \tag{12.18}$$

where m and q are the electron mass and charge, h is Planck's constant, and V is the potential difference through which electrons are accelerated. For example, electrons of 100 keV energy will have wavelengths of 0.037 Å. The higher the operating voltage of a TEM instrument, the greater is its lateral spatial resolution. A TEM with 400 keV has point-to-point resolutions better than 0.2 nm [20,21]. TEM can offer magnification up to 5 million times at maximum. TEM is used to determine directly the size of nanoparticles or nanocrystallites and the interfaces in nanocomposites. For example, the interface between the nickel (Ni) with size of ca. 30–50 nm and titanium nitride (TiN) has been clearly observed. HRTEM is used to determine the size of the extremely small nanoparticle and the distance between the atomic planes. For example, cubic titanium carbide ($Ti_{44}C_{56}$) nanoparticle with size of ca. 12–18 nm has the atomic planes distance of 0.25 nm for the (111) planes. The TEM can provide information on the particle size, size distribution, and morphology of the nanoparticles. This is the only technique in which the individual particles are directly observed and measured. TEM can also be used to see degree of dispersion or aggregation of the nanoparticles (or particulates) in the matrix. HRTEM can provide thickness of the shell on the core-shell NPs, due to the contrast differ form the core and shell. However, TEM requires elaborate sample preparation. It is slow and few particles are examined at once. A typical histogram is generally plotted between the frequency of occurrence and the size range. About 500–1000 particles or grains should be measured to present statistically reliable mean size data for an optimum sample size.

It is to be noted that the electron microscopy is a local method and provides information on the size of the object only in the field of observation. However, the observed area may not be true representative or entire volume of the substance. Electron microscopy is used to examine a section with a size of 10 μm × 10 μm (i.e. 10^{-6} cm^2) of a specimen with a surface area of 1 cm^2. In this case, the examined area represents only one millionth parts (i.e., 10^{-6}) of the entire surface of the specimen, which is probably not a true representation of the actual sample surface. Therefore, electron microscopy examination must be carried out on several areas (or sections) in order to obtain statistically averaged out information for the entire substance. Low-voltage TEM is more commonly used to visualize biomolecules attached to a nanomaterial core. The attached layer of biomolecule appears to be much smaller than expected from a DLS analysis. It is due to the difference in nature of sample preparation for characterization. In TEM, a dried sample is taken, while suspended nanoparticles are taken in DLS technique. Moreover, staining the biomolecules with contrast agents may also alter the sizes of the nanomaterial bioconjugates [22,23].

Selected-area electron diffraction (SAD) technique is used to obtain the diffraction information from a small area of the specimen. For SAD, the sample is first examined in image mode until a feature of interest is found. Then, the intermediate aperture (objective aperture is removed) is inserted and positioned around this feature. The microscope is then switched into diffraction mode. SAD can be performed on regions of few hundred nanometers to 1 μm in diameter. The SAD of a region smaller than this size range is difficult to study due to the spherical aberration of the objective lens. It is useful to select a single crystalline region for analysis at a time and to examine the crystallographic orientation between two crystals. As discussed, electron diffraction uses monochromatic wavelengths between 0.0025 and 0.0037 nm. In contrast, XRD uses monochromatic wavelengths between 0.07 and 0.2 nm. Additionally, electron diffraction is based on the registration with photo plates or CCD devices, whereas XRD applies goniometer readings. However, for the nanomaterials with size <100 nm, the SAD has the problems of high-effective probe size of approximately few 100 nm and low signal-to-noise ratio from the nanostructures. These difficulties can be overcome by using nanobeam electron diffraction (NBD) technique, which uses a nanometer-sized parallel beam. In NBD, the electron beam is positioned exactly on the nanostructure of interest and then the diffraction pattern is obtained. There are several modes in TEM. The conventional TEM uses a broad electron beam, while scanning-TEM (STEM) uses a focused electron beam across the sample surface in a raster manner. A focused electron beam is produced from a LaB6 source which has a high-current density and extremely small diameter. STEM has both the features of SEM and TEM. In this, high-energy electron beam is scanned across the sample allowing the samples of higher thicknesses. Staining is generally not necessary for low-Z elements due to a higher sensitivity to sample density/composition. The majority of STEM instruments are simply conventional TEMs with the addition of scanning coils. The advanced STEMs with lens aberration correction have the resolution limits to as low as 0.1 nm. The STEM attached with EDX detector makes it suitable for analytical techniques including elemental mapping. HRTEM images the structures with atomic-level resolution or better than this (i.e., sub-Å-level resolution) by using aberration-corrected HRTEM. In brief, both TEM and SEM are widely used due to their very high spatial resolution and sensitivity to composition.

12.5.2.2 Specimen Preparation for TEM

Sample preparation for TEM analysis is the most important but tedious job. The thickness of the sample should be such that the sample should be transparent to the primary electron beam and there should be sufficient interaction between the volume of sample and electron beam. In general, the higher the accelerating voltage, the thicker may be the sample cross section. For example, epoxy-based sections of up to 250 nm thick may be examined using TEM operated at 100 keV. However, sample thicknesses in the range of 60–100 nm are typically used for TEM analysis in order to improve the resolution through reduced electron scattering. For higher atomic weight material, the specimen should be thinner. Such thin sections can be prepared by using crushing, mechanical polishing, ion milling, chemical etching, ultramicrotome, focused ion beam (FIB), electropolishing, etc. Specimens for the TEM analysis are placed on special micromesh grids of a conductive metal such as Cu, Au, or Ni. Most commonly used grids are made of Cu. The typical size of grids is 3 mm diameter and 10–25 μm thick. The mesh number of a grid indicates the number of grid openings per linear inch. The smaller the grid number, the larger is the aperture size. In general, the grids are coated with light elements such as C or Be, which act as support film that holds the sample in place. The support film must be as transparent as possible while providing the support for the sample. The light elements are used to prevent the interference from the primary electron beam. Some of the selected methods of sample preparation for TEM analysis are discussed below.

Powder suspension method: In this, a stable dispersion of the powder with dimensions small enough (i.e., electron transparent) is prepared in a suitable liquid medium. The concentration of powder in solvent should be <100 parts per million (ppm) to avoid the reaggregation of the nanoparticles. Then, a few drops of the dispersion are placed on a carbon-coated copper grid using a pipette or dropper. The grids are dried before examining. Sometimes, surfactant or surface-modifying agent is also added into the suspension to get stable suspension. The suspension is agitated using ultrasonic bath for ca. 10–15 min. Once the stable suspension is prepared, a drop of dilute suspension containing nanoparticles is deposited on carbon-coated copper grid. One has to be careful about coating of surfactant on nanoparticle particularly when the size is few nanometers and irregular aggregates left on the grid due to meniscus as the drop dries.

Ultramicrotomy: Ultramicrotomy has been used for producing thin (thickness: <100 nm) sections of biological or soft tissues and polymer samples for a long time. In addition, thicker sample is embedded within a polymeric matrix such as methacrylates, epoxy, and polyester resins, and cut into thin cross sections. Thin films deposited from the liquid or vapor phase are generally stripped from the substrate. If the films are thin enough, they are examined directly by TEM. To examine the cross section of a thin film, the samples is mounted in an acrylic resin and then sectioned to get a thin slice. The cross-section of the sample is trimmed to have a tip of about 1 mm² for diamond knife cutting. The sample holder gradually moves toward the knife while it repeatedly moves up and down. The firmly mounted specimen is sectioned as it passes the edge of the knife blade. Ultramicrotomy method is good to get a uniform thickness of multiphase materials. However, ultramicrotomy introduces severe mechanical stresses and artifacts.

Mechanical polishing: This method involves reducing the specimen thickness (also called pre-thinning) to about 100 μm using grinder and/or dimple grinder followed by final thinning. For this, a sample of around <1 mm thick is prepared by cutting with a diamond saw. Then, a 3-mm-diameter disc is cut from it using a specially designed punch. Now, the thickness is further reduced by grinding both sides of a disc using a dimple grinder. The final thinning can be done by electrolytic thinning (for metal), ion milling (for metal and ceramic), and ultramicrotome cutting (for polymeric and biological) of the samples to a thickness of <100 nm. Ion milling is a sputtering process that can remove very fine quantities of material. Typically ion milling conditions are 4–5 kV, 8–12° with beam currents of 0.5 mA. Final polishing with low voltage is extremely important to remove thin amorphous layers on the specimen.

Electropolishing/Electrolytic Thinning: Electropolishing is a process in which the metallic sample (anode) is polished by electrochemical action, that is, by dissolving the sample in an electrolyte and applying a suitable potential. A sample (as anode) is placed in an electrochemical cell. A suitable electrolyte is used to electrochemically reduce sample thickness. A sample is placed in a holder with two sides facing guides of electrolyte jet streams (cathode). The electrolyte jet polishes both sides of the sample. Electrolytic thinning is very efficient and fast process. It generally takes 3–15 minutes. The electrolyte, its temperature, and bias voltage are some of the important parameters in controlling the rate of dissolution of the sample.

Focused ion beam (FIB) milling: FIB is used to etch or mill away undesired portions of the sample. FIB uses beam of energetic gallium ions from a liquid gallium source. An ion beam with energy of 1-10 keV bombards the sample. The sample is placed in the center at an optimized angle (of about 5–30°) to the ion beam in order to get a high yield of sputtering. It creates a thick amorphous layer on both sides of thin sections through gallium implantation. These amorphous layers limit high-spatial resolution in TEM analysis and it is necessary to cool the chamber in order to prevent sample heating by the high energy ion beam. By combining FIB with low-angle and low-energy ion polishing, high-quality TEM samples can be obtained. Ion milling does not need electrically conductive samples. Thus, this technique is suitable for metal, ceramics and polymers, as long as they are thermally stable. This technique is especially useful for the study of polymer films deposited on the substrates. In contrast to ultramicrotomy, the FIB does not induce mechanical stress on the sample surface. However, this technique results in amorphization, shrinkage, and bond cleavage, particularly for soft polymers [6].

12.6 Scanning Probe Microscopy

SPMs provide 3-D images of specimen surfaces with atomic resolution. There are two basic types of SPMs: STM and AFM, which can be used to study the surface, surface modification (nanolithography), and atom-by-atom manipulation.

12.6.1 Basic Principle of STM

In an STM, a bias voltage is applied between a sharp metal tip and a conducting sample (metal or semiconductor) to be investigated and then, the metal tip (or probe) is scanned over the surface in a raster pattern. When the separation between the tip and sample surface is of $\sim$0.1 nm, the current (I) tunnel from the sample surface to the tip or vice versa, depending on the sign of the bias voltage. The exponential dependence of the separation (s) (i.e., $I \propto e^{-1/s}$) between the tip and target surface gives STM its remarkable sensitivity, that is, for a typical value of average barrier height of several eV, I changes by an order of magnitude for every 1 Å change of the separation. For a gap of $\sim$5 Å, the current with 1 V applied will be $\sim$1 nA.

STM mainly consists of a scanning metal tip over the surface to be investigated, as shown in Figure 12.15a. The metal tip is fixed to a rectangular piezodrive made of a piezoceramic. The tips are typically made of tungsten or molybdenum wires with tips radii <1.0 µm.

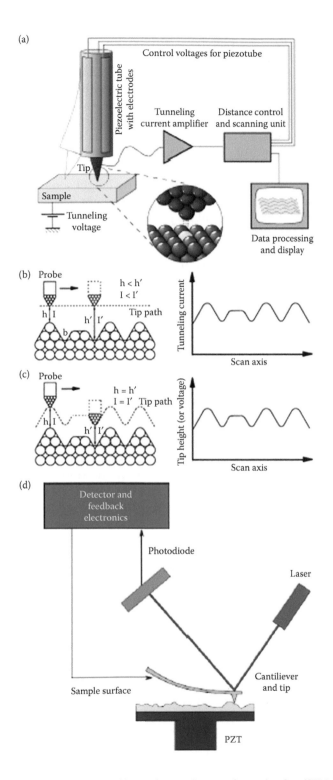

FIGURE 12.15
Schematic diagram of (a) STM (From http://www.iap.tuwien.ac.at/www/surface/STM Gallery, courtesy of Michael Schmid, TU Wien.), (b) constant height mode, (c) constant current mode, and (d) AFM (From http:// en.wikipedia.org/wiki/Image:Atomic force microscope block diagram.png.)

A servo system records the current flowing through the probe and adjusts the height of the probe so as to keep the tunnel current constant. Due to a strong dependence of current on distance, even a very simple servo control system can hold the tip height constant to within a tiny fraction of an Angstrom, if the work function remains constant over the surface.

There are two modes of operation of STM, that is, constant height and constant current modes. In a constant height mode (Figure 12.15b), the tip travels in a horizontal plane above the sample. The induced tunneling current depending on the topography and the local surface electronic properties constitutes the STM image. In a constant current mode (Figure 12.15c), the tip height is adjusted by feedback signals to keep the tunneling current constant during scanning, and the variation of the tip height constitutes the STM image. In other words, the image is formed by plotting the tip height (strictly, the voltage applied to the z-piezo) as a function of the lateral tip position. A constant current mode is generally used for atomic-scale images. Hence, this mode is not practical for rough surfaces. In STM, the electron energies are even smaller than typical energies of chemical covalent bonds, allowing the nondestructive atomic resolution imaging. An electron energy of ~1 eV is typically used for STM imaging; however, it is not high enough to resolve individual atoms because the corresponding electron wavelength is ~1.2 nm. This wavelength is larger than typical interatomic distances (~0.3 nm) in solids. Hence, for "atom by atom manipulation" or nanolithography, the STM uses energies in the range of 4–12 eV. The STM provides local information with atomic resolution directly in a real space, while XRD provides indirectly. Real-space information is particularly important for the study of nonperiodic features such as vacancies, interstitial, impurity sites, steps, dislocations and grain boundaries, and other physical and chemical inhomogeneities.

As discussed above, the tunnel current is extremely sensitive to the vacuum gap, and only to a thin layer at a sample surface. The lateral resolution is given by the width of the tunnel channel, which is extremely narrow. At present, the resolution of STM reaches 0.05 Å vertically and well below 2 Å laterally. The application of STM is limited to conductive samples, and requires ultrahigh vacuum. In addition to the gap, the tunnel current is sensitive to material composition and strain. Atomic resolution in both lateral and vertical directions makes STM an ideal tool for the characterization. The STM allows to image atomic structures directly in a real space including irregularity at the atomic level. In addition, it can be used to measure the physical properties of materials on a nanoscale [6,12,20].

12.6.2 Atomic Force Microscopy

An AFM operates by measuring the attractive or repulsive forces between the fine tip with size of the order of nanometers (20–50 nm or less) and sample. These forces vary with the spacing between the tip and sample and cause the cantilever to deflect when the cantilever drags the tip over the sample. An image of the surface topography is generated by measuring the deflection of cantilever. The tip made of silicon or silicon nitride is attached to a cantilever. As the tip is brought into close proximity of a sample surface, van der Waals force between the tip and the sample leads to a deflection of the cantilever. The force is measured either by recording the bending of a cantilever on which the tip is mounted (i.e., contact mode) or by measuring the change in resonance frequency due to the force (i.e., tapping mode). The AFM is used to characterize the shapes of nanostructures with a typical resolution of several nanometers laterally and several Å vertically. For two electrically neutral and nonmagnetic bodies held at a distance of one to several tens of nanometers, van der Waals forces usually dominate the interaction force between them. Van der Waals forces (*F*) are usually attractive and rapidly increase as atoms, molecules, or bodies approach one another. As a first approximation, the force can be described by using the following equation:

$$F = \frac{-H \times R}{(6 \times d^2)} \tag{12.19}$$

where H is the material-dependent Hamaker constant, R is the radius of the tip, and d is the distance between the tip and the specimen surface. The forces can be derived from the deflection Δz of a cantilever (or spring), on which one of the material sheets is mounted. According to Hooke's law, the force can be represented by the following equation:

$$F = c\Delta z$$
$$(c = E \times w \times t^3 / 4l) \tag{12.20}$$

where E is Young's modulus and w, t, and l are the width, thickness, and length of the cantilever, respectively. For example, a cantilever made of aluminum foil with size of 4 mm (length) $\times$ 1 mm (width) $\times$ 10 μm (thick) would have a spring constant of ~1 N/m. This cantilever can measure a force as small as 10^{-10} N and thus can measure the deflection of ~0.1 nm easily. Precise control of the tip/sample interactions is now possible with PicoForce AFM. It is good instrument to investigate the mechanical, electrical, and thermal forces of the materials. In contrast to STM, AFM can also analyze the electrically insulating samples. It hardly needs any sample preparation. However, it has a drawback of investigating only the surface structures. For example, in case of oxidized semiconductor surface, the AFM shows the image of oxide. As discussed, STM monitors electric tunneling current between the sample top surface and bottom of the probe tip while AFM monitors the force exerted between top surface of the sample and the bottom of the probe tip.

Primary modes of operation for an AFM are contact mode, noncontact mode, and dynamic contact mode, depending on the difference in the extent of tip–sample interaction during the measurement. In a contact (repulsive)-mode operation, either the constant height or the constant force mode can be used. The force of interaction between the tip and the sample lies in the repulsive regime in the intermolecular force curve. The force is kept constant during scanning by maintaining a constant deflection. Contact-mode AFM provides 3-D information of the sample nondestructively with 1.5 nm lateral and 0.05 nm vertical resolutions. In a noncontact operation, the cantilever tip is made to vibrate near the sample surface with spacing on the order of a few nm or intermittently touches the surface at lowest deflection, also known as the tapping mode. In the noncontact mode, the tip is oscillated at a distance from the sample so that the two are no longer in contact. The constant height (deflection mode) is useful in high-resolution analysis of the samples that are extremely flat. Tapping mode is generally used for imaging soft and poorly immobilized samples. The tip is oscillated at its resonating frequency and positioned over the sample so that it contacts the sample for a short-time interval during oscillation. If the tip is scanned at constant height, there would be a risk that the tip would collide with the surface, causing the damage to the sample. Therefore, in most cases, a feedback mechanism is employed to adjust the tip-to-sample distance to keep the force between the tip and the sample constant.

As discussed above, AFM relies on the mechanical defection of a cantilever to relay information about the contour of a specimen (Figure 12.15d). In AFM, a laser beam is focused on the top of the cantilever, which is reflected into photodiode. Differences in the reflected beam are measured by split photodiode and recorded as change in topography. The photodiode detector is able to resolve the probe motion <1 nm. The bending of the cantilever obeys Hooke's law for small displacements, and the force between the tip and the sample can be calculated. If the whole apparatus is raster-scanned across the surface, then an image of the specimen can be generated.

AFM provides a 3-D surface profile of a sample without any special surface preparation and high vacuum. It is a nondestructive technique, whereas electron microscopy uses high-energy electron beams which may destroy the organic parts of a hybrid sample. However, AFM shows a maximum height on the order of micrometers and a maximum area of around $100 \times 100 \, \mu m$. Its scanning speed is quite low compared to SEM. AFM provides atomic-level resolution because of the force of interaction between the atoms of the tip and the sample. In contrast to TEM and SEM, AFM reveals a range of information about the biomolecule, and the nanomaterial–biomolecule interaction on a single particle basis. AFM can examine nonconductive, wet, and soft samples in physiological environments. The sensitivity and versatility of AFM allows a 3-D mapping of ligands attached to a nanoparticle surface. AFM can also determine how strongly the ligands are bound to that nanoparticle surface and the structure of the underlying nanomaterial core [21,22].

12.7 X-ray Photoelectron Spectroscopy

X-ray photoelectron spectroscopy (XPS) is a surface-sensitive analytical tool used to examine the chemical compositions, empirical formula, chemical state, and electronic state of the surface of a specimen placed under ultrahigh vacuum. It is based on the measurement of the kinetic energy of photoelectrons generated when the specimen is illuminated with soft x-ray (1.5 keV) radiation. The interaction of x-rays with the top layer of (few nanometers thick) the specimen excites electrons (referred to as photoelectrons) of the specimen. Some of the electrons from the upper layer ~5 nm are emitted from the specimen and can be detected. Since the energy of x-ray radiation (i.e., photon energy, hv) used is known and the kinetic energy (E_k) can be measured by an analyzer, the binding energy (E_b) can be determined using the following equation:

$$E_b = hv - E_k - \Phi, \tag{12.21}$$

where Φ is the work function of the material. Shifts in the binding energies provide additional chemical information (e.g., the oxidation state of the element). The binding energy of the ejected electron is the characteristic of an element. Figure 12.16 shows the XPS spectrum of gold nanoparticles vapor-deposited (PVD) onto a silicon substrate containing ~4 nm oxide layer, in the region corresponding to the binding energy range of 110–70 eV. The XPS spectrum includes the Si 2p (Si^{4+} and Si^0) and Au 4f peaks recorded at 90° and 30° electron takeoff angles. From the binding energies measured, the chemical elements can be easily identified as corresponding to Au^0, Si^0, and Si^{4+}. It can be seen that the XPS intensity ratio of peaks from elements of the core and the shell is independent of the electron takeoff angles [26]. In contrast to XPS where a relatively low-energy x-rays are used to eject the photoelectrons from an atom via the photoelectric effect, secondary ion mass spectroscopy (SIMS) bombards a source of ions on the sample surface and sputters mainly neutral atoms.

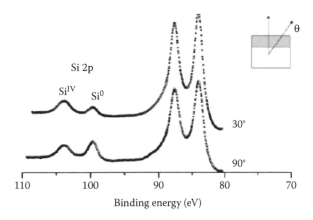

FIGURE 12.16
The XPS spectrum of gold nanoparticles deposited onto a silicon substrate containing ~4 nm oxide layer, recorded at 90° and 30° electron takeoff angles. (Reproduced with permission from Tunc I. et al. 2005. XPS characterization of Au (core)/SiO₂ (shell) nanoparticles. *J. Phys. Chem. B*, 109(16): 7597–7600. Copyright 2005, American Chemical Society.)

12.8 Thermal Analyzer

Thermal analyzers help in determining the amount of conjugated biomolecule and the thermal stability of nanomaterials. Thermogravimetric analysis (TGA) is the method that utilizes a high-precision electronic balance to determine the changes in the weight of a sample with increasing temperatures. It is widely used to characterize a variety of nanomaterials functionalized with coupling agent, surfactant, biomolecules, and drug molecules. It determines the thermal stability (i.e., degradation temperatures) and the amount of inorganic constituents, which usually stay until the end of the measurement due to its high-thermal resistance. It also determines the level of absorbed moisture or organic volatiles in the materials. Typically TGA plots show the weight lost as a function of temperature or time (at constant temperature). The typical temperature ranges that can be distinguished are for the loss of moisture and absorbed solvents up to 150°C and the decomposition of organic constituents between 300°C and 500°C. Generally, TGA analysis is carried out under air or an inert gas. TGA in combination with nuclear magnetic resonance (NMR) can also be used to reveal information about the average number of ligands attached per nanoparticle or QD and the extent of surface functionalization. TGA can also be reliably used to evaluate the purity of nanomaterials. However, conventional TGA consume several milligrams of the nanomaterials and thus, measuring small samples and thin films with minor modifications in surface are below the limits of conventional TGA. Microscale TGA (μ-TGA) uses samples on the order of 1 μg and can detect mass changes less than a nanogram, and thus significantly improving the detection limits of conventional TGA. Differential scanning calorimetry (DSC) is used to study the transitions temperatures such as melting, crystallization, glass transition, and decomposition temperature. In addition to the activation energies and grain growth exponents, isothermal DSC can distinguish between normal and abnormal grain growth. DSC is a technique that compares the difference in the amount of heat required to increase the temperature of a sample and a reference. Both the sample and the reference are maintained at the same temperature under controlled atmosphere (oxidative or inert) throughout the experiment. The basic principle is that when the sample undergoes a physical transformation (i.e., phase transition or thermal decomposition), more or less heat is required compared to the reference to maintain both at the same temperature. The gain or loss of the heat from the sample depends on whether the process is exothermic or endothermic. For example, melting points of solids require more heat flow to the sample; therefore, these processes are endothermic, while thermal decompositions (e.g., oxidation processes) are mostly exothermic events. An exothermic or endothermic event in the sample results in a deviation in the difference between the two heat flows to the reference and unknown sample, and this results in a peak in the DSC curve. The DSC plots the heat flow (or heat capacity) to the sample as a function of temperature (or time) during dynamic heating. The difference in heat flow between the sample and reference also delivers the quantitative amount of energy absorbed or released during such transitions. This information can be obtained by integrating the peak and comparing it to a given transition of a known sample. During the heating of a polycrystalline sample in a DSC, there is a release of the interfacial enthalpy of the grain due to grain growth. In contrast to the conventional micron-sized grains, the grain growth of nanoscale grains (ca. 10–20 nm) releases measurable amount of enthalpy due to the high interface-to-volume ratio in nanostructured materials [23,27,28]. Moreover, DSC can be used to find out change in Curie temperature of a ferroelectric ceramic [29] and the melting points of nanoparticles and nanowires as the size is reduced. DSC also provides melting temperature and transition enthalpy of the alkane thiols with chain lengths containing carbon atoms >12, which are coated on metallic nanoparticles. Both melting temperatures as well as the enthalpy were found to increase with an increase in the chain length of the thiols [30].

12.9 Zeta Potential

Measurement of the zeta (ζ) potential of a nanopowder in solution provides an information on the net charge a nanoparticle or modified nanoparticle has. It provides insight on nanoparticle stability. The ζ potential is commonly determined by applying an electric field across a sample and measuring the velocity at which charged species move toward the electrode. This is proportional to the ζ potential. The ζ potential can also be used to infer particle stability; the values >30 mV indicate stability, while the values <30 mV represent particles with a tendency toward agglomeration or instability. There are many factors including pH, concentration, ionic strength of the solution, temperature, radiation, and the nature of the surface ligands which can influence nanoparticles stability and hence the ζ potential [27,28].

References

1. Zhang J. Z. 2009. *Optical Properties and Spectroscopy of Nanomaterials*. World Scientific, Singapore.
2. Gusev A. I., Rempel A. A. 2008. *Nanocrystalline Materials*. Cambridge International Science Publishing, United Kingdom .
3. Guzelian A. A., Katari J. E. B., Kadavanich A. V., Banin U., Hamad K., Juban E., Alivisatos A. P., Wolters R. H., Arnold C. C., Heath J. R. 1996. Synthesis of size-selected, surface-passivated InP nanocrystals. *J. Phys. Chem.* 100: 7212.

4. Cao G. 2004. *Nanostructures & Nanomaterials: Synthesis, Properties & Applications.* Imperial College Press, London.

5. Akbari B., Tavandashti P. M., Zandrahimi M. 2011. Particle size characterization of nanoparticles—A practical approach. *Iranian J. Mater. Sci. Eng.* 8(2): 48–56.

6. Wen J. G. 2014. Transmission electron microscopy. In Mauro S. (Ed.), *Practical Materials Characterization.* Springer, New York.

7. Ciszewski M., Mianowski A., Nawrat G. 2013. Preparation and electrochemical properties of sodium-reduced graphene oxide. *J. Mater. Sci.: Mater. Electron.* 24: 3382–3386.

8. Valiev R. Z., Islamgaliev R. K., Alexandrov I. V. 2000. Bulk nanostructured materials from severe plastic deformation. *Prog. Mater. Sci.* 45: 103–189.

9. Spanier J. E., Robinson R. D., Zhang F., Chan S. W., Herman I. P. 2001. Size-dependent properties of CeO_{2-y} nanoparticles as studied by Raman scattering. *Phys. Rev. B* 64: 245407.

10. https://en.wikipedia.org/wiki/Ultraviolet%E2%80%93visible_spectroscopy.

11. Kulkarni S. K. 2015. *Nanotechnology: Principles and Practices.* 3rd ed. Co-published by Springer, with Capital Publishing Company, New Delhi, India.

12. Zhang J. Z. 2009. *Optical Properties and Spectroscopy of Nanomaterials.* World Scientific Publishing Co. Pte. Ltd., Singapore.

13. Vossmeyer T., Katsikas L., Gienig M., Popovic I. G., Diesner K., Chemseddine A., Eychmiiller A., Weller H. 1994. CdS nanoclusters: Synthesis, characterization, size dependent oscillator strength, temperature shift of the excitonic transition energy, and reversible absorbance shift. *J. Phys. Chem.* 98: 7665–7673.

14. Smith A. M., Nie S. 2004. Chemical analysis and cellular imaging with quantum dots. *Analyst* 129: 672–677.

15. Brunauer S., Emmett P. H., Teller E. 1938. Adsorption of gases in multimolecular layers. *J. Am. Chem. Soc.* 60: 309.

16. Vollath D. 2013. *Nanomaterials: An Introduction to Synthesis, Properties and Applications.* 2nd ed. Wiley-VCH Verlag GmbH & Co. KGaA, Germany.

17. Goyal R. K., Kapadia A. S. 2013. Study on phenyltrimethoxysilane treated nano-silica filled high performance poly(etheretherketone) nanocomposites. *Compos. B* 50: 135–143.

18. Powers K. W., Brown S. C., Krishna V. B., Wasdo S. C., Moudgil B. M., Roberts S. M. 2006. Research strategies for safety evaluation of nanomaterials. Part VI. Characterization of nanoscale particles for toxicological evaluation. *Toxicol. Sci.* 90(2): 296–303.

19. Mitin V. V., Kochelap V. A., Stroscio M. A. (Eds.). 2008. *Introduction to Nanoelectronics: Science, Nanotechnology, Engineering, and Applications.* Cambridge University Press, New York.

20. Leng Y. 2008. *Materials Characterization: Introduction to Microscopic and Spectroscopic Methods.* John Wiley & Sons (Asia) Pte Ltd., Singapore.

21. Kickelbick G. (Ed.). 2007. *Hybrid Materials: Synthesis, Characterization, and Applications.* Wiley-VCH Verlag GmbH & Co. KGaA, Betz-Druck GmbH, Darmstadt.

22. Matija L. 2004. Reviewing paper: Nanotechnology: Artificial versus natural self-assembly. *FME Trans.* 32: 1–14.

23. Sakka Y., Ohno S. 1996. Hydrogen sorption-desorption characteristics of mixed and composite Ni–TiN nanoparticles. *Nanostruct. Mater.* 7: 341–353.

24. http://www.iap.tuwien.ac.at/www/surface/STM Gallery

25. http://en.wikipedia.org/wiki/Image:Atomic force microscope block diagram.png

26. Tunc I., Suzer S., Correa-Duarte M. A., Liz-Marzán L. M. 2005. XPS characterization of Au (core)/SiO_2 (shell) nanoparticles. *J. Phys. Chem. B* 109(16): 7597–7600.

27. Shah M., Fawcett D., Sharma S., Tripathy S. K., Poinern G. E. J. 2015. Green synthesis of metallic nanoparticles via biological entities. *Materials* 8: 7278–7308.

28. Sapsford K. E., Tyner K. M., Dair B. J., Deschamps J. R., Medintz I. L. 2011. Analyzing nanomaterial bioconjugates: A review of current and emerging purification and characterization techniques. *Anal. Chem.* 83: 4453–4488.

29. Pendse S., Goyal R. K. 2016. Disappearance of Curie temperature of BaTiO$_3$ nanopowder synthesized by high energy ball mill. *J. Mater. Sci. Surf. Eng.* 4(3): 383–385.
30. Pradeep T. 2007. *Nano: The Essential.* Tata McGraw-Hill, New Delhi.

Further Reading

1. Fultz B., Howe J. M. 2008. *Transmission Electron Microscopy and Diffractometry of Materials.* 3rd Edn. Springer, New York.
2. Leng Y. 2008. *Materials Characterization: Introduction to Microscopic and Spectroscopic Methods.* John Wiley & Sons (Asia) Pte Ltd, Singapore.

13

Corrosion Behavior of Nanomaterials

13.1 Introduction

Corrosion can be defined as the destruction or degradation of a material due to its reaction with environment. Corrosion of materials is a continuing problem which results in economic losses of the order of billions of dollars to nations. Corrosion involves two important reactions: oxidation reaction and reduction reaction. As represented by Equations 13.1 and 13.2 for a metal in deaerated acid solution, at anode the oxidation reactions occur, while the reduction reactions occur at cathode in the presence of an electrolyte for the exchange of ions and electrons between the electrodes.

$$M = M^{2+} + 2e^- \text{ (Oxidation reaction)} \tag{13.1}$$

$$2H^+ + 2e^- = H_2 \text{ (Reduction reaction)} \tag{13.2}$$

There are several factors such as temperature, pH, humidity, corrosive species, salinity (in sea water), composition and microstructure of materials which affect their corrosion rate. Metallic corrosion is a major problem in oil production, shipping and aviation industries, power plants, pipelines, drinking water systems, bridges, and public buildings. The corrosion typically occurs under acidic conditions, but microbial corrosion may occur even under ambient temperatures and neutral pH conditions. The role of pH in corrosion science has been well explained using the Pourbaix diagrams, also known as E_H-pH diagrams. The Pourbaix diagram indicates which phase is stable at a particular value of potential and pH. Quantitative corrosion rates are calculated using the Tafel analysis. In this, a plot of logarithm of the current density (I) versus the electrode potential (V) is drawn. The corrosion rate can be calculated by extrapolating a linear fit to the data of the Tafel plot by excluding the part of the curve at large over potentials. The lowest value of current (I_{corr}) is obtained from the point of intersection. The corrosion rate (CR) is calculated using the following equation:

$$CR = \frac{I_{corr} \times K \times EW}{\rho \times A} \tag{13.3}$$

where K is the corrosion rate constant, EW is the equivalent weight, ρ is the material density, and A is the sample area.

Electrochemical impedance spectroscopy (EIS) is used to interpret the mechanism of corrosion protection by the metals or alloys. In this, a small sinusoidal perturbation is applied to the sample under examination, and the impedance modulus is recorded as a function of frequency. The analysis of the frequency behavior of the impedance allows the determination of the corrosion mechanism and the robustness of the coating. Corrosion characteristics such as corrosion potential (E_{corr}), corrosion current (i_{corr}), anodic Tafel slope (β_a), and cathodic Tafel slope (β_c) are obtained from the intersection of cathodic and anodic Tafel curve tangents using the Tafel extrapolation method. From these values, the polarization resistance (R_p) and the corrosion rate (r_{corr}) can be calculated using the below equations [1]:

$$R_P = \frac{\beta_a \beta_c}{2.303 \, i_{corr} \, (\beta_a + \beta_c)} \tag{13.4}$$

$$r_{corr} = 0.00327 \frac{i_{corr} M}{nd} \tag{13.5}$$

where M, n, and d are the molar mass, charge number, and density of tested metal, respectively. The corrosion rate is determined in millimeter per year (mpy).

The failure of an engineering material is strongly influenced by its structure and properties. In many cases, failure starts from the material surface, particularly during wear and corrosion processes.

In recent years, nanomaterials have attracted great interest because of their excellent hardness, better microstructure, and improved toughness. It has also been reported that a nanostructured surface layer of a low-carbon steel exhibited greatly improved wear and friction properties in selected corrosive environments. A high-volume fraction of intergranular defects, primarily free volumes and microvoids, is often associated with the ultrafine grain size of the nanomaterials. These defects could lead to poor corrosion performance since localized corrosion initiates at surface heterogeneities and weak structural sites. On the other hand, some of the alloying elements of the alloys can diffuse easily through the grain boundaries to the surface of the alloy to form the protective passive layer. In the nanocrystalline alloys, the grain boundary volume could be up to 50% of the total volume. Therefore, nanomaterials may be assumed to have a combination of a crystalline structure with long-range order for the atoms far from the grain boundaries and a disordered structure with some short-range order at the grain boundaries or interface region. Thus, in a medium, where a nanostructured alloy could passivate oxide film formed on the alloy surface would be more uniform, and thus more protective than that on the same microcrystalline alloy. However, there are some exceptional cases of nanomaterials which have more corrosion rate than their bulk microcrystalline counterparts of the alloys. In this chapter, the corrosion behavior of selected nanocrystalline metals and alloys has been discussed in brief for the understanding.

13.2 Corrosion Resistance of Nanocrystalline Metals

The corrosion behavior of electrodeposited nanocrystalline Ni of various grain sizes has been investigated. Figure 13.1 shows the active–passive polarization behavior of nanograined Ni

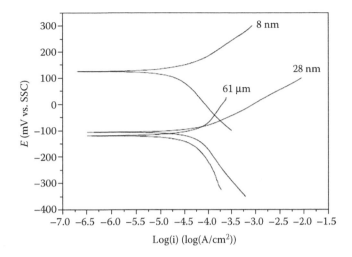

FIGURE 13.1
The Tafel plots for bulk Ni (61 μm), 28 and 8 nm grain-sized nanocrystalline Ni conducted in 1 mol/l H_2SO_4 at a scan rate of 0.5 mV/s. (Reprinted from *Corros. Sci.*, 46, Mishra R., Balasubramaniam R., Effect of nanocrystalline grain size on the electrochemical and corrosion behavior of nickel, 3019–3029, Copyright 2004, with permission from Elsevier.)

as compared to coarse-grained polycrystalline Ni. There is a progressive shift in the zero current potential (ZCP) toward the noble direction for the nanograined Ni as compared to bulk Ni. In nanocrystalline Ni, a large volume fraction of atoms lie in the intercrystalline region of nanocrystalline nickel; hence, the positive shift in the ZCP may be due to the changes in the hydrogen reduction process. It can also be seen that the nanograined Ni exhibits higher passive current density than the micrometer-grained Ni. This is due to easier diffusion of Ni cations through a more defective Ni film, resulting in higher passive current densities in the passive region of nanograined Ni. The lower tendency for localized grain boundary corrosion in nanocrystalline Ni may contribute to an increase in breakdown potential. Relatively poor quality of passive film on nanocrystalline Ni compared to bulk Ni has shown increased corrosion resistance due to the higher-activation barrier to dissolution from nanocrystalline Ni surfaces [2].

In case of cobalt (Co), the overall corrosion behavior is not greatly affected by reducing the average grain size from micrometer to nanometer. It has been widely reported that an increase in the phosphorus (P) content leads to a transition in the alloy structure from crystalline to amorphous. This in turn exhibits better resistance to aqueous corrosion than unalloyed pure metals in acidic media (Figure 13.2). The limited dissolution of P-rich amorphous alloys is mainly associated with an anodic film formation and thus, inhibits the anodic process. In comparison to that of nano-Co, the anodic polarization curve of nano-Co/1.1 P shifts to a more positive value of potential and decreases the anodic dissolution rate to −0.1 V SCE [3].

13.3 Corrosion Resistance of Nanocomposite Coating

Magnetic nanocomposites or alloys (i.e., $Nd_2Fe_{14}B$ phase) are gaining importance in many industrial sectors due to their high-energy density $(BH)_{max}$. The corrosion behavior of

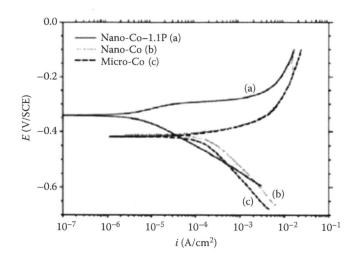

FIGURE 13.2
Corrosion curves of micro-Co, nano-Co, and nano-Co/1.1 P alloys in deaerated 0.1 H_2SO_4 solution. (Reprinted from *Electrochemical Acta*, 51, Jung H., Alfantazi A., 1806–1814, Copyright 2006, with permission from Elsevier.)

microcrystalline NdFeB-type (sintered) magnets has been studied extensively in different media. This has low-corrosion resistance due to the high content of neodymium- (Nd) and boron-rich phases in the grain boundary region. The corrosion rates of NdFeB magnets decrease with an increase in the grain size from 100 to 600 nm of the matrix phase. This is attributed to the changes, heterogeneity, and reduction in the volume fraction of Nd-rich intergranular phase in larger grain-sized sample. Figure 13.3 shows potentiodynamic polarization curves of NdFeB magnets with different grain sizes in N_2-purged 0.1 M H_2SO_4 solution at 25°C and 720 rpm using a scan rate of 0.2 mV/s. A decrease in both anodic and cathodic activities of NdFeB magnets occurs with an increase in the grain size

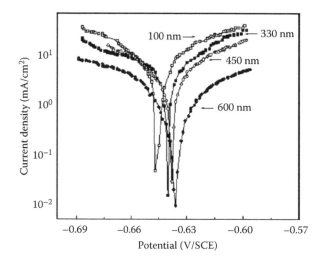

FIGURE 13.3
Potentiodynamic polarization curves of NdFeB magnets with various grain sizes in N_2-purged 0.1 M H_2SO_4 at 25°C and 720 rpm. (Reprinted from *Corros. Sci.*, 44, El-Moneim A. A. et al., 1097–1112, Copyright 2002, with permission from Elsevier.)

of the magnet. The corrosion current density decreases with increasing grain size. The abnormal dissolution behavior of magnets with cathodic polarization indicates that applying any kind of cathodic protection can lead to catastrophic corrosion of nanocrystalline NdFeB magnets [4,5].

The atomically thin layers of graphene as a protective coating inhibit corrosion of underlying metals. Cyclic voltammetry measurements reveal that the graphene coating effectively suppresses metal oxidation and oxygen reduction. The Tafel analysis (Figure 13.4) indicates that copper coated with graphene is corroded seven times slower in an aerated Na_2SO_4 solution as compared to the pure copper. Similarly, Ni coated with a multilayer graphene film corrodes 4–20 times slower than pure Ni, depending upon the type of method used for coating. So far, graphene is the thinnest known corrosion-protecting coating [6].

The use of graphene, as a passive layer, delays microbially induced galvanic corrosion (MIC) of metals for periods more than 100 days. The rate of Ni dissolution in the graphene-coated Ni anode is reduced approximately more than an order of magnitude than the uncoated Ni electrode and the MIC of graphene-coated Ni is inhibited by over 40-fold. The graphene coatings combat MIC by preventing the access of microbes to the Ni surface and by forming a protective passive film between the Ni surface and the anolyte. This film minimizes charge (Ni^{2+}) transport into the solution and protects the Ni surface from microbial by-products (e.g. H^+) that enhance Ni dissolution. Due to high inertness, the graphene coating provides a better and stable barrier than the oxide film [7].

The use of polyaniline (PANI)-functionalized–graphene nanocomposite coatings on steel exhibits outstanding barrier properties against O_2 and H_2O as compared to pure PANI and PANI/clay composites. Figure 13.5 shows the corrosion protection of bare steel and PANI, PANI/graphene- (0.1 wt%), PANI/clay- (0.5 wt%), PANI/graphene- (0.25 wt%), and PANI/graphene (0.50 wt%)-coated steel in a corrosive medium of 3.5 wt%

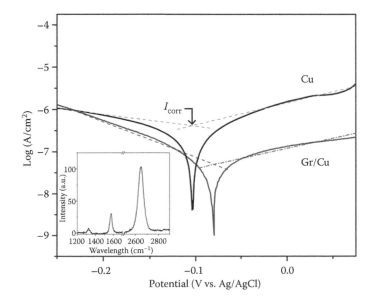

FIGURE 13.4
The Tafel plots of Cu and graphene (Gr)-coated Cu samples. Best fits are shown by dotted lines. (Reprinted with permission from Prasai D. et al., Graphene: Corrosion-inhibiting coating. *ACS Nano* 6(2): 1102–1108. Copyright 2012, American Chemical Society.)

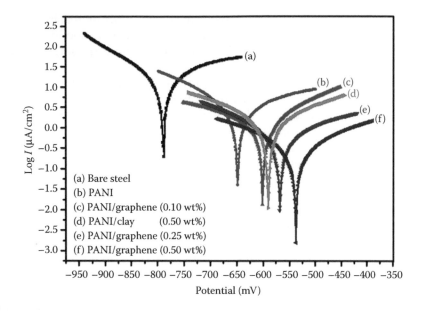

FIGURE 13.5
The Tafel plots for bare steel and PANI-nanocomposites-coated steel electrodes measured in 3.5 wt% NaCl aqueous solution. (Reprinted from *Carbon*, 50(14), Chang C. H. et al., Novel anticorrosion coatings prepared from polyaniline/graphene composites, 5044–5051, Copyright 2012, with permission from Elsevier.)

aqueous NaCl electrolyte. The PANI/graphene (0.50 wt%) coating exhibits the highest and lowest values of E_{corr} and I_{corr}, respectively, compared to bare steels and PANI coatings.

The excellent corrosion protection of PANI/graphene (0.50 wt%) coating is due to the well-dispersed 4-aminobenzoyl group-functionalized graphene-like sheets in the PANI matrix that increases the tortuosity of the diffusion pathways for oxygen and water vapor. It can also be seen that PANI/graphene (0.50 wt%) nanocomposite has better corrosion resistance than PANI/clay for a given weight fraction (i.e., 0.50 wt%). This is due to well-dispersed functionalized graphene sheets, with a relatively high-aspect ratio (~500) compared to clay (~220), in a polymer matrix. This enhances the gas barrier against O_2 and H_2 and hence, makes PANI/graphene much more effective than PANI/clay nanocomposites [8].

Aal et al. [9] studied the effect of Ni/TiO$_2$ nanocomposite coatings prepared by codeposition of Ni and TiO$_2$ nanoparticles powder on the surface of A356 Al alloy in 3.5% NaCl solution at room temperature. Figure 13.6 shows the potentiodynamic polarization curves of pure Ni (average crystallite size: ~46 nm) and its nanocomposites containing TiO$_2$ nanoparticles (average crystallite size: ~34 nm) coatings in nondeaerated 3.5% NaCl solution. The thickness of the coating was ~15 μm. It can be seen that the corrosion potential of Ni/TiO$_2$ nanocomposite coating is more positive than that of the pure Ni coating. The corrosion potentials (E_{corr}) of the coated samples increased gradually as the content of TiO$_2$ nanoparticles increased in Ni matrix. For example, E_{corr} increased from −0.415 V for pure Ni-coated sample to −0.38 V for 9 wt% TiO$_2$-filled-Ni-coated sample. The Ni/9% TiO$_2$ coating shows three times higher corrosion resistance compared to pure Ni coating under the similar testing conditions. It shows that the addition of TiO$_2$ in the nickel matrix coating on the Al alloy surface significantly

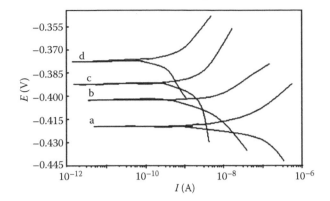

FIGURE 13.6
Potentiodynamic polarization curves of (a) Ni, (b) Ni/3 wt% TiO_2, (c) Ni/5 wt% TiO_2, and (d) Ni/9 wt% TiO_2 coated A356 Al alloy. (Reprinted from *Mater. Sci. Eng. Part A*, 474, Aal A. A., Hard and corrosion resistant nanocomposite coating for Al alloy, 181–187, Copyright 2008, with permission from Elsevier.)

increases the corrosion resistance due to two main reasons [9]. Ni/TiO_2 nanocomposite coatings (0–8.3 wt% TiO_2 particle with size of ~25 nm) deposited on low-carbon steel by electrodeposition process in a Ni plating bath containing TiO_2 nanoparticles increases the corrosion performance of coatings in salty, acidic, and alkaline solutions. As the content of TiO_2 nanoparticle in the Ni/TiO_2 matrix coating increases, the polarization resistance increases, the corrosion current decreases, and the corrosion potential shifts to more positive values (Figure 13.6).

Figure 13.6 clearly shows an increase of corrosion protection by TiO_2 nanoparticles due to the following reasons. First, TiO_2 nanoparticles act as inert physical barriers to the initiation and development of defects, thus modifying the microstructure of the Ni layer. Second, TiO_2 nanoparticles in the Ni layer act as corrosion microcells, where TiO_2 nanoparticles act as cathode and Ni matrix acts as anode because TiO_2 has more positive standard potential than Ni. These microcells facilitate the anode polarization and, hence localized corrosion is inhibited, and mainly homogeneous corrosion occurs [9,10]. The microhardness of Ni/TiO_2 nanocomposite coatings is gradually increased with increasing TiO_2 nanoparticles and it is almost doubled at 8.3 wt% TiO_2 nanoparticle content. A significant increase in microhardness for the nanocomposite coatings may be attributed to higher hardness of TiO_2 particles compared to pure Ni matrix, and additional strengthening effect due to the grain refinement and the dispersion. The coefficient of frictions of the Ni/TiO_2 nanocomposite coatings are found more stable and decrease to approximately >65% compared to that of pure Ni matrix coating. Both increase in microhardness and a decrease in coefficient of friction lead to increase in wear resistance of the nanocomposite coatings by more than two times compared to pure Ni coating [10]. Similarly, Ni/SiC (size: 30 nm) nanocomposite coatings deposited on pure Ni substrate by electrodeposition process from a nickel sulfate bath containing SiC nanoparticles exhibit enhanced microhardness and wear resistance compared to pure Ni coating. The Ni/SiC nanocomposite coatings exhibit a coefficient of friction approximately more than four times lower than the pure Ni coating on Ni substrate under similar wear test conditions. In addition, the coefficient of friction of Ni/SiC nanocomposite is found more stable during full range of sliding distance than that of the pure Ni coating [11].

References

1. Ahmad Z. 2006. *Principles of Corrosion Engineering and Corrosion Control.* 1st ed. Elsevier, London.
2. Mishra R., Balasubramaniam R. 2004. Effect of nanocrystalline grain size on the electrochemical and corrosion behavior of nickel. *Corros. Sci.* 46: 3019–3029.
3. Jung H., Alfantazi A. 2006. An electrochemical impedance spectroscopy and polarization study of nanocrystalline Co and Co–P alloy in 0.1 M H_2SO_4 solution. *Electrochem. Acta* 51: 1806–1814.
4. El-Moneim A. A., Gebert A., Schneider F., Guteisch O., Schultz L. 2002. Grain growth effects on the corrosion behavior of nanocrystalline NdFeB magnets. *Corros. Sci.* 44: 1097–1112.
5. Koch C. C., Lya I., Ovid'ko A., Seal S., Veprek S. 2007. *Structural Nanocrystalline Materials: Fundamentals and Applications.* Cambridge University Press, New York.
6. Prasai D., Tuberquia J. C., Harl R. R., Jennings G. K., Bolotin K.I. 2012. Graphene: Corrosion-inhibiting coating. *ACS Nano* 6(2): 1102–1108.
7. Krishnamurthy A., Gadhamshetty V., Mukherjee R., Chen Z., Ren W., Cheng H.-M., Koratkar N. 2013. Passivation of microbial corrosion using a graphene coating. *Carbon* 56: 45–49.
8. Chang C.-H., Huang T.-C., Peng C.-W., Yeh T.-C., Lu H.-I., Hung W.-I., Weng C.-J., Yang T.-I., Yeh J.-M. 2012. Novel anticorrosion coatings prepared from polyaniline/graphene composites. *Carbon* 50(14): 5044–5051.
9. Aal A. A. 2008. Hard and corrosion resistant nanocomposite coating for Al alloy. *Mater. Sci. Eng. Part A* 474: 181–187.
10. Baghery P., Farzam M., Mousavi A. B., Hosseini M. 2010. Ni/TiO_2 nanocomposite coating with high resistance to corrosion and wear. *Surf. Coat. Technol.* 204: 3804–3810.
11. Zhou Y., Zhang H., Qian B. 2007. Friction and wear properties of the co-deposited Ni/SiC nanocomposite coating. *Appl. Surf. Sci.* 253: 8335–8339.

14

Applications of Nanomaterials

Owing to unique thermal, mechanical, and electrical properties, nanomaterials have been exploited for various applications such as nanofluids/ferrofluids, hydrogen storage devices, solar energy, antibacterial coating, Giant magnetic resistant sensors, single-electron transistor (SET), construction industry, self-cleaning coating, nanotextile, biomedical field, nanopore filters, water treatment, automotive sector, etc.

14.1 Nanofluids

14.1.1 Introduction

Nanofluids are a new class of fluids engineered by dispersing nanometer-sized nanoparticles (NPs), nanofibers, nanotubes, nanowires, nanorods, nanosheets, or droplets in base fluids like oil or water. Nanofluids are two-phase systems with one phase (i.e., solid phase) in another (i.e., liquid phase). They have been found to possess enhanced thermophysical properties such as thermal conductivity, thermal diffusivity, viscosity, and convective heat transfer coefficients compared to base fluids. For a two-phase system, the stability of nanofluids is a big challenge. In this chapter, the stability of nanofluids, properties and applications for various fields including energy, mechanical, and biomedical fields are discussed.

14.1.2 Stabilization of Nanofluids

Nanofluids are stable suspensions of NPs in a liquid, generally water or oil. Due to large surface area and continual NP–NP collisions caused by the Brownian agitation, the NPs in a host fluid interact strongly through van der Waals interactions, which lead to the formation of agglomerations. The size of these agglomerates varies up to few micrometers. Due to gravity, such particles settle out of the suspension, and create the problem of chocking and abrasion of conduit walls. This reduces heat extraction efficiency of the resultant fluids. Such problems can be minimized by stabilizing the NPs by steric stabilization and electrostatic stabilization, as shown in Figure 14.1. In steric stabilization (Figure 14.1a), long-organic molecules are attached at the surface of the NPs which act as distance holders. These molecules avoid direct contact in-between the particles and, therefore, the tendency of agglomeration prevents. The hydrophilic end of the organic molecule ensures compatibility with a polar solvent. In electrostatic stabilization (Figure 14.1b), organic molecules are added into the nanofluids, which exhibit an electrostatic dipole moment at the surface of NPs. The molecules form a kind of electrostatic double layer. As the surface of the NPs suspended in liquid is now covered with electric charges of the same sign, the particles repel each other. However, the net charge of each particle remains zero. In both cases, the particles are kept at a certain distance; this thwarts the formation of *van der Waals* bond clusters. Therefore,

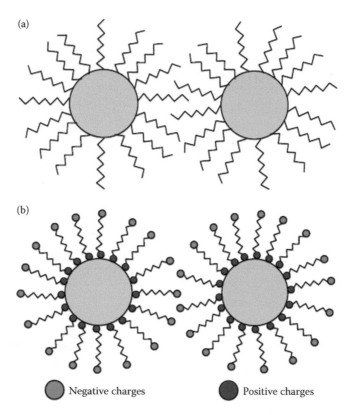

FIGURE 14.1
Stabilization of nanofluids by (a) steric stabilization and (b) electrostatic stabilization. (Vollath D. (Ed.): *Nanoparticles: Nanocomposites-Nanomaterials,* 2013. Copyright Wiley-VCH Verlag GmbH & Co. KGaA. Reproduced with permission.)

the *Brownian* molecular movements of the NPs prevent sedimentation of the particles. In general, a nanofluid contains up to 10 vol% particles and ca. 5–15 vol% surfactants or stabilizers. Conventional fluids have the problem of settling out of micrometer-sized particles due to gravity, while this problem has been minimized by adding nanosized particles and stabilizers in the nanofluids [1].

14.1.3 Synthesis of Nanofluids

In general, the heat capacity and thermal conductivity of the NPs are larger than that of liquids. Therefore, it is worthy to add NPs to fluids to improve their performance. Nanofluids are synthesized by mixing a thermally conducting nanopowder such as carbon nanotube (CNT), metal oxide, carbide, nitride, etc. in a suitable liquid. This NP/liquid mixture is sonicated at high intensity over a specific duration (i.e., 15–30 min) to break the agglomerated powders and form a well-dispersed suspension. Then, a stabilizing agent or surfactant is added to this suspension to avoid the reagglomeration of the NPs in the nanofluid. Moreover, the addition of surfactant increases the compatibility between the NPs and the fluid, which make the nanofluid stable for a long time. However, the addition of stabilizing agent may be restricted to some applications. Due to high-surface reactivity, metallic NPs oxidize or form an interfacial layer which is not desirable for the nanofluids. Besides heat capacity and thermal conductivity, an effect of NPs addition on the rheological properties of nanofluid is also very important. For example, the dynamic viscosity (η) of a nanofluid consisting of CuO NPs, higher than 0.1 vol% in ethylene glycol, increases exponentially with a power of three. Such dramatic increase of the dynamic viscosity may impair long-term stability of the coolant used for the cooling systems in the automotive industry. This is due to the tendency of sedimentation or thermophoresis of the NPs present in nano-fluids. This factor must be kept in mind while synthesizing nanofluids or ferrofluids for specific applications.

Ferrofluids are a special type of nanofluids consisting of superparamagnetic particles (size: ~10 nm) in a solvent. The stabilization of ferrofluids is more difficult as compared to the other nanofluids, as in-between the ferromagnetic particles, there is a dipole–dipole interaction, with the tendency to form lumps. Furthermore, besides segregation due to gravity and thermophoresis, one must take note of segregation of the particles in an external magnetic field. Magnetic unmixing is avoided by limiting the size of the magnetic particles. Therefore, magnetic phenomena depend strongly on the particle size. Commercial ferrofluids consist of ca. 3–8 vol% magnetic NPs and usually >10 vol% sur-factant. Depending on the application, water or oil is used as carrier liquid. The NPs in the ferrofluids are individual permanent magnets. In the absence of an external magnetic field, the net magnetization of the ferrofluid is zero. When the magnetic field is applied, the particles adjust within milliseconds in the direction of the magnetic field and after remov-ing the external field, the orientation of the magnetic dipoles (the particles) randomize again. In a magnetic field gradient, the whole fluid moves to the region of the highest flux. In contrast to other nanofluids, ferrofluids give rise to unusual fluid mechanical response when magnetic fields are applied to them [1].

14.1.4 Applications of Nanofluids

14.1.4.1 Automotive Applications

Engine oils, automatic transmission fluids, coolants, lubricants, and other high-tempera-ture heat transfer fluids used in conventional truck thermal systems have inherently poor heat transfer efficiency. This shortcoming could be minimized by using appropriate nano-fluids in the radiator and brake of vehicles.

Radiator: A radiator is a type of heat exchanger used to transfer heat from the hot coolant that flows through it to the air blown through it by the fan. Most of the modern cars use aluminum radiators made by brazing thin aluminum fins to flattened aluminum tubes. The coolant flows from the inlet to the outlet through many tubes mounted in a parallel arrangement. The fins conduct the heat from the tubes and transfer it to the air flowing through the radiator. Nanofluids enable the potential to allow higher temperature cool-ants and higher heat rejection in the automotive engines. The higher temperature radiator could reduce the radiator size by ~30%. This translates ~10% fuel savings due to reduced aerodynamic drag and fluid pumping and fan requirements. A nanofluid based on water with 0.65 vol% Fe_2O_3 NPs enhances the heat transfer of ~9% in comparison with pure water in a car radiator [2]. The use of nanofluids as coolants would allow smaller size and better positioning of the radiators. In addition, due to better efficiency of the nanofluids, coolant pumps could be shrunk and truck engines could be operated at higher tempera-tures, allowing for more horsepower while still meeting stringent emission standards. For example, the application of nanofluids in radiators could reduce its size up to 10%, which in turn saves ~5% fuel. The use of nanofluid also reduces the coefficient of friction and wear rate of pumps and compressors, which leads to >6% fuel savings. It is envisaged that further improvement in nanofluids could save more energy in the future. The aluminum NPs with thin layer of alumina produce hydrogen by splitting water during the combustion process. In this case, the aluminum acts as a catalyst and the aluminum NPs serve to decompose the water to yield more hydrogen. It has been studied that the combustion of diesel fuel mixed with aqueous aluminum nanofluid increases the total combustion heat while decreasing the concentration of smoke and nitrous oxide in the exhaust emission from the diesel engine.

Brakes: A vehicle's kinetic energy is dispersed through the heat produced during the process of braking, and this is transmitted throughout the brake fluid in the hydraulic braking system. If the heat causes the brake fluid to reach its boiling point, a vapor lock is created that stops the hydraulic system from dispersing the heat. This occurrence will cause a malfunctioning of the brake. Under such circumstances, nanofluids can maximize the performance by transferring more amount of heat compared to pure oil.

14.1.4.2 Biomedical Applications

Cancer therapeutics: Nanofluids containing magnetic NPs in liquid (also called as ferrofluids) are used in cancer imaging and drug delivery. This involves the use of ferrofluid as delivery vehicles for drugs or radiation in cancer patients. Ferrofluids are injected into the bloodstream to reach a tumor with magnets. It will allow doctors to deliver high local doses of drugs or radiation without damaging nearby healthy tissue. It is true that the magnetic NPs are more adhesive to tumor cells than nonmalignant cells; hence, they absorb much more power than the particles in alternating current magnetic fields tolerable in humans. They make excellent candidates for cancer therapy. This combination of targeted delivery and controlled release will also decrease the likelihood of toxicity.

Imaging: Colloidal gold, the most stable known colloid, has been used for several centuries as colorant of glass and silk. In medieval medicine, gold has been used for the diagnosis of syphilis. Nowadays, it is used in chemical catalysis, nonlinear optics, supramolecular chemistry, molecular recognition, and the biosciences. Magnetic resonance imaging (MRI) is one of the most important noninvasive imaging techniques in clinical diagnostics. MRI has the ability to image tissues with high resolutions in three dimensions, down to ~50 μm at high fields. Ferrofluids are used to improve the contrast in NMR imaging in medical diagnostics. In NMR, the spin resonance of protons (hydrogen nuclei) is determined. As this resonance frequency is proportional to the magnetic field, any local variation of the magnetic susceptibility leads to a variation in the resonance signal. MRI produces images by measuring the radio frequency (RF) signals arising from the magnetic moments of lipid and mainly water protons in living tissues. The use of water proton has the highest sensitivity among biologically relevant nuclei and water is present in high concentrations in most of the tissues. The magnetic properties of water are sensitive to the local microstructure and composition of biological tissues. The magnetic properties of the hydrogen nuclei in water can be modulated by interactions with magnetic entities (endogenous or exogenous) [3]. The efficiency of iron oxide as probes is size dependent and increases with increasing particle crystallinity. The inherent negative contrast associated with iron oxide NPs limits their use in low-signal regions of the body or in organs with intrinsically high-magnetic susceptibilities (e.g., lungs) [4]. Iron oxide magnetic NPs are used routinely as contrast agent in MRI and other applications such as cancer thermal treatment, magnetic targeting, and remotely triggered drug release [5]. As the ferrofluid is injected into the blood of the patient, blood vessels will become visible. If the surface of the magnetic particles is covered with proteins or enzymes, which are characteristic for a specific organ or tumor, the particles are collected at these places. This greatly improves the contrast.

Dampers: Electrorheological (ER) and magnetorheological (MR) fluids have the ability to undergo changes in yield stress when an electric or magnetic field is applied, respectively; this property can be used as dampers with smart capabilities. For a constant field, as the force increases, the plug thickness decreases and the equivalent viscous-damping constant also decreases. In advanced prosthetic devices, MR fluid is used as real-time controlled dampers. In this, a small MR fluid damper is used to control, in real time, the motion of an artificial limb based on inputs from a group of sensors [1].

Nano cryosurgery: Cryosurgery is a popular nonroutine therapy which uses freezing to destroy undesired tissues. The introduction of NPs enhances freezing and could also make conventional cyrosurgery more flexible. The applications of magnetite (Fe_3O_4) and diamond are perhaps most popular and appropriate because of their good biological compatibility.

14.1.4.3 Coolants

Conventional fluids have a limitation of low-thermal conductivity for the energy-efficient heat transfer fluids required in several industries. This limitation can be overcome by suspending thermally conductive NPs including CNTs in the fluids. These nanofluids possess significantly higher-thermal conductivity compared to pure fluids. An increase in thermal conductivity of nanofluids depends on the volume fraction, size, type, and aspect ratio of the NPs [6]. Nanofluids as coolants in industries could save good amount of energy and reduce emissions. The use of smart nanofluids in heat valve controls the flow of heat poorly or efficiently as per the requirements for a particular application [7].

Coolant for pressurized water reactor (PWR): In a PWR nuclear power plant system, the vapor bubbles formed on the surface of the fuel rods conduct very little heat compared to liquid water. Interestingly, when the nanofluids with alumina suspension are used in place of water, the fuel rods become coated with NPs which push newly formed bubbles away and thus prevent the formation of a layer of vapor around the rods, and subsequently increase the critical heat flux significantly.

Coolant for machining: Life of the cutting tool is decreased due to the heat liberated and the friction associated with the cutting process. This is overcome by using cutting fluids. However, environmental hazards posed by the fluids have limited their usage, thus users are forced to use minimum quantity of fluids. Nevertheless, its capability to carry away the heat and to provide adequate lubrication is limited. Nanofluids, with their better cooling and lubricating properties, have emerged as a promising solution. Cutting nanofluids possess heat transfer capacity higher than ∼6% compared to fluid without NPs. Moreover, the highest nodal temperature of nanofluid is <50°C, which is significantly lower than a value of >160°C for fluid without NPs [8].

Coolant for diesel electric generators: The efficiency of waste heat recovery heat exchanger is increased when nanofluids are used. This is due to the higher heat transfer coefficient of the nanofluids compared to the base fluid. For example, the heat exchanger efficiency of a base fluid based on ethylene glycol–water mixture is 78.1% and increased to 81.1% for nanofluid containing 6% Al_2O_3 NPs. This is mainly attributed to a decrease in the specific heat associated with an increase in Al_2O_3 NPs concentration [8].

Coolant for solar water heater: Solar water heating (SWH) is the conversion of sunlight into renewable energy for water heating using a solar thermal collector. SWH systems mainly comprise a water storage tank and solar collectors. The solar collector efficiency increases from ~10% for a base fluid to a limiting value of ~80% for a nanofluid containing 2 vol% NPs of about 5 nm in size and thereafter the increment in efficiency gets saturated on further increasing the volume fraction of the NP. Increasing the particle volume fraction leads to a corresponding increase in attenuation of sunlight passing through the collector, and this, in turn, increases the collector efficiency [9]. For a given volume fraction, the collector efficiency increases slightly (3%–5%) with a decrease in the size of silver NPs from >50 to 4 nm [10].

Coolant as lubricant: Lubrication is a very important part of the automobile. The tiny spherical NPs (of nanofluid) dispersed in liquid act as a rolling medium between two moving/sliding parts and this system exhibits powerful rolling–sliding and provides a very low coefficient of friction compared to fluid without NPs. This nanofluid also reduces the contact temperature between the sliding parts. For example, nanofluid based on paraffin oil with TiO_2 (~50 nm) NPs maximally reduces the value of coefficient of friction at an oil temperature of 45–96°C [11].

Coolant to extract geothermal power: Nanofluids can be used to extract the Earth's geothermal energy. The Earth's crust temperature *varies* between 500°C and 1000°C. Nanofluids can be employed to cool the pipes exposed to such high temperatures. When drilling, nanofluids can serve in cooling the machinery and equipment working in high-friction and high-temperature environment. As a fluid superconductor, nanofluids could be used as a working fluid to extract energy from the earth core and processed in a PWR power plant system, producing large amounts of work energy.

Coolant in high-performance loudspeakers: One of the most important parts of the loudspeaker is the magnetic system. Among many parts of the speaker, the voice coil is an important part to make a qualified sound with the input signals of the current. As the current is supplied to the voice coils, the voice coils twist around the bobbin, moving up and down along the bobbin. Then, the sound of the speaker is generated using the formation of a magnetic field, and heat is also generated by the electrical resistance of the voice coil [12]. The higher the magnetic field in the gap, where the voice coil is located, the more is the possibility to reduce the electric current in this coil, and, therefore, the ohmic losses. Placing a ferrofluid in the gap instead of air increases the magnetic field (~0.8–1.8 T) around the coil (Figure 14.2a). The presence of ferrofluid in this gap improves the loudspeaker's performance significantly by increasing the damping capacity and by reducing the temperature of voice coil. The high-thermal conductivity (0.160 W/mK) of ferrofluid compared to air

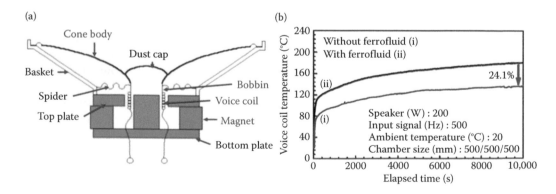

FIGURE 14.2
(a) Schematic diagram of a general speaker and (b) voice coil versus elapsed time for a speaker. (From Lee M. 2014. *Entropy* 16: 5891–5900. Open access.)

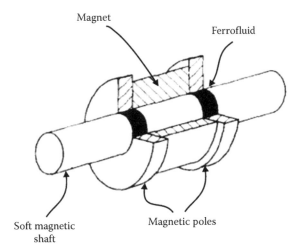

FIGURE 14.3
Basic elements of ferrofluid dynamic seal. (Reprinted from *J. Magn. Magn. Mater.*, 85, Raj K., Moskowitz R., Commercial applications of ferrofluids, 233–245, Copyright 1990, with permission from Elsevier.)

(i.e., <0.04 W/mK) increases the cooling of the coil, and thus, reduces its temperature. The damping capacity is improved by using ferrofluid of appropriate viscosity. In short, the use of ferrofluid may double the efficiency of loudspeakers due to decrease of the coil temperature by about 24% compared to that of fluid without NPs, as shown in Figure 14.2b [13].

14.1.4.4 Dynamic Seal

A dynamic seal retains fluid by keeping out contaminants from the system. For example, it is necessary to seal hermetically the hole through which the axle passes. This is achieved by making the hole inside a magnet (Figure 14.3) and the shaft made of soft magnetic material. A groove in the shaft is filled up with ferrofluid, which is kept in place by the magnetic field. This obstructs the entry of any foreign particles, leaving the axle free to rotate [13].

14.2 Hydrogen Storage

Hydrogen is the most abundant element in the universe. It has great potential as an energy source. It is also nonpolluting, and forms water as a harmless by-product. However, it is difficult to store for its use as a fuel. For example, 1 g of hydrogen gas occupies about 11 L of space at atmospheric pressure. To store it in a less space, the gas must be pressurized to several hundred atmospheres and stored in a pressure vessel. In liquid form, it can only be stored under cryogenic temperatures. The best solution is to storage it in hydride form. This method uses an alloy that can absorb and hold large amounts of hydrogen by bonding with hydrogen and forming hydrides. A hydrogen storage alloy must be capable of absorbing and releasing hydrogen without compromising its own structure. Some alloys such as MgH_2, Mg_2NiH_4, $FeTiH_2$, and $LaNi_5H_6$ store hydrogen at a higher density. Among these alloy, MgH_2 has the highest (theoretical) hydrogen storage density (i.e., 6.5×10^{22} H atoms/cm^3 or 7.6 wt% hydrogen). However, in MgH_2, both hydrogenation and dehydrogenation reactions are very slow and, hence, relatively high temperatures are required. The addition of CNT significantly promotes hydrogen diffusion in the host metal lattice of MgH_2 due to the reduced pathway length and creation of fast diffusion channels. Mechanical milling also helps to reduce the particles of MgH_2 into micro- or nanocrystalline phases and thus leads to lowering the activation energy of desorption. The activation energies of the H_2 sorption for the bulk MgH_2, mechanically milled MgH_2, and nanocatalyst-doped MgH_2 are 162 kJ/mol H_2, 144 kJ/mol H_2, and 71 kJ/mol H_2, respectively [14]. Table 14.1 shows hydrogen storage density that is stored by various metals, alloys and CNTs. A best hydrogen storage system has weight efficiency (the ratio of H_2 weight to system weight) of 6.5 wt% hydrogen and a volumetric density of 63 kg H_2/m^3. The most common application for hydrogen storage alloys (i.e., nickel–metal hydride) is in batteries which usually last over 500–700 cycles. They have 40% higher electrical capacity than traditional Ni–Cd battery and are environment-friendly.

The ideal hydrogen storage system needs to be light, compact, relatively inexpensive, safe, easy to use, and reusable without the need for regeneration. Metal hydride systems store hydrogen as a solid in combination with other materials. The metal splits the hydrogen gas molecules and binds the hydrogen atoms to the metal until released by heating. However, metals have the drawback of high weight. Complete hydrogen desorption occurs only at very high temperature (i.e., at 1000°C in an inert gas), which is a critical procedure in order to remove the chemisorbed gases [15].

Nanomaterials are the best choice for increasing the kinetics of hydrogen uptake and release. The absorption and desorption characteristics can be fine-tuned by controlling the particle sizes. Figure 14.4 shows higher amount of hydrogen absorption by the Mg_2Ni nanomaterials compared to macroscopic bulk counterpart. It is due to the fact that NPs have much smaller size and very high-specific surface area compared to bulk materials, which promote fast hydrogen exchange by decreasing diffusion paths for hydrogen. In addition, bulk NPs have high-volume fraction of open pores which also lead to increased diffusion rates and thus, absorption increases [16]. In recent years, nanomaterials have attracted great interest because of their higher-surface/volume ratio, which is the key attribute for hydrogen storage devices. Nanostructured materials such as CNTs, nanomagnesium-based hydrides, complex hydride/CNTs nanocomposites, boron nitride nanotubes, TiS_2/MoS_2 nanotubes, Alanates, polymer nanocomposites, and metal organic frameworks (MOFs) are considered to be potential candidates for storing large quantities of hydrogen (in atomic

TABLE 14.1

Materials and Their Hydrogen Storage Density

Material	Hydrogen storage density by wt% (conditions)	Ref.
MgH_2	7.6	[14]
Mg_2NiH_4	<4 (>523 K)	[20]
$FeTiH_2$	<2 (>263 K)	[20]
Pure carbon nanotubes (CNT)	0.6 (80 bar/300 K)	[63]
Bulk boron containing carbon material	0.2 (80 bar/300 K)	[63]
Boron containing CNT	1.13 (1 atm./77 K)	[63]
(specific surface area: 523 m^2/g)	2 (80 bar/300 K)	
SWNTs (dia.: 1.85 nm)	4.2 (100 bar/room temperature)	[63]
Platinum dispersed SWNTs	3.03 (78 bar/125 K)	[63]
MOFs (zinc)-derived carbon (MDC)	3.25 (1 bar/77 K)	[63]

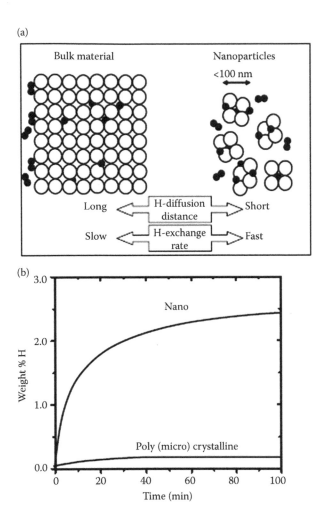

FIGURE 14.4
(a) Schematic diagram showing the effect of size on the H diffusion distance and (b) adsorption kinetics in nano and bulk polycrystalline Mg_2Ni at 200°C. (With kind permission from Springer Science+Business Media: *Resonance*, Nanomaterials for hydrogen storage, 12, 2007, 31–36, Sunandana C. S.)

or molecular form). The storage efficiency is affected by the crystallinity and nanocatalyst doping on the metal or complex hydrides [14]. MOFs (zinc)-derived carbon (MDC) exhibits an extremely high-ultramicropore volume (0.63 cm^3/g) and yielded an H_2 storage capacity of 3.25 wt% at 77 K and 1 bar [17].

The interaction between hydrogen and carbonaceous materials is based on physisorption or chemisorption. Physisorption is due to the van der Waals attractive forces, while chemisorption is due to the overlap of the highest occupied molecular orbitals of carbon with the hydrogen electron, which overcome the activation energy barrier for hydrogen dissociation. Physisorption of hydrogen limits the hydrogen-to-carbon ratio to less than one hydrogen atom per two carbon atoms (i.e., 4.2 mass %). In chemisorption, a ratio of two hydrogen atoms per one carbon atom is realized. It is worthy to note that the materials with large hydrogen-to-carbon ratios liberate the hydrogen at elevated temperature. However, physically adsorbed hydrogen molecules are weakly bound to a surface and, hence, are easily released. The curvature of CNTs and fullerenes increases the reactivity of these materials

TABLE 14.2

Comparison of H_2 Storage of CNTs with Graphite and Metal Organic Frameworks (MOFs)

Materials	Absorption Temperature (Pressure)	Hydrogen Storage Capacity (wt%)	Ref.
Metal-organic frameworks	78 K (−195°C)	4	[14]
Metal-organic frameworks	Room temperature	1	[14]
Metal-organic frameworks (MOF-177) (sp. surface area: 5250 m²/g)	196°C (6 MPa)	7.5	[15,21]
CNT	398–773 (1 atm.)	0.4	[20]
K-doped CNT	<313 (1 atm.)	14.0	[20]
Li-doped CNT	473–673 (1 atm.)	20.0	[20]
Li-doped graphite	473–673 (1 atm.)	14.0	[20]
K-doped graphite	<313 (1 atm.)	5.0	[20]
Pure carbon nanotubes (CNTs)	300 K (80 bar)	0.6	[63]
Bulk boron containing carbon material	300 K (80 bar)	0.2	[63]
Boron containing CNT (sp. surface area: 523 m²/g)	77 K (1 atm.)	1.13	[63]
	300 K (80 bar)	2	

with hydrogen [18]. Compared to metal hydrides, 78% of the adsorbed hydrogen could be released from SWNTs under ambient pressure at room temperature, while the remaining residual stored hydrogen requires heating to ~200°C. SWNTs have shown two chemisorption sites: the exterior and the interior of the tube wall. The SWNTs with average diameter of 1.85 nm have a hydrogen storage capacity of 4.2 wt% at room temperature and 100 bar. The platinum-dispersed SWNTs have shown a hydrogen uptake up to 3.03 wt% at 125 K and at 78 bar. Polyaniline could store as much as 6–8 wt% of hydrogen. The maximum hydrogen storage capacity of SWNTs can exceed 14 wt% (160 kgH²/m³) but reproducibility is a challenge. An excellent review on storage of hydrogen in nanostructured carbon materials is suggested to readers to refer [19]. As shown in Table 14.2, lithium- or potassium-doped carbon nanotubes (MWNT with diameter ca. 25–35 nm and specific surface area ~130 m²/g) can absorb ~20 or ~14 wt% of hydrogen at moderate (200–400°C) or room temperatures, respectively, under ambient pressure. These values are greater than those of metal hydride and graphite systems. For example, H_2 uptake of Li- and K-doped CNTs is ~40% and 180%, respectively, greater than those of graphite powder (specific surface area ~8.6 m²/g). The high uptake of alkali-doped CNTs is due to the hollow cylindrical tube of CNTs with much more open edge and greater interplanar distance (0.347 nm for CNT vs. 0.335 nm for graphite) [20].

As an energy carrier in vehicle, hydrogen has several advantages such as highest energy density by weight, a renewable source of energy, and pollution-free proton exchange membrane (PEM) fuel cell. Vehicle can have fuel economy of 80.5 km/kg H_2 with zero emissions. However, hydrogen storage in metals leads to volume change [21].

14.3 Solar Energy

14.3.1 Introduction

Increasing demand for energy has forced us to seek environmentally clean alternative energy resources. Recent efforts to design ordered assemblies of semiconductor and metal

NPs as well as carbon nanostructures provide innovative strategies for designing the next-generation energy conversion devices. Renewable energy such as solar radiation is ideal to meet the projected demand. The majority of solar panels (~90%) are based on silicon technology, which is matured. Moreover, silicon does not absorb sunlight strongly; hence, silicon cells contain a relatively thick layer of silicon, which is brittle and therefore must be supported on a rigid and heavy piece of glass. This adds cost and limits applications. The widespread use of inorganic solar cells remains limited due to the high costs imposed by fabrication procedures involving high-vacuum, elevated temperature (400–1400°C), and numerous lithographic steps. Organic solar cells that use polymers can be processed from solution at a low cost with solar power efficiencies of up to 2.5%. Nowadays, there is a focus on organic photovoltaics, dye-sensitized solar cells (DSSCs), perovskite photovoltaics, and inorganic quantum dot (QD) solar cells. These emerging cells are expected to be less expensive, thinner, flexible, and amenable to a broad range of wavelengths, making them suitable for applications beyond rooftop and solar farm panels [22].

14.3.2 Photoelectrochemical Cells

The photoelectrochemical cells, also called photovoltaic cells or solar cells, are used for a higher-conversion efficiency of solar energy to electrical power. Solar cells are usually made of the semiconductor materials. Figure 14.5 shows the operation of a basic solar cell. Photoelectrochemical devices made up of silicon-based p–n junction material and other heterojunction material such as indium gallium phosphide, gallium arsenide, and cadmium telluride/cadmium sulfide have shown highest efficiency of ~20%. However, the high cost of production, expensive equipment, and necessary clean-room facilities associated with the development of these devices have directed exploration of solar energy conversion to cheaper materials and devices. Nanostructures are advantageous for photoelectrochemical cell devices for efficient and higher conversion of light to electrical power due to their large surface area at which photoelectrochemical processes take place. There are several nanostructured materials such as TiO_2, SnO_z, ZnO, and Nb_2O_5, which have been studied for solar cell devices but the highest overall light conversion efficiency of these devices has achieved hardly up to 10% [23].

Photovoltaics are the direct conversion of light into electricity at the atomic level. The materials exhibiting photoelectric effect absorb photons of light and release electrons. On capturing these free electrons, an electric current is generated which can be used as electricity. In solar cells, a thin semiconductor wafer (i.e., Si) is specially treated to

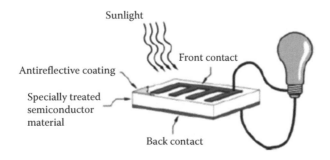

FIGURE 14.5
Basic setup of photovoltaic cell or solar cell. (Kickelbick (Ed.): *Hybrid Materials: Synthesis*, Characterization, and Applications, 2007, Copyright Wiley-VCH Verlag GmbH & Co. KGaA. Reproduced with permission.)

form an electric field, positive on one side and negative on the other. When light energy strikes the solar cell, the electrons are knocked out from the atoms in the semiconductor material. If electrical conductors are attached to the positive and negative sides, forming an electrical circuit, the electrons can be captured in the form of an electric current. In a single-junction photovoltaic cell, only photons whose energy is equal to or greater than the bandgap of the cell material can free an electron for an electric circuit. Thus, the photovoltaic response of single-junction cells is limited to the portion of the sun's spectrum whose energy is above the bandgap of the absorbing material. This limitation can be overcome by using two or more different cells (i.e., multijunction or cascade or tandem cells) having more than one bandgap and more than one junction to generate a voltage. Multijunction devices can achieve higher total conversion efficiency due to conversion of more of the energy spectrum of light to electricity, for example, gallium-arsenide-based multijunction cells have shown efficiencies of ~35% under concentrated sunlight [23].

Inspired from natural solar cell (chlorophyll), electrochemical DSSC based on nano-structured TiO_2 having a wide bandgap is made light-sensitive through a monolayer of an organic dye on its surface. The high-surface-to-volume ratio of TiO_2 can adsorb larger amounts of the dye to ensure higher-efficient absorption of the solar light. In contrast to a thick layer of the silicon solar cells, a nanometer-thin sensitizer film (of DSSC) is enough for total light capture. The TiO_2 nanocrystals carry the received photoexcited electrons away from electron donors, in order to absorb another photon by the donor molecules. These cells have an efficiency of 10% under direct sunlight, but they suffer from some limitations of required morphology of the nanoporous oxide layers, use of volatile solvent as electrolyte, low stability of TiO_2 (by UV light), and weak stability of the oxide-sensitizer bridge. These limitations have been minimized by covalent grafting of organic dyes, which leads to stable dye-oxide cross-linkages and by designing quasi-solid or solid electrolytes to replace the organic liquid [23].

In solar cell, compared to dye molecules, the use of semiconductor QDs as sensitizers has unique advantages, including tailored energy gaps (i.e., by controlling their size), large extinction coefficients (due to quantum confinement effect), and large intrinsic dipole moments, which may lead to rapid charge separation.

QD-sensitized solar cell (QD-SSC): DSSCs are considered to be a low-cost alternative to conventional solid-state solar cells. DSSCs are based on photosensitization of mesoporous TiO_2 films by adsorbed sensitizers. Ruthenium complexes and organic dyes are commonly used as sensitizers of DSSCs and power conversion efficiency up to 11% has been achieved. In addition to the organic sensitizers, semiconductor QDs such as CdS, CdSe, PbS, PbSe, and InP, which absorb light in the visible region, can also serve as sensitizers of DSSCs. The specific advantages of QDs over organic materials in light harvest come from the quantum confinement effect. By using a QD, it is possible to utilize hot electrons to generate multiple electron–hole pairs per photon through the impact ionization effect. Furthermore, QD sensitizer has a higher-extinction coefficient than conventional dyes, which is known to reduce the dark current and increase the overall efficiency of a solar cell. Therefore, the maximum efficiency theoretically predicted for a QD-sensitized solar cell is higher (i.e., 44%) than that for DSSCs using organic sensitizers. QD-SSCs based on CdS-QDs-sensitized nanocrystalline TiO_2 photoelectrode have shown an energy conversion efficiency up to 2.8% under 1 sun illumination. This efficiency is presently the highest reported for QD-SSCs. The use of CdSe-QDs along with CdS-QDs increases the absorption range of the wavelength from 550 (of CdS bulk) to 720 nm (of CdSe bulk). Hence, if two QDs are sequentially assembled onto a TiO_2 film, forming a cascade cosensitized structure, an energy conversion efficiency as high as 4.22% under 1 sun illumination can be achieved. However, the efficiency of a CdS/CdSe-sensitized solar cell is strongly dependent on the position of CdS and CdSe with respect to TiO_2. When CdS is located between CdSe and TiO_2 (TiO_2/CdS/CdSe), both the conduction and valence bands edges of the three materials increase in the order: TiO_2<CdS<CdSe, which is advantageous to the electron injection and hole recovery of CdS and CdSe [24].

A photovoltaic device consisting of 7×60 nm CdSe nanorods and the conjugated polymer poly-3(hexylthiophene) has been assembled from solution with an external quantum efficiency of over 54% and a monochromatic power conversion efficiency of 6.9% under 0.1 mW/cm^2 illumination at 515 nm. Under Air Mass (AM) 1.5 Global solar conditions, a power conversion efficiency of 1.7% has been obtained. By changing the nanorod radius, the bandgap of QDs can be tuned, which in turn optimize the overlap between the absorption spectrum of the cell and the solar emission spectrum. The increasing aspect ratio of rods, due to quantum confinement, enhances the absorption coefficient compared to the bulk. Such 1-D nanorods provide a directed path for electrical transport and hence are preferred over QDs or sintered nanocrystals [25]. The maximum incident-photon-to-charge carrier efficiency (IPCE) is more for tubular TiO_2 than that of particulate TiO_2. The maximum IPCE observed at the excitonic band is found to increase with decreasing particle size [26]. The use of bifunctional surface modifiers containing carboxylate and thiol functional groups facilitates binding of CdSe QDs to the TiO_2 NPs. This exhibits a photon-to-charge carrier generation efficiency of 9%–12% depending upon the chain length of linker molecules. These lower IPCE values indicate that a larger fraction of carriers or electrons are lost due to scattering as well as charge recombination at TiO_2/CdSe interfaces and internal TiO_2 grain boundaries. The TiO_2 films show photocurrent responses in the UV region (<380 nm) while TiO_2/CdSe QDs-films show responses below 600 nm indicating that the TiO_2/CdSe QDs-films are useful as light harvesting antenna [27].

Solar cells based on a nanostructured TiO_2 and colloidal CdSe QDs capped with trioctylphosphine (TOP) have shown that the open structure of the TiO_2 provides higher photovoltaic conversion efficiencies. In addition, the use of ZnS treatment and Cu_2S counter-electrode leads to a cell performance efficiency of ~1.83% under full 1 sun illumination intensity, which is considered the highest efficiency ever reported for the solar cells containing presynthesized colloidal CdSe-QDs. The ZnS passivation treatment on nanostructured TiO_2 sensitized with CdSe-QDs has shown an increase in solar cell performance due to an increase in the recombination resistance between TiO_2 and the electrolyte [28]. Figure 14.6a shows the dependences of the photovoltaic conversion efficiencies of the CdSe-QD SSCs on the CdSe adsorption time without and with the ZnS coating. The maximum value of the photovoltaic conversion efficiency (after 30 h absorption time) increases from 1.16% (without ZnS coating) to 2.02% after the ZnS coating. This is due to the fact that the ZnS coating passivates the surface states of CdSe, which results in the suppression of the surface trapping of photoexcited electrons and holes in the CdSe-QDs. Thus, the photoexcited electrons can efficiently transfer into the TiO_2 conduction band, and a higher photocurrent density is obtained. This is confirmed by the large increase in the IPCE values of the CdSe-QD-SSCs with the ZnS coating compared to those without it (Figure 14.6b) [29].

14.3.3 Thermoelectric Devices

Heat is carried both by free electrons (and holes in semiconductors) and thermal vibrations. Thermal vibrations are the only mechanism of heat transport in electrical insulators.

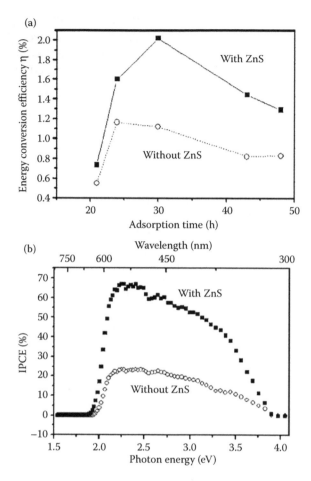

FIGURE 14.6
Dependences of the photovoltaic conversion efficiencies of CdSe-QD-sensitized solar cells on the (a) CdSe adsorption time and (b) the incident photon to current conversion efficiency (IPCE) versus wavelength of illumination of photon. (Reprinted with permission from Shen Q. et al. 2008. Effect of ZnS coating on the photovoltaic properties of CdSe quantum dotsensitized solar cells. *J. Appl. Phys.* 103: 084304. Copyright 2008, American Institute of Physics.)

In a solid state, these vibrations are quantized; such quantized lattice vibrations are called as phonons. The heat transport in solids is dominated by long-wavelength thermal vibrations, and there is a rapid increase in the Boltzmann occupation of these low-energy states that gives rise to the T^3 dependence of the specific heat at low temperature for a dielectric solid. The thermoelectric cooler consists of a large area junction between p-type and n-type semiconductors. Typically, the semiconductors are arranged as vertical bars connected by a horizontal metal bridge. Passing a current across the junction in the forward direction sweeps holes away from the junction in one direction and electrons away from the junction in the opposite direction. These carriers take the heat away from the junction, with the result that the junction cools. The opposite side of the device gets hot. Excellent electrical conductivity is required for good thermoelectric cooling but low-thermal conductivity is also required. The low-thermal conductivity is required to prevent the leakage of the heat from the hot side to the cold side via thermal vibrations. For this, semiconductors made of heavy atoms are used because the speed of sound is lower in heavy materials.

However, it is also true that bulk materials that are good electron conductors are also good heat conductors. A figure of merit (i.e., ZT) for thermoelectric devices is given by the following equation:

$$ZT = \frac{S^2 \sigma T}{K} \tag{14.1}$$

where S is the Seebeck coefficient (volts per degree), σ is the electrical conductivity, T is the absolute temperature, and K is the thermal conductivity. The thermoelectric performance increases with S and σ, while the corresponding temperature gradient falls with thermal conductivity. As seen from Equation 14.1, the performance is directly proportional to the ratio of electrical conductivity to thermal conductivity. Since the scattering path lengths for phonons and electrons are different, it is possible to design nanostructures with super-lattices that are phonon blocking but electron transmitting. The interaction of the Fe, Co, and Ni NPs may play a role of enhancing the thermoelectric efficiency of SWNTs. Thermal conductivity of silicon nanowires can be dropped by a factor of 4–5 after the chemically etching of the surfaces compared to that of without etching, and a nanowire can have a thermal conductivity 100 times lower than the bulk material and a nanotube has higher-thermal conductivity than diamond [30–32].

14.4 Antibacterial Coating

Silver has been used for the treatment of medical ailments for over 100 years due to its antibacterial, antifungal, and antiviral properties. Its NPs, due to a large surface-area-to-volume ratio provide a more efficient means for antibacterial activity even at very low concentration. Silver NP is an effective killing agent against a broad spectrum of Gram-negative and Gram-positive bacteria. It is also an effective and a fast-acting fungicide against a broad spectrum of common fungi including genera such as *Aspergillus*, *Candida*, and *Saccharomyces*. Its NPs with an average size of 5–20 nm inhibit HIV-1 virus replication (also called antiviral property). Moreover, silver inhibits the formation of biofilms. Biofilms are complex communities of surface attached aggregates of microorganisms embedded in a self-secreted extracellular polysaccharide matrix. Biofilm-forming bacteria act as efficient barriers against antimicrobial agents and the host immune. Owing to these properties, silver NPs have been used significantly for consumer products, food technology, textiles/fabrics, and medical applications including wound care products and implantable medical devices. The silver NPs with size of <25 nm show a very large relative surface area, which increase the contact between the silver NPs and bacteria or fungi. This enormously improves the bactericidal and fungicidal effectiveness. The use of silver NPs in shocks helps to maintain healthy and bacteria-free feet. A number of products use silver NPs in medicine and water purification. Silver-impregnated bandages and dressings are used for the treatment of serious burns, local treatment of wounds, and elimination of pathogenic bacteria. Ceramic filters with a coating of silver NPs are suggested to be used for drinking water purification. Nonliving surfaces that penetrate the body or are implanted within the body are prone to the growth of microbial biofilms. Silver coatings on the surfaces of artificial joints, pacemakers, artificial heart valves, and Teflon sleeves have great potential to prevent these deadly microbial growths. Recently, a number of companies are marketing urinary, dialysis, and other catheters with such coatings [33]. Both natural and synthetic textile fibers do not have resistance to bacteria or pathogenic fungi. Thus, various antibacterial finishes and disinfection techniques have been used to control microorganisms for these textiles for a long time. Silver has been known as nontoxic and skin-friendly (i.e., does not cause skin irritation) antibacterial agent. Cotton- and polyester-woven fabrics, coated with silver NPs (size ca. 2–5 nm), show excellent antibacterial effect against *Escherichia coli* (*E. coli*) (Gram-negative) and *Staphylococcus aureus* (*S. aureus*) (Gram-positive) bacteria. The fabrics padded through 50 ppm silver colloidal solution show better bacteriostasis than the samples treated with 25 ppm solution. The concentration of silver NPs decreases rapidly after laundering five times [34]. The laundering durability can be increased by using amino-treated silver NPs. The Ag NP with an average crystallite size of ~11 nm treated with amino-terminated hyperbranched polymer (HBP-NH$_2$) and grafted on the oxidized cotton fabric shows >99.4% bacterial reduction rate against *S. aureus* and *E. coli*. Such high-antibacterial efficiency is obtained from the cotton fabric consisting of ~150 ppm Ag NPs only and can retain ~116 ppm Ag NPs even after 50 washing cycles with its antibacterial activity at over 96% reduction level [35]. The rate of killing bacteria increases with increasing silver ion concentration of the fiber. Similarly, zinc oxide (ZnO) shows excellent antibacterial effect without toxic by-products to human being. Thus, ZnO NP with an average crystallite size of ~12 nm is very efficient in impart of antibacterial effect to cotton fabric against *E. coli* and *S. aureus*. The coated cotton sample displays high activity with a great reduction of bacteria [36]. The needle-shaped-ZnO-nanorod-coated cotton fabric has shown an excellent UV-blocking property with UV protection factor value over 105. After 20 washings, the cotton fabric still has UV protection factor value of 36.2, which is still considered good UV protection [37].

   In case of wounds, the disruption of any of the numerous healing processes can lead to problems in the time-sensitive healing actions of the dermal and epidermal layers. Bacterial infection is one of the major obstacles to proper wound healing. Keeping the wound free of bacteria is imperative to the proper repair of dermal wounds. Silver has been widely used to treat wounds for its bactericidal properties. The ability of silver NPs to reduce the growth of bacteria at the area of interest without causing any serious damage to the adjacent skin cells is the main feature of silver's clinical safety and antibacterial activity. The average healing times of animals inoculated with silver NPs are significantly reduced and the appearances of the healed areas of the Ag NP-treated wound are almost similar to that of normal skin. The silver NPs can be used to prepare antibacterial cotton fabrics as wound-treatment kits [38]. Skin generally needs to be covered with a dressing immediately after it is damaged. At present, there are three categories of wound dressing: biologic, synthetic, and biologic–synthetic, which are commonly used clinically, but they have some disadvantages. An ideal dressing should maintain a moist environment at the wound interface, allow gaseous exchange, act as a barrier to microorganisms, and remove the excess exudates. It should also be nontoxic, nonallergenic, nonadherent, and easily removable without trauma. It should be made from a readily available biomaterial that requires minimal processing. These treated fabrics possess antimicrobial properties, and promotes wound healing. Currently, a novel wound dressing composed of nanosilver and chitosan is found to have better rate of wound healing and is associated with silver levels in blood and tissues lower than levels associated with the silver sulfadiazine dressing. Sterility and pyrogen tests of the silver nanocrystalline chitosan dressing were negative. Thus, this dressing may find wide application in clinical settings [39].

14.5 Giant Magnetoresistance

The ferromagnetic bits are written into (and read from) the disk surface, which is uniformly coated with a ferromagnetic film having a small coercive field. Both writing and reading operations are accomplished by the read head. The read heads in magnetic hard drives are based on a giant magnetoresistance (GMR) sensor. The density of information that can be stored in a magnetic disk is primarily limited by the domain size. Below this domain size, the individual atomic magnetic moments remain independent of each other. In present technology, the linear bits are about 100 nm in length and 0.3–1.0 μm in widths. The ferromagnetic domain magnetization is parallel or antiparallel to the linear track. The localized, perpendicular magnetic fields that appear at the junctions between parallel and antiparallel bits are sensed by the read head. The sensor consists of a narrow electrically conducting layer (e.g., Cu) sandwiched between two magnetic films. The conducting layers have a width that is smaller than the electron scattering length in the conducting material (100 nm or so). Current in the conducting layer may be sum of two currents, one coming from spin-up electrons, the other mediated by spin-down electrons. Electron scattering in magnetic materials is strongly dependent on the orientation of the electron spin with respect to the magnetization of the material surrounding the conductor. GMR occurs when the magnetic layers above and below the conductor are magnetized in opposite directions, that is, in the absence of the magnetic field the magnetizations of the ferromagnetic layers are antiparallel and this results in a substantial (~20%) increase in the resistance (R) of the material as shown in Figure 14.7. Increasing the external magnetic field (H) aligns the magnetic moments and saturates the magnetization of the multilayer, which leads to a drop in the electrical resistance of the multilayer. This is due to the fact that when the top and bottom magnetic layers are magnetized in the same direction, electrons of one particular spin polarization will be strongly scattered in both layers. However, electrons of the opposite spin polarization will hardly be scattered at all, resulting in a low-resistance path across the device. When the magnetic materials are magnetized in opposite directions, electrons of one spin will be strongly scattered in one of the magnetic layers, while electrons of the opposite spin will be strongly scattered in the other magnetic layer. This results in a significant increase in the resistance of the device. In a magnetic hard drive reading head, the upper magnetic layer is made of cobalt and it remains magnetized in one direction. The lower magnetic layer is made from a NiFe alloy (i.e., soft material), for which the magnetization is readily realigned by the magnetic bits on the hard disk that spins just below the read head. Thus,

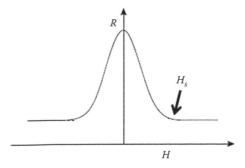

FIGURE 14.7
Variation in Giant magnetoresistance as a function of external magnetic field.

the magnetization of the bits on the surface of the disk can be read out as fluctuations in the resistance of the conducting layer. The width of the transition region between adjacent bits, in which the localized magnetic field is present, is between 10 and 100 nm. The localized fields extend linearly across the track and point upward (or down) from the disk surface. The total thickness of the sensor sandwich, along the direction of the track, is presently ∼80 nm. In the future, this thickness may fall to 20 nm. In this read head sandwich, the Cu layer is ∼15 atoms in thickness. For example, laptops and high-speed computers use a GMR nanosensor. The sensitivity of the GMR magnetoresistive sensor is ∼1%/Oe [40,41].

14.6 Single-Electron Transistor (SET)

The existence of a potential-dependent blockaded region in the current–voltage (I–V) curve of double-junction devices is the basis of a SET. The SET consists of an isolated metal particle coupled by tunnel junctions to two microscopic electrodes. The isolated metal particle is capacitively coupled with a gate electrode that is used to control the potential of the metal particle independently. When the potential is adjusted to values that correspond to high conductance through the small metal particle, the transistor is on. The particle is generally made of semiconducting particle like n-type (i.e., GaAs), which is insulated from the source and drain contacts by thin films of the insulating alloy (i.e., AlGaAs). The potential of the dot is controlled by a gate insulated from the dot by a layer of InGaAs. This finite source–drain voltage SET opens a window of potential for tunneling via the QD [41].

14.7 Construction Industry

The use of NPs in constructions possesses superior structural properties, functional paints and coatings, and high-resolution sensing/actuating devices. CNTs, as filler for polymeric chemical admixtures, can remarkably improve the mechanical durability by gluing the concrete mixtures and prevent the crack propagation. The addition of CNTs as crack-bridging agents into nondecorative ceramics can enhance their mechanical strength and reduce their fragility, as well as improve their thermal properties. Nano- and microscale sensors and actuators are implanted in construction structures for accurate real-time monitoring of material/structural damage and health and environmental conditions. For example, CNT/polycarbonate nanocomposite produces momentary changes in the electrical resistance when the device senses strain inputs, providing an early indication on the possible structural damage.

The SiO_2 and Fe_2O_3 NPs can be used as filling agents to pack the pores that concrete develops due to cement hydration. This will prevent propagation of crack and can enhance the mechanical properties of concrete. The addition of SiO_2 nanolayers between two glass panels can make fireproof windows. Silica NPs on windows control exterior light as an antireflection coating, thus contributing to energy conservation (or as air conditioning). TiO_2 coated on pavements, walls, and roofs also functions as an antifouling agent to keep the surfaces dirt-free under solar irradiation. In addition to bacterial/viral inactivation, the photoinduced superhydrophilic property of TiO_2 prevents the hydrophobic dust accumulation on windows. Light-mediated TiO_2 surface hydroxylation endows window glass with antifogging properties by decreasing the contact angle between water droplets and the glass surface. Flexible solar cells (of silicon or TiO_2-based photovoltaic cells) placed on roofs and windows produce electricity under sunlight illumination.

The addition of magnetic nickel NPs during concrete formation increases the compressive strength by over 15% as the magnetic interaction enhances the mechanical properties of cement mortars. Copper NPs decrease the surface roughness of steel to promote the weldability and impart the steel surface corrosion resistant. Uniform dispersion of metal carbides, oxides, and nitrides NPs in the steel matrix strengthens steel against creep approximately two orders of magnitude. Silver NPs can be embedded in paint to inactivate pathogenic microbes and provide antimicrobial properties to surfaces, particularly hospital walls [42].

14.8 Self-Cleaning Coatings

Self-cleaning surfaces have become a reality due to photocatalytic coatings of TiO_2 NPs. These NPs initiate photocatalysis, a process by which dirt is broken down by exposure to the sun's UV rays and washed away by rain. Volatile organic compounds are oxidized into carbon dioxide and water. There are mainly two types of self-cleaning coatings: hydrophobic and hydrophilic. Both of them clean themselves through the action of water, the former by rolling droplets and the latter by sheeting water that carries away dirt. Hydrophilic coatings have an additional property of chemically breaking down the adsorbed dirt in the presence of sunlight.

14.8.1 Hydrophobic coating

The self-cleaning action of hydrophobic coatings stems from their high-water contact angles. Water on these surfaces forms almost spherical droplets, which are rolled away carrying dust and dirt particles with them. The rolling motion more efficiently cleans the surface and is less likely to leave behind dirt than the sliding motion that occurs at lower contact angles. The requirements for a self-cleaning hydrophobic surface are a very high-water contact angle and a very low roll-off angle. A surface with these two properties is known as superhydrophobic or ultraphobic. For example, a 2 μm thick film of vertically aligned CNTs coated with a very thin conformal layer of polytetrafluoroethylene results in a highly textured surface. This hydrophobic coating exhibits a water contact angles as high as 170°.

14.8.2 Hydrophilic coatings

These coatings chemically break down dirt when exposed to light, a process known as "photocatalysis," although it is the coating not the incident light that acts as a catalyst. Several major glass-manufacturing companies have been producing self-cleaning windows or glasses. These windows are coated with a thin transparent layer of TiO_2. This coating acts to clean the window in sunlight through two distinct properties: photocatalysis and hydrophilicity. The photocatalysis causes the coating to chemically break down the organic dirt adsorbed onto the window, while hydrophilicity causes water to form the sheets. In this case, water contact angles are reduced to very low values in sunlight and dirt is washed away. This coating is also known as superhydrophilic. In general, TiO_2 has been used for self-cleaning windows and hydrophilic self-cleaning surfaces because of its good photocatalysis and superhydrophilic state. In addition, it is a nontoxic, chemically inert in the absence of light, inexpensive, relatively easy to handle and deposit into thin films. It is used as a pigment in cosmetics and paint as additives.

A semiconductor under normal conditions, for example TiO_2 absorbs light with energy equal to or greater than its bandgap energy, resulting in excited charge carriers: an electron (e^-) and a hole (h^+). Although the fate of most of these charge carriers is rapid recombination, some migrate to the surface. There, holes cause the oxidization of adsorbed organic molecules, while the electrons eventually combine with atmospheric oxygen to give the superoxide radical, which quickly attacks nearby organic molecules. The result is a cleaning of the surface by "cold combustion," the conversion of organic molecules to carbon dioxide and water (and other products if heteroatoms are present) at ambient temperatures. This process is remarkably effective and clean because it does not give any detectable by-products on a TiO_2 surface. Superhydrophilicity in TiO_2 is also a light-induced property. Holes produced by photoexcitation of the semiconductor oxidize lattice oxygen at the surface of the material, resulting in oxygen vacancies. These can be filled by adsorbed water, resulting in surface hydroxide groups that make the wetted surface more favorable compared to the dry surface, lowering the static contact angle to almost $0°$ after irradiation. Both self-cleaning properties of TiO_2 are therefore governed by the absorption of ultrabandgap light and the generation of electron/hole pairs. The photocatalytic properties of TiO_2 depend greatly on its form, particle size, morphology, and the method of preparing the sample. In recent years, the self-cleaning coating products based on TiO_2 have been developed and marketed by multinational glazing companies, for self-cleaning windows and painting buildings. The self-cleaning surfaces are made by applying a thin nanocoating film, painting a nanocoating, or integrating NPs into the surface layer of a substrate material [43].

14.9 Nanotextiles

14.9.1 Textile Fabrics

Textile fabrics are one of the best platforms for deploying the nanotechnology. Nano-Pel is a water- and oil-repellent treatment that can be applied to all major apparel fabrics, including cotton, wool, polyester, nylon, rayon, and blends. Nano-Care is a product for 100% cotton that imparts wrinkle resistance in addition to water and oil repellency. Synthetic textile materials including nylon and polyester have hydrophobicity, poor wicking, and permeability properties. These properties make them uncomfortable to wear in hot weather. Water does not readily spread out over surfaces made of these materials. Nylon and polyester also often exhibit static cling and stain retention. These drawbacks can be minimized by attaching the hydrophilic materials to the hydrophobic fibers. This treatment eliminates static cling and enables the release of stains during laundering. A treatment that involves a 3-D molecular network surrounding a fiber is called Nano-Dry. This hydrophilic treatment is applied to polyester and nylon fabrics, which attaches a hydrophilic network to a hydrophobic substrate without altering the strength, color, fastness, and tactile feel. A typical solid content for Nano-Dry treatment on to the fabric surface is ca. 0.1–0.15 wt%

of the fabric, which is significantly smaller than that of (i.e., 0.8–4.0 wt%) conventional film-forming hydrophilic treatments. The performance of water absorbency of the Nano-Dry fabric is durable for the life of the garment, while traditional commercially available treated fabrics have durability only up to 5–10 home launderings. One more development of Nano-Touch-treated fabric regains four times the moisture content compared to normal polyester, that is, up to 1.2% moisture regain. This moisture absorption improves the "natural" feel of the fabric. Since Nano-Touch has carbohydrate sheath wrapping on to the fiber, the sheath is naturally hydrophilic and will wick moisture in a similar manner to the Nano-Dry treatment [44].

14.9.2 Intelligent Textiles

Intelligent (or smart) textiles are the textiles which can sense and react to environmental conditions or stimuli. Nanoscale manipulation results in new functionalities for the smart textiles such as self-cleaning, sensing, actuating, and communicating. There are "passive smart textiles" and "active smart textiles," which are capable of sensing environmental conditions and, both actuators and sensors, respectively. The basic components within smart textiles are sensors, actuators, and control units. These components are integrated seamlessly into the textile without compromising the original tactile, flexibility, and comfort properties of clothing. This is made possible by incorporating inherently conducting polymers (ICPs), CNTs, and antimicrobial nanocoatings to the textile fabric at various production stages; at the fiber-spinning level, during yarn/fabric formation, or at the finishing stage. The ICPs such as polypyrrole (PPy) and polyaniline (PANi) conduct electricity and have the ability to sense and actuate. Actuators based on ICPs can generate much higher stresses with a strain comparable to natural skeletal muscle. Sensors based on ICPs can change their resistivity or generate an electrical signal in response to external stimuli. Thus, ICP-based intelligent polymer systems have the ability to sense, process information, and actuate. For example, ICP-based mechanical actuators can achieve average stresses ca. 10–20 times those generated from natural muscle, realize strains (>20%) comparable to natural muscle, and achieve fast-freestanding beam actuation with an operational frequency of up to 40 Hz. Moreover, by incorporating CNTs, the electrical and mechanical properties of ICPs can also be improved. For example, a PANi-CNT composite nanofiber produced by a wet spinning technique has shown the ultimate tensile strength and elastic modulus higher by 50%–120% compared to pure PANI, with an electronic conductivity of up to $\sim$750 S/cm. Thus, the unique properties of composites nanofibers make them potentially useful in electronic textile applications. The addition of TiO_2 NPs into fabric provides biological protection.

Membrane of expanded poly(tetrafluoroethylene) (PTFE) with pores of <1 µm diameter allows water vapor (or sweat) to penetrate the material but prevents the passage of liquid. This makes them more comfortable to wear by maintaining the body's natural thermoregulatory function. Similarly, high-performance moisture-wicking fabrics worn next to the skin transport perspiration away from the body to the outside of the garment where it can more quickly evaporate. It is even possible to maintain constant body temperature using phase change materials (PCMs) which absorb, store, and release heat as the material changes phase from solid to liquid and back to solid. The microencapsulated PCMs can be applied as a finishing on fabrics or infused into fibers during the manufacturing process. Several companies have commercialized smart textiles such as shock-absorbing shocks, Knee Sleeve, keyboard, light-emitting diodes integrated fabric, etc. [45].

14.10 Biomedical Field

14.10.1 Medical Prostheses

Polymer nanofibers have been suggested for making vascular prostheses and breast prostheses. Protein nanofibers can be deposited as a thin porous film onto a prosthetic device for implantation in the central nervous system. The film with a gradient fibrous structure is expected to reduce the stiffness mismatch at the tissue/device interface, thereby preventing the fracture or fatigue failure of the device after implantation. Some of the biodegradable polymeric nanofibers have been demonstrated for tissue engineering scaffolds.

14.10.2 Drug Delivery

Polymeric nanofibers have been found to exhibit promising potential in drug delivery. Due to a very high-specific surface area, polymeric nanofibers have high-dissolution rate of a drug. The drug and the carrier (i.e., nanofibers) can be mixed together in the resulting electrospun nanofiber. One of the attractive applications in nanomedicine is the use of nanorobots or nanorobots devices for improved therapy and diagnostics, which have the potential to serve as the vehicles for delivery of therapeutic agents, detectors, or guardians against early disease and perhaps repair of metabolic or genetic defects.

14.10.3 Cancer Treatment

Novel nanodevices are capable of detecting cancer at its earliest stages, pinpointing its location within the human body, and delivering chemotherapeutic drugs against malignant cells. The nanomedicine is being developed for early detection of tumor and its treatment. Tumor diagnosis and prevention is the best cure for cancer. Nanodevices, especially nanowires, can detect the cancer-related molecules, contributing to the early diagnosis of tumor. Nanowires having the unique properties of selectivity can be designed to sense molecular markers of malignant cells. Nanowires coated with a probe can be used to recognize the specific RNA sequences. Proteins that bind to the antibody will change the nanowire's electrical conductance, and this can be measured by a detector. As a result, proteins produced by cancer cells can be detected and earlier diagnosis of tumor can be achieved.

14.10.4 Wound Dressing

For wound and burn treatment, polymeric nanofibers show promising application. Biodegradable polymeric nanofibers have been sprayed on the skin wounds using an electrical field to produce a fibrous mat dressing. This dressing enables the formation of normal skin growth and prevents the formation of scar tissue that results from conventional treatment. Nonwoven nanofibrous membranes for wound dressing can be made with pore size ranging from 500 to 1000 nm in order to shield the wound against bacterial penetration.

14.10.5 Cosmetics

Polymer nanofibers have been tested as cosmetic skincare masks for the treatment of skin healing or cleansing and other therapeutic purposes with or without various additives.

The small interstices and high-surface area of nanofibrous skin masks facilitate higher utilization and transfer rate of the additives to the skin for the fullest use of the additive potential. Such cosmetic skin masks made from electrospun nanofibers can be gently and painlessly applied to the 3-D topography of the skin for healing and skincare. In addition, nanofibrous membranes are highly efficient for fluid absorption and dermal delivery and hence, polymeric nanofibers are promising in the development of high-efficiency hemostatic devices [44]. The haemostatic devices can be used to support hemostasis (i.e. stop bleeding), thus reduce the operative risks and length.

14.11 Nanopore Filters

There are mainly two types of nanopore filters, which are commonly used: Nuclepore and Nanopore. The nuclepore filters are sheets of polycarbonate of 4.7 cm diameter and thickness of 10 µm with closely spaced arrays of parallel holes running through the sheet. The filters have pore sizes in the range of 15–12,000 nm. The pores are made by exposing the polycarbonate sheets to perpendicular flux of ionizing particles, followed by controlled chemical etching to enlarge the parallel holes to the desired diameter. The actual pore size is ~165 nm. The filters are robust and the pore size distribution is relatively narrow. The pores density is about 3×10^8/cm and the porosity is ~12%. The smallest filters will block the passage of bacteria and perhaps even some viruses, and are used in many applications including water filters. The Anapore filters are made of alumina grown by anodic oxidation of aluminum metal. These filters have porosity up to 40%, and are stronger and more temperature resistant than the Nuclepore filters. The Anapore filters have been used to synthesize the nanowires. Anopore filters possess a larger and more uniform pore size distribution than Nuclepore filters, and appear to have a smoother cavity surface. They also have a much larger surface area [46,47].

14.12 Water Treatment

Wastewater is contaminated by metal ions, radionuclides, organic and inorganic compounds, and pathogenic bacteria and viruses. Due to extended droughts, population growth, and more stringent health-based regulations, there is an increasing demand for clean water for a healthy human life. Water treatment generally involves adsorption and/or photocatalysis of contaminants and their reduction by NPs and bioremediation. Remediation is the process of pollutant transformation from toxic to less toxic in water and soil. Hence, one of the most promising applications of the NPs has been in water treatment. There are several nanomaterials such as TiO_2, CNTs, zerovalent iron, dendrimers, and silver, which can purify water through: adsorption of heavy metals and other pollutants, removal and inactivation of pathogens, and transformation of toxic materials into less toxic compounds. These nanomaterials are incorporated in nanostructured catalytic membranes to purify the waste water. For example, nanoporous activated carbon fibers with an average pore size of 1.16 nm are used to absorb benzene, toluene, xylene, and ethylbenzene.

Conventional water treatment techniques including reverse osmosis, distillation, biosand, coagulation–flocculation, and filtration are not capable of removing all heavy metal ions. Fortunately, NPs can be used to treat water contaminated by dye, papermaking wastes, pesticide, and oily wastes. Some nanomaterials like zerovalent iron possess excellent reductive capabilities and can render toxic pollutants less toxic without producing intermediate by-products. Owing to high-surface area, some metal oxides are considered better candidates for the removal of heavy metal ions such as As(V) and Cr(VI). Similarly, flowerlike hierarchically superstructured γ-AlOOH with BET surface area of $\sim$146 m²/g quickly removes the Pb(II) and Hg(II) ions from aqueous solutions. The use of some nanomaterials in water purification is discussed below in brief.

14.12.1 TiO₂ Nanomaterials

TiO₂ powder is most widely used in the degradation of organic and inorganic compounds. It has comparatively low cost, high photosensitivity, easy availability, minimal toxicity, and environment-friendly characteristics. TiO₂ nanopowders are suspended to increase the contact between TiO₂ NPs and target pollutants in water, and hence improve photocatalytic efficiency. TiO₂ membranes are multifunctional in providing liquid separation of both contaminated and treated water and offer photocatalytic degradation of pollutants. The use of TiO₂ nanowires for membranes exhibits greater photocatalytic efficiency compared to conventional bulk materials due to the higher-surface area and percent surface atoms. These TiO₂ membranes can have multifunctional activities of filtration, self-cleaning, photocatalytic degradation, and disinfection. However, the separation of TiO₂ powders from treated wastewater prior to discharge is a time-consuming task. This challenge limits the application of TiO₂ photocatalysis in water treatment. This limitation can be minimized by immobilizing TiO₂ on a substrate. However, the use of immobilization decreases its photocatalytic ability.

14.12.2 CNT Powder

Due to outstanding adsorbance of organic and inorganic pollutants, CNTs can be used in devices as adsorbents and detection matrices. Compared to activated carbon, functionalized CNTs can adsorb higher amounts of low-molecular weight and polar compounds. CNTs also show some intrinsic antimicrobial activity.

14.12.3 Zerovalent Metals

Due to large surface area and reactivity, zerovalent metals are favorable compounds for an effective detoxification of organic and inorganic pollutants in aqueous solutions. Fe⁰ and its bimetallic NPs can reduce Cr(VI) to less toxic species. Silver compounds and silver ions have antimicrobial ability which disinfects medical devices and water. As discussed in section 14.4, they are effective biocides against a number of Gram-positive and Gram-positive bacteria. Silver NPs can directly damage bacterial cell membrane. In general, the release of silver ions from silver nanomaterials leads to bactericidal activity by increasing membrane permeability. Silver NP–alginate composite beads have been used for drinking water disinfection. Similarly, MgO NPs have been reported as effective biocides against Gram-positive and Gram-negative bacteria. CuO NPs are effective for As(III) and As(V) adsorption.

14.12.4 Dendrimers

Dendrimers are relatively monodispersed hyperbranched-well-defined macromolecules (or polymers) with controlled composition and 3-D architectures. Dendritic polymers possess high capacity and recyclable water-soluble ligands for grabbing toxic metal ions, radionuclides, and inorganic anions. It is known that nanofiltration and reverse osmosis membranes are very effective in filtration of solutes of size 1000 Da and 1000–3000 Da, respectively, but they require high pressures to operate. In contrast, ultrafiltration membranes require lower pressure but do not efficiently discard dissolved organic and inorganic solute with molar mass <3000 Da. The use of dendritic polymers in membranes effectively removes the toxic metal ions, radionuclides, organic and inorganic solutes, and living entities (bacteria and viruses) from contaminated water. For example, cross-linked, polystyrene-supported, low-generation diethanolamine-typed dendrimers exhibit good adsorption capacities for Cu^{2+}, Ag^+, and Hg^{2+} ions. These dendritic polymers can be recycled after the filtration.

14.12.5 Graphene

Graphene nanosheets (GNS) and graphene oxides (GO) have been known to exhibit high-specific surface area, high porosity, high thermal stability, and ease of chemical modification. Therefore, GNS and GO have emerged as ideal materials for adsorbents to eliminate the heavy metal ions, inorganic ions, as well as organic dyes. These graphene-based materials have been found to be suitable for air remediation through the removal of toxic gases, emitted due to the large-scale expansion of industries, including SO_x, CO, NH_3, and others. In other words, this can help to avoid natural calamities on account of global warming. The adsorption by graphene does not produce any harmful by-products during adsorption process. GO and GNS have shown a very strong adsorption tendency toward fluorides, phosphates, sulfides, nitrates, heavy metal ions, and organic compounds which are present in water. The adsorption efficiency of GNS and GO is higher in magnitude than activated carbon or CNT.

14.12.6 Polymeric Nanofibers

Polymeric nanofibers have been used as filter media in industrial filtration over the past two decades. They provide higher filter efficiency at equal pressure drop. They can also be electrostatically charged to modify the ability of electrostatic attraction of particles without an increase in the pressure drop to further improve the filtration efficiency. Protective clothing must have the ability to react with chemical agents such as sarin, soman, tabun, mustard gas, and biological and chemical warfare. Current protective clothing containing charcoal absorbents has high weight and poor water permeability. It is desirable for protective clothing to possess the lightweight, breathable fabric, permeable to air and water vapor, insoluble in all solvents, and highly reactive with nerve gases and other deadly chemical agents. For example, insoluble linear poly(ethylenimine) nanofibers are capable of neutralizing the chemical agents without hindering the air and water vapor permeability to the clothing. Electrospun nanofibers show lower impedance to moisture vapor diffusion and maximum efficiency in trapping the aerosol particles as compared to conventional textiles [48–50].

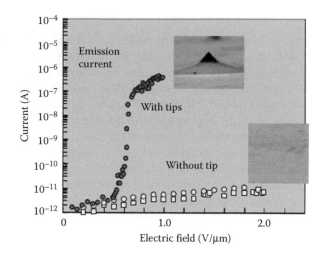

FIGURE 14.8
Electron emission characteristics. (Schulte J: *Nanotechnology: Global Strategies, Industry Trends and   Applications.*
2005. Copyright Wiley-VCH Verlag GmbH & Co. KGaA. Reproduced with permission.)

14.13 Nanodiamond

Diamond is an excellent semiconductor, which exhibits an extremely high hardness, very
high-thermal conductivity, and high-acoustic velocity. It can be precisely machined on a
nanometer scale compared to other materials. The nanostructured diamond (or diamond
nanoemitter) is a steeple diamond single crystal made by reactive ion etching using patterned aluminum (Al) sacrificial masks. The Al mask disappears after precisely guiding
the position of the steeple on the diamond surface. Diamond nanoemitter has a radius of
2 nm at the top of the steeple, which is made by top-down nanotechnology. Figure 14.8
shows the comparison of electron emitted from the nanodiamond tips and flat diamond.
At the applied electric field of 1.0 V/μm, the electron emitted from the tip is approximately
six orders of magnitude higher than as compared to that from the flat diamond surface.
This is called the nanosize effect. The development of diamond nanoemitter could bring
down the required temperature of over 2000°C for reasonable electron emission from tri-
ode vacuum tube to almost room temperature (i.e., 30°C). This can squeeze down the size
significantly [44].

14.14 Automotive Sector

14.14.1 Introduction

The automotive industry in Europe, has been taking efforts in exploiting the nanotechnology for the applications in power train, lightweight construction, energy conversion,
pollution sensing and reduction, interior climate, wear reduction, CO_2-free engines, safe
driving, quiet cars, self-cleaning body, windscreens, and tailor-made outer body color.

Nanotechnology is already in use for automobile components as antireflection coating based on multiple nanolayers on glass (trade name Schott Conturan) by Audi and DaimlerChrysler trucks. Sun protecting glazing with infrared reflecting nanolayers embedded into sheets of glass (trade name Sekurit Thermocontrol) is being used in buses (e.g., Evobus). Thermoplastic nanocomposite with nanoflakes (trade name Basell TPO-Nano) is used for stiff and light exterior parts [51]. The nanotechnology can be exploited for automotive sector by: (i) converting the energy of sunlight into electricity using the solar cells, (ii) converting solar energy into hydrogen fuel by splitting water into its constituents, (iii) storing hydrogen in solid-state forms, (iv) utilizing hydrogen to generate electricity through the use of fuel cells, and (v) using nanofluids in radiators to increase the fuel efficiency. The addition of CNTs to plastics or metals may make airplanes and vehicles lighter and therefore reduce the fuel consumption. The addition of carbon black nanopowder to tyres increases strength and reduces the rolling resistance (or rolling drag), and thus it results in a fuel savings of up to 10%. The use of self-cleaning coating onto the glass saves both energy and water in cleaning process since such surfaces need not be cleaned continuously as in conventional cars. The fuel or motor oil (nano) additives can reduce the fuel consumption of vehicles and extend the engine life. Nanoceramic-based coatings can replace environmentally harmful or hazardous chromium (VI) layers. Moreover, titanium dioxide and silica nanopowders can replace the environmentally damaging bromine in flame retardants [52].

14.14.2 Solar Energy

The use of nanostructured layers in thin film (thickness ~150 nm) solar cells exhibits several advantages including increased multiple reflections, reduced travel path for light-generated electrons and holes, and tailored energy bandgap of layered materials. Compared to conventional materials, a significantly shorter path length of electrons and holes reduces the recombination losses. Thin film allows better design flexibility in the absorber of solar cells and increases solar cells efficiency. The use of carbon black aqueous nanofluids could effectively enhance the solar absorption efficiency in the wavelength range from 200 to 2500 nm. Car manufacturers have an option to put solar cells on the car roof to provide the ventilation in the interior when the engine is off. However, low efficiency of standard cells and small area of the covered car roof limits use. One can deposit triple cells of nanocomposites containing semiconductor nanodots (i.e., Ge or Si), which can use a wider part of the solar spectrum [53–55].

14.14.3 Fuel Cell

The fuel cell can be defined as an electrochemical energy scavenger that converts the chemical energy of a fuel such as hydrogen or methanol into electricity through a chemical reaction with an oxidizing agent (i.e., oxygen or air). The net reaction that occurs in a fuel cell can be written as: $H_2 + \frac{1}{2}O_2 \rightarrow H_2O$. This reaction shows that the reaction between hydrogen and oxygen produces electricity and heat along with water as a by-product, which is environment-friendly with no harmful carbon dioxide (CO_2) emissions. Conventional fuel cells have several drawbacks including high cost, durability, and operation issues. To a certain extent, these limitations can be reduced by using nanofuel cells which have better efficiency, nanosize, and safe storage of hydrogen. Hydrogen energy is used extensively in the space vehicles since it has excellent energy-to-weight ratio for any fuel. Nanostructured materials are effectively utilized to improve the transformation of hydrogen energy into electricity by using the fuel cells. Nanoblades could hold large volumes of hydrogen. Recently, CNT fuel cells have potential to store hydrogen and are considered environmentally friendly. Hydrogen fuel cell (HFC) has been suggested in automotive industry as a possible replacement for fossil fuels in passenger cars and public transportation. The use of hydrogen as fuel in an internal combustion engine liberates clean water vapor, which is better than CO_2 emissions (from fossil fuels) [52].

14.14.4 Vehicle Radiator

Engine manufacturers have been trying to produce high-efficiency engine at low cost. However, engine overheating has been a primary concern for many manufacturers. Vehicle engine cooling system takes care of excess heat produced during engine operation. However, low-thermal conductivity of engine coolant limits the cooling efficiency of a vehicle radiator, which makes it difficult in maintaining the compact size of the cooling system. In addition, increasing the cooling rate by fins and microchannel has reached their limits. Therefore, one of the innovative ways is to use nanocoolant (or nanofluid) to enhance the heat transfer in an automotive car radiator.

The average heat transfer coefficient is directly proportional to the volume fraction of the thermally conducting particles added to the fluid and the Reynolds number. For example, the addition of 0.5 vol% MWNTs increases the average heat transfer coefficient ~200% compared to base fluid (i.e., water/ethylene glycol) [56].

14.14.5 Diesel Particulate Filter

Diesel particulate filter is a device that remove diesel particulates (solid organic fraction of hydrocarbon, soot etc.) to prevent their release to the atmosphere. These diesel particulates cause serious health concerns due to their carcinogenity. Nano-Pt-CeO$_x$ particles have been used to optimize the contact between the NPs and the diesel soot in diesel particulate filters. Due to the smaller particle size, a relatively lower filter regeneration temperature is achieved (<450°C). Such nanocomposites are currently used in Toyota or General Motors (GM) cars instead of regular plastic. These nanocomposite filters are lighter, stronger, and longer-lasting than conventional plastics. Owing to lighter filters, cars use less fuel, and hence, less expensive for people to own.

14.14.6 Other Applications

Toyota has been making new bumpers for cars, which are 60% lighter than the old bumpers and twice as hard to scratch or dent. The 2002 Chevrolet Astro and General Motors Corp. Safari midsize vans are the first vehicles to use an advanced polymeric nanocomposite on an exterior application and there are many more examples of using polymeric nanocomposites in automotive sectors by other manufacturers. Compared to a conventional polymeric composites (i.e., talc-filled polypropylene), the new nanocomposite can provide up to a 20% weight reduction with a similar stiffness of 1–1.2 GPa, and improved paint adhesion and quality. In addition, the new nanocomposites are cost-effective and more recyclable due to less clay content [57]. Scratch-resistant, dirt-repellent, self-healing paints, and ultrathin coatings for mirrors and reflectors have been studied by Daimler Chrysler. The addition of spherical silica NPs (mean diameter: ca. 7–40 nm) in paint, after drying and hardening process, provides a very dense and ordered matrix on the paint surface. This paint provides significantly better scratch resistance and paint brilliance. The addition of CNTs to styrene-butadiene rubber (SBR) matrices improves the tensile strength, tear strength, and hardness of the nanocomposites by ~600%, 250%, and 70%, respectively, compared to that of pure SBR matrix. A nanoclay containing brominated isobutylene-co-para-methyl styrene elastomer (BIMSM) has been commercialized by ExxonMobil, which shows better (by ~50%) air retention properties compared to those of halobutyl rubbers. The addition of o-clay to the tyre provides an isotropic performance and exhibit higher stiffness, better thermoplastic stability, and reduced decay [57]. The nanoscale additives of carbon black, silica, and clay are used as pigments and reinforcing agents during tyre manufacturing. The additives are incorporated into the rubber before the vulcanization process. The dispersion of nanoscale carbon black inclusions increases the tensile strength and wear resistance of tyres because of the chemical double bond between the carbon black and the rubber. The addition of silica NPs can enhance rolling resistance and wet traction of tyres. The crystalline silica particles are used to increase the friction between the surface of the tyre and the road for improved rolling resistance and wet-tyre traction. Nanoadditives also improve the lifetime of tyres considerably. Rubber tyres modified by nanographene platelets (NGPs) considerably improve the abrasion resistance of tyres when they are added to tyre treads. This reduce noise, tyre erosion, and air release rates, and enhance tyre life expectancy [58].

14.15 Catalysts

It has been studied well that the catalytic activity and selectivity of some nanomaterials are significantly higher than those of their bulk counterparts, primarily due to their large surface area. For example, the chemical reactivity of nanocrystalline TiO_2 is much higher than that of commercial TiO_2 in removing S from H_2S by decomposition. The higher activity of the nanocrystalline TiO_2 can be attributed to its rutile crystal structure, high-surface area, and its oxygen-deficient composition. Similarly, small Pt particles with size in the range of 3–6 nm has shown an unusual catalytic activity in the hydrogenation of p-chloronitrobenzene. Commercial catalysts with ~20 nm Pt particles only promote the catalytic reduction of the NO_2 group, while the 3–6-nm-sized Pt particles catalyze the unusual formation of dicyclohexylamine 8 in large quantity [59].

Incompletely burnt hydrocarbons (HCs), CO, and nitrogen oxides (NO_x, including NO and NO_2) produced by gasoline and diesel internal combustion engine vehicles are major air pollutants in cities. To comply with current environmental protection legislation, highly efficient catalysts for the complete elimination of these compounds are needed. Commercial catalytic convertors based on the use of platinum (Pt), palladium (Pd), and rhodium (Rh) play a major role in the cleaning of automobile emissions. Cerium, iron, manganese, and nickel are also used but each has its own limitations. Pure gold NPs and its mixture with oxides have demonstrated significant activity in the conversion of toxic compounds including carbon monoxide, unburnt HCs, and nitrogen oxides into CO_2, CO_2/H_2O, and N_2, respectively, in engine emissions and they have advantages over the Pt, Pd, and Rh metals. The conversion of harmful pollutants into harmless by-products after the reaction in the presence of catalysts in catalytic convertor can be represented by the following equations [60].

Reduction of nitrogen oxides to nitrogen:.

$$2NO_x \rightarrow xO_2 + N_2 \tag{14.2}$$

Oxidation of carbon monoxide to carbon dioxide:

$$2CO + O_2 \rightarrow 2CO_2 \tag{14.3}$$

Oxidation of HCs to CO_2 and water:

$$HC + O_2 \rightarrow CO_2 + H_2O \tag{14.4}$$

In contrast to Pt, Pd, and Rh metals, Au is more readily available and has a lower-market price. As a consequence, there has been an increasing interest in the application of Au in automobile emission control due to its excellent catalytic activity when dispersed as fine particles. Bulk gold is known for chemically inertness but when the gold is small enough with particle size <10 nm, it shows surprisingly active catalytic behavior for CO oxidation and propylene epoxidation at low temperatures. More importantly, the efficiency of mixed precious metal (i.e., gold) with oxides can have better catalytic activities than the use of Au alone. These catalysts will therefore be effective in reducing the running costs of chemical plants and could increase the selectivity of the reactions involved in various applications such as air cleaning, low light-off autocatalysts, and purification of hydrogen streams used for fuel cells [61]. The most active catalyst in CO oxidation is the multicomponent catalyst $Au/MgO/MnO_x/Al_2O_3$ with MgO being a stabilizer for the Au particle size and MnO_x being the cocatalyst. This catalyst also exhibits good performance in selective oxidation of CO in a hydrogen atmosphere. Figure 14.9a shows the conversion of propene over three types of alumina-supported catalysts, for example, CeO_x/Al_2O_3, Au/Al_2O_3, and the multicomponent catalyst consisting of both Au (average size of 3 nm) and CeO_x, $Au/CeO_x/Al_2O_3$. It can be seen clearly that the multicomponent $Au/CeO_x/Al_2O_3$ catalyst is much

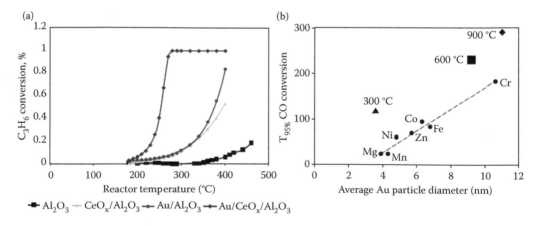

FIGURE 14.9
(a) Synergistic effects of Au, oxides, and mixture of (Au/oxides) catalysts on oxidation of propene and (b) temperature needed for 95% CO conversion versus the average gold particle size. The Au/Al_2O_3 and $Au/MO_x/Al_2O_3$ catalysts were calcined at different temperatures in order to change the average Au particle size. (From Grisel R. et al. 2002. *Gold Bull.* 35: 39–45. Open access.)

more active at low temperatures than the monocomponent ones. Figure 14.9b shows the temperatures needed for 95% CO conversion ($T_{95\%}$) by monocomponents and multicomponents catalysts as a function of average gold particle size (4–11 nm). The gold NPs with size <5 nm show high activity. The addition of MO_x (i.e., MgO or MnO) in multicomponents (i.e., in $Au/MO_x/Al_2O_3$) has a much more beneficial effect on the activity at low temperatures compared to that of Au NPs alone. For example, the $Au/MnO_x/Al_2O_3$ catalyst with an average particle size of 4.2 ± 1.4 nm has a $T_{95\%}$ that is 100° lower than that found for Au/Al_2O_3 with an average particle size of 3.6 ± 1.4 nm. In addition, the CO conversion over $Au/MO_x/Al_2O_3$ is higher at ambient temperature than the CO conversion over the Au/Al_2O_3 catalyst at 150°C. The addition of MO_x probably stabilizes the gold NPs and acts as cocatalyst [62].

References

1. Vollath D. (Ed.). 2013. *Nanoparticles: Nanocomposites-Nanomaterials*. Wiley-VCH, Germany.
2. Peyghambarzadeh S. 2013. Experimental study of overall heat transfer coefficient in the application of dilute nanofluids in the car radiator. *Appl. Therm. Eng.* 52: 8–16.
3. Xie J. 2010. PET/NIRF/MRI triple functional iron oxide nanoparticles. *Biomaterials* 31: 3016–3022.
4. Hahn M. 2011. Nanoparticles as contrast agents for in-vivo bioimaging: current status and future perspectives. *Anal. Bioanal. Chem.* 399: 3–27.
5. Duguet E., Vasseur S., Mornet S., Devoisselle J. M. 2006. Magnetic nanoparticles and their applications in medicine. *Nanomedicine* 1: 157–168.
6. Marquis F. D. S., Chibante L. P. F. 2005. Improving the heat transfer of nanofluids and nanolubricants with carbon nanotubes. *JOM* 57: 32–43.
7. Donzelli G., Cerbino R., Vailati A. 2009. Bistable heat transfer in a nanofluid. *Phys. Rev. Lett.* 102: 104503.
8. Saidur R. 2010. A review on applications and challenges of nanofluids. *Renew. Sust. Energ. Rev.* 15: 1646–1668.
9. Tyagi H. 2009. Predicted efficiency of a low-temperature nanofluid-based direct absorption solar collector. *J. Solar Energ. Eng.* 131: 1–7.
10. Mahian O. 2012. A review of the applications of nanofluids in solar energy. *Int. J. Heat Mass Transf.* 57: 582–594.
11. Kao M. 2007. Evaluating the role of spherical titanium oxide nanoparticles in reducing friction between two pieces of cast iron. *J. Alloys Comp.* 483: 456–459.
12. Lee M. 2014. Heat transfer characteristics of a speaker using nano-sized ferrofluid. *Entropy* 16: 5891–5900.
13. Raj K., Moskowitz R. 1990. Commercial applications of ferrofluids. *J. Magn. Magn. Mater.* 85: 233–245.
14. Niemann M. U., Srinivasan S. S., Phani A. R., Kumar A., Goswami D. Y., Stefanakos E. K. 2008. Nanomaterials for hydrogen storage applications: A review. *J. Nanomater.* Article ID 950967. doi: 10.1155/2008/950967.
15. Yang J., Sudik A., Wolverton C., Siegelw D. J. 2010. High capacity hydrogen storage materials: Attributes for automotive applications and techniques for materials discovery. *Chem. Soc. Rev.* 39: 656–675.
16. Sunandana C. S. 2007. Nanomaterials for hydrogen storage. *Resonance* 12: 31–36.
17. Yang S. J., Jung H., Kim T., Park C. R. 2012. Recent advances in hydrogen storage technologies based on nanoporous carbon materials. *Progress Nat Sci. Mater. Int.* 22(6): 631–638.

18. Griessen R., Feenstra R. 1985. Volume changes during hydrogen absorption in metals. *J. Phys. F: Met. Phys.* 15: 1013.

19. Züttel A., Orimo S. 2002. Hydrogen in nanostructured, carbon-related, and metallic materials. *MRS Bull.* 27: 705–711.

20. Chen P., Wu X., Lin J., Tan K. L. 1999. High H_2 uptake by alkali-doped carbon nanotubes under ambient pressure and moderate temperatures. *Science* 285: 91–93.

21. Yurum Y., Taralp A., Veziroglu T. N. 2009. Storage of hydrogen in nanostructured carbon materials. *Int. J. Hydrog. Energ.* 34: 3784–3798.

22. http://cen.acs.org/articles/94/i18/future-low-cost-solar-cells.html

23. Kickelbick (Ed.). 2007. *Hybrid Materials: Synthesis, Characterization, and Applications.* Wiley-VCH Verlag GmbH & Co. KGaA, Betz-Druck GmbH, Darmstadt.

24. Lee Y. L., Lo Y. S. 2009. Highly efficient quantum-dot sensitized solar cell based on co-sensitization of CdS/CdSe. *Adv. Funct. Mater.* 19: 604–609.

25. Huynh W. U., Dittmer J. J., Alivisatos A. P. 2002. Hybrid nanorod-polymer solar cells. *Science* 295: 2425–2427.

26. Kongkanand A., Tvrdy K., Takechi K., Kuno M., Kamat P. V. 2008. Quantum dot solar cells. Tuning photoresponse through size and shape control of CdSe-TiO_2 architecture. *J. Am. Chem. Soc.* 130: 4007–4015.

27. Robel I., Subramanian V., Kuno M., Kamat P. V. 2006. Quantum dot solar cells. Harvesting light energy with CdSe nanocrystals molecularly linked to mesoscopic TiO_2 films. *J. Am. Chem. Soc.* 128: 2385–2393.

28. Gimenez S., Mora-Sero I., Macor L., Guijarro N., Lana-Villarreal T., Gomez R., Diguna L. J., Shen Q., Toyoda T., Bisquert J. 2009. Improving the performance of colloidal quantum-dot-sensitized solar cells. *Nanotechnology* 20: 295204.

29. Shen Q., Kobayashi J., Diguna L. J., Toyoda T. 2008. Effect of ZnS coating on the photovoltaic properties of CdSe quantum dotsensitized solar cells. *J. Appl. Phys.* 103: 084304.

30. Volz S., Chen G. 1999. Molecular dynamics simulation of thermal conductivity of silicon nanowires. *Appl. Phys. Lett.* 75: 2056–2058.

31. Mingo N., Broido D. A. 2005. Carbon nanotube ballistic thermal conductance and its limits. *Phys. Rev. Lett.* 95: 096105.

32. Hochbaum A. I., Chen R., Delgado R. D., Liang W., Garnett E. C., Najarian M., Majumdar A., Yang P. 2008. Enhanced thermoelectric performance of rough silicon nanowires. *Nature* 451: 163–167.

33. Luoma S. N. 2008. *Silver nanotechnologies and the environment: Old problems or new challenges? PEN 15.* Woodrow Wilson International Center for Scholars Project on Emerging Nanotechnologies.

34. Lee H. J., Yeo S. Y., Jeong S. H. 2003. Antibacterial effect of nanosized silver colloidal solution on textile fabrics. *J. Mater. Sci.* 38: 2199–2204.

35. Zhanga D., Chenb L., Zanga C., Chena Y., Lina H. 2013. Antibacterial cotton fabric grafted with silver nanoparticles and its excellent laundering durability. *Carbohyd. Polym.* 92: 2088–2094.

36. El-Nahhal I. M., Zourab S. M., Kodeh F. S., Elmanama A. A., Selmane M., Genois I., Babonneau F. 2013. Nano-structured zinc oxide–cotton fibers: Synthesis, characterization and applications. *J. Mater. Sci.: Mater. Electron.* 24: 3970–3975.

37. Mao Z., Shi Q., Zhang L., Cao H. 2009. The formation and UV-blocking property of needle-shaped ZnO nanorod on cotton fabric. *Thin Solid Films* 517: 2681–2686.

38. Nam G., Rangasamy S., Purushothaman B., Song J. M. 2015. The application of bactericidal silver nanoparticles in wound treatment. *Nanomater. Nanotechnol.* 5: 23.

39. Lu S., Gao W., Gu H. Y. 2008. Construction, application and biosafety of silver nanocrystalline chitosan wound dressing. *Burns* 34: 623–628.

40. Wolf E. L. 2006. *Nanophysics and Nanotechnology.* Wiley-VCH, Germany.

41. Lindsay S. M. (Ed.). 2010. *Introduction to Nanoscience.* Oxford University Press, New York.

42. Lee J., Mahendra S., Alvarez P. J. J. 2010. Nanomaterials in the construction industry: A review of their applications and environmental health and safety considerations. *ACS Nano* 4: 3580–3590.

43. Parkin I. P., Palgrave R. G. 2005. Self-cleaning coatings. *J. Mater. Chem.* 15: 1689–1695.

44. Schulte J. 2005. *Nanotechnology: Global Strategies, Industry Trends and Applications.* John Wiley & Sons Ltd., England.

45. Coyle S., Wu Y., Lau K. T., Rossi D. D., Wallace G., Diamond D. 2007. Smart nanotextiles: A review of materials and applications. *MRS Bull.* 32: 434–442.

46. Crawford G. P., Steele L. M., Ondris-Crawford R., Iannocchione G. S., Yeager C. J., Doane J. W., Finotello D. 1992. Characterization of the cylindrical cavities of Anopore and Nuclepore membranes. *J. Chem. Phys.* 96: 7788.

47. Wolf E. L. (Ed.). 2006. *Nanophysics and Nanotechnology: An Introduction to Modern Concepts in Nanoscience.* 2nd ed. Wiley-VCH Verlag GmbH &Co. KGaA, Weinheim.

48. Hu A., Apblett A. (Eds.). 2014. *Nanotechnology for Water Treatment and Purification.* Springer International Publishing, Switzerland.

49. Ghasemzadeh G., Momenpour M., Omidi F., Hosseini M. R., Ahani M., Barzegari A. 2014. Applications of nanomaterials in water treatment and environmental remediation. *Front. Environ. Sci. Eng.* 8(4): 471–482.

50. Schulte J. 2005. *Nanotechnology: Global Strategies, Industry Trends and Applications.* John Wiley & Sons Ltd., England.

51. Presting H., König U. 2003. Future nanotechnology developments for automotive applications. *Mater. Sci. Eng. C* 23: 737–741.

52. Hussein A. K. 2015. Applications of nanotechnology in renewable energies-a comprehensive overview and understanding. *Renew. Sust. Energ. Rev.* 42: 460–476.

53. Sethi V., Pandey M., Shukla P. 2011. Use of nanotechnology in solar PV cell. *Int. J. Chem. Eng. Appl.* 2: 77–80.

54. Hussein A. K. 2015. Applications of nanotechnology in renewable energies: A comprehensive overview and understanding. *Renew. Sust. Energ. Rev.* 42: 460–476.

55. Presting H., König U. 2003. Future nanotechnology developments for automotive applications. *Mater. Sci. Eng. C* 23: 737–741.

56. M'hamed B., Sidik N. A. C., Akhbar M. F. A., Mamat R., Najafi G. 2016. Experimental study on thermal performance of MWCNT nanocoolant in Perodua Kelisa 1000cc radiator system. *Int. Commun. Heat Mass Transf.* 76: 156–161.

57. Koch C. C., Ovid'ko I. A., Seal S., Veprek S. 2007. *Structural Nanocrystalline Materials: Fundamentals and Applications.* Cambridge University Press, New York.

58. Asmatulu R., Nguyen P., Asmatulu E. 2013. Nanotechnology safety in the automotive industry. In *Nanotechnology Safety.* Asmatulu R. (Ed.), 1st Ed., Elsevier, USA. pp. 57–72.

59. Suryanarayana C., Koch C. C. 2000. Nanocrystalline materials—Current research and future directions. *Hyperfine Interact.* 130: 5–44.

60. Zhang Y., Cattrall R. W., McKelvie I. D., Kolev S. D. 2011. Gold, an alternative to platinum group metals in automobile catalytic converters. *Gold Bull.* 44: 145–153.

61. Thompson D. T. 2002. Catalysis of gold nanoparticles deposited on metal oxides. *Cattech* 6: 102–115.

62. Grisel R., Weststrate K. J., Gluhoi A., Nieuwenhuys B. E. 2002. Catalysis by gold nanoparticles. *Gold Bull.* 35: 39–45.

63. Sankaran M., Viswanathan B. 2007. Hydrogen storage in boron substituted carbon nanotubes. *Carbon* 45: 1628–1635.

15

Risks, Toxicity, and Challenges of Nanomaterials

As discussed in previous chapters, nanomaterials are mainly in the forms of particles, tubes, wires, films, thin sheets, nanofibers, and nanocomposites. They exhibit interesting physical, chemical, physicochemical, and biological properties. The engineered nanomaterials with size <100 nm are used increasingly in engineering applications, medical imaging, disease diagnoses, drug delivery, cancer treatment, gene therapy, cosmetic, and other areas. The potential for nanoparticles (NPs) in these areas appears to be almost infinite. Nanomaterials have been considered for the most powerful computers and satellites, faster cars and planes, and better microchips and batteries due to their outstanding mechanical, electrical, magnetic, optical, and thermal properties. They can also be used to produce military armor, artificial muscles, and drug delivery systems. However, there are concerns on the toxicological, health, and environmental effects of direct and indirect exposure to nanomaterials. These toxic nanomaterials can be found in air, water, soil, plants, and animal and human bodies. Moreover, these invisible nanomaterials can be suspended in air and water for days or even weeks in work areas, which can bring huge risk factors throughout the fabrication, transportation, handling, usage, waste disposal, and recycling processes. Nanomaterials can enter the human body through various ways, including inhalation, ingestion, and/or contact through skin. These NPs can reside in a body for a long period of time, thus causing asthma, lung, liver cancer, bronchitis, Parkinson's and Alzheimer's diseases, heart disease, Crohn's disease, birth defects, and many other potential diseases. The extent of toxicity of nanomaterials depends upon the specific surface area, composition, size, and shape of the nanomaterials. The NPs with a higher-surface area have a higher toxicity to human and animal cells than NPs with a lower surface area. The NPs can diffuse into the body faster than the larger particles and reach very sensitive organs of the body and can easily react with the human body, stay inert, and/or interact with the human system. Surface charge can cause an electrostatic interaction between NPs and cells and enhance the penetration rate of these NPs in sensitive parts of the organs and other tissues in the body, thus causing the toxicity of nanomaterials. The possible risks and toxic health effects of these NPs associated with human exposure are not studied in detail so far. However, some risks and toxic health effects from these NPs to human body reported in literature have been summarized in this chapter [1–6].

15.1 Introduction to Nanoparticle Risks and Toxicity

The risks and toxicity by the NPs can arise during fabrication, transportation, handling, usage, waste disposal, and recycling. There are three main routes for NPs to enter the human body: inhalation into the pulmonary system, absorption through the dermal (or skin) system, and ingestion through the gastrointestinal system. This is due to constant contact of human lungs, skin, and the gastrointestinal tract with the environment. While the skin is generally an effective barrier to foreign substances, the lungs and gastrointestinal tract are more vulnerable. Injections and engineered implants are other possible routes of exposure. Due to their small size, NPs can translocate from these entry portals into the circulatory and lymphatic systems, and ultimately to body tissues and organs. Some NPs, depending on their composition and size, can produce irreversible damage to cells by oxidative stress and/or organelle injury. The NPs may also cause tissue inflammation and cell death. Figure 15.1 shows macrophage cell of a rat and its organelles compared to NPs of various sizes varying from 100 to 1000 nm. It can be clearly seen that the NPs enter cells and interact with various cell components, nucleus, mitochondria, etc. Human macrophages are up to two times larger than rat macrophages [6].

It is interesting to see that the computer processors, thin-film coatings, microchip electronics, and nanocomposites are not likely to cause harm and adverse health effects. In addition, there are several NPs which have positive health effects in medical diagnostics and treatments. For example, the magnetic NPs are used to improve MRI image quality and to kill the cancer cells; silver NPs have antibacterial and antifungal properties, and functionalized fullerene is used as antioxidants. However, there are some NPs which adversely affect the health depending on their size, concentration, aggregation, composition, crystallinity, surface functionalization, etc. For example, shorter MWNTs are much safer compared to the longer MWNTs that are highly toxic and can cause asbestos-like effects in the human body. In addition, when CNTs are incorporated with matrix materials to make nanocomposites or nanodevices, their toxicity and risk factors are reduced. Figure 15.2 illustrates the possible adverse health effects due to inhalation, ingestion, and contact of the human skin with NPs. The extent of toxicity of any NP to an organism depends on the individual's genetic complement [6].

15.1.1 Dermal Exposure

Skin with 1.5 m^2 surface area is the first defense barrier against the outside environment and typically the first place on the body that is exposed to the toxicity of nanomaterials. The skin is structured in three layers: epidermis, dermis, and subcutaneous. It is difficult for ionic compounds as well as water-soluble molecules to pass through this first barrier.

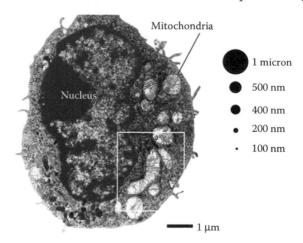

FIGURE 15.1
Rat macrophage cell size is compared to the NP's size. (From Buzea C., Pacheco I. I., Robbie K. 2007. *Biointerphases* 2(4): MR17–71. With kind permission.)

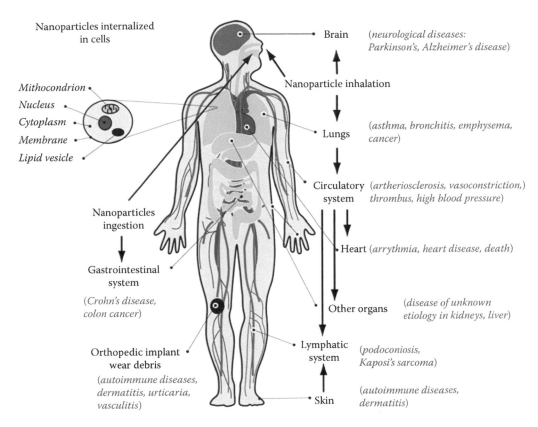

Nanoparticles internalized in cells

Mithocondrion
Nucleus
Cytoplasm
Membrane
Lipid vesicle

Nanoparticles ingestion

Nanoparticle inhalation

Brain (*neurological diseases: Parkinson's, Alzheimer's disease*)

Lungs (*asthma, bronchitis, emphysema, cancer*)

Circulatory system (*artheriosclerosis, vasoconstriction, thrombus, high blood pressure*)

Heart (*arrythmia, heart disease, death*)

Other organs (*disease of unknown etiology in kidneys, liver*)

Gastrointestinal system
(*Crohn's disease, colon cancer*)

Lymphatic system (*podoconiosis, Kaposi's sarcoma*)

Orthopedic implant wear debris
(*autoimmune diseases, dermatitis, urticaria, vasculitis*)

Skin (*autoimmune diseases, dermatitis*)

FIGURE 15.2
Illustration of human body with pathways of exposure to nanoparticles. (From Buzea C., Pacheco I. I., Robbie K. 2007. *Biointerphases* 2(4): MR17–71. With kind permission.)

Nanomaterials can penetrate the skin through the pores and go deeper under the skin. The surface of damaged skin is an ineffective particle barrier, and therefore, eczema, sunburn, acne, and shaving wounds may accelerate skin uptake of NPs. Then, once in the bloodstream, nanomaterials can be transported throughout the body into sensitive organs and tissues, including the brain, nervous system, kidneys, bone marrow, heart, liver, and spleen. Nanomaterials have been proven toxic to tissues and cell DNA, causing the higher oxidative stress, inflammation of cytokine production, and even cell death. Furthermore, cutaneous absorption could be another major exposure route for workers handling NPs, since these particles can end up in the circulatory system after passing through all the layers of the skin [2].

15.1.2 Ingestion Exposure

NP's entry by way of oral ingestion is similar to the dermal route, but the surface area of the digestive system is more than two orders of magnitude (200 m²) higher compared to the skin surface area (1.5 m²). Once in the stomach, the epithelium absorbs only small particles, which are taken in by the villi (finger-like structures) and covered by numerous microvilli. It has been found from various studies that the particles with a diameter of 14 nm permeated the system within 2 min but particles with a diameter of 415 nm took 30 min, and particles with a diameter of 1000 nm are practically unable to pass the barrier system. Ingestion can occur from unintentional hand-to-mouth transfer of unwanted materials and from contaminated food and drinks [1–3].

15.1.3 Inhalation Exposure

The respiratory system has about 2300 km of airways in and out of the body and 300 million alveoli for a total area of 140 m^2. The airways are strongly guarded by the immune system with a mucus layer that covers the pulmonary system. The wall between capillaries and alveolar is as thin as 0.5 μm. Nanomaterials suspended in the air (or airborne) pose the biggest threat to the body's respiratory system because they can break down into respirable-sized particles. The first line of contact of the NPs would be with the lungs, which provide an entryway into the body, and from there they travel to the organs, where they accumulate. NPs can deposit on the lower airways (pulmonary deposition mode), whereas microparticles are mainly found in the upper airways (trachea and nose) [1–3].

15.2 Risks And Toxicity from Metallic Nanoparticle

Silver NPs are one of the fastest growing products in the nanotechnology industry. Pure silver has the highest electrical and thermal conductivity of all metals and the lowest contact resistance. Ancient civilizations were aware of the bactericidal properties of silver. Metallic silver was used for surgical prosthesis and splints, fungicides, and coinage. Soluble silver compounds such as silver salts have been used in treating mental illness, epilepsy, nicotine addiction, gastroenteritis, and infectious diseases, including syphilis and gonorrhea. It has been used for jewelry, silverware, and photography in large amount since a long time. However, the rise of digital photography has resulted in an enormous reduction of silver emissions. The widest use of silver in medicine is in combination with sulfadiazine, as antibacterial agent, for the treatment of burns. Colloidal silver proteins, which were used to fight cold, are gaining popularity as a dietary supplement for treating certain diseases such as allergy prophylaxis.

Now, silver NPs are widely used as antibacterial and/or antifungal agents for various applications such as air sanitizer sprays, socks, slippers, pillows, wet wipes, face masks, detergent, shampoo, soap, toothpaste, coatings of refrigerators, air filters, vacuum cleaners, washing machines, food storage containers, cellular phones, and even in liquid condoms. In addition, coatings of NPs are widely used for stain- and wrinkle-free fabrics and clothes with built-in sunscreen and moisture management technology. Nanocoatings are also used in wetsuits for higher performance of athletes or self-cleaning surfaces. Textiles with 30-nm-embedded NPs help in preventing the pollen from entering gaps in the fabric.

Silver may be released into the air and water through natural processes such as weathering of rocks or by human activities. Rain may wash silver out of soil into the groundwater. Silver can be present in four different oxidation states: Ag^0, Ag^+, Ag^{2+}, and Ag^{3+}. The former two are the most abundant ones; the latter two are unstable in the aquatic environment. The free silver ion is Ag^+. In the environment, silver is found as a monovalent ion together with sulfide, bicarbonate, sulfate, or more complex forms with chlorides and sulfates adsorbed onto particulate matter in the aqueous phase. Metallic silver itself is insoluble in water, but metallic salts such as silver nitrate ($AgNO_3$) and silver chloride ($AgCl$) are readily soluble in water. Metallic silver appears to pose minimal risk to health, whereas soluble silver compounds are more readily absorbed and produce adverse effects. The acute toxicity of silver is dependent on its chemical form and the availability of free silver ions. For an aqueous concentration of only 1–5 mg/L, sensitive aquatic organisms and insects, trout, and flounder can be killed. Furthermore, accumulation of silver in species exposed to a slightly lower concentration of silver has led to adverse effects on growth. Silver is usually not available in concentrations high enough to pose a risk to human health and the environment. If nanosilver is in large concentration, then it could pose a threat to human health and environment. Nanosilver is also used in washing machines because of its antimicrobial activity; however, wastewater containing silver ions is drained into the ecosystem. This may affect the beneficial bacteria in soil, which are essential for the soil used for farming. The addition of silver NPs to socks kills the bacteria associated with foot odor. Unfortunately, the silver may easily leak into wastewater during two to four washings depending upon the manufacturers of the socks. Thus, these silver NPs disturb helpful bacteria used in wastewater treatment facilities or endanger aquatic organisms in lakes and streams. Such incidents have raised the concern of governments and the public. There are Ag-resistant bacteria in our mouths, which may be possibly related to the Ag in amalgam fillings. Ag-resistant bacteria have also been found in nature, food, and intestinal bacteria from different geographic locations and hospitals [1,2].

Gold NPs have been widely used in various fields due to their unique electronic, optical, thermal, chemical, inertness, biological, and catalytic properties. Owing to nontoxic and nonirritating properties, gold is sometimes used as a food decoration. In Asia, including India, gold has been used in the form of ayurvedic medicine for rejuvenation and revitalization during old age (as Swarna Bhasma) for a long time. Gold has been used as a drug (called Nervin) for the revitalization of people who suffer from nervous disorders. Biocompatible gold NPs labeled with specific targeting biomolecules/drugs are playing a key role in the diagnosis and therapy of several incurable diseases including cancer treatment. The oxidation state of gold varies from -1 to $+5$. Aurous ion (Au^{1+}) is the most common oxidation state, which reacts with soft ligands such as thioethers, thiolates, and tertiary phosphines. Some adverse effects such as skin irritation, dermatosis, stomatitis, contact allergy, and hypersensitivity reactions are associated with over exposure to gold and gold compounds. For example, gold sodium thiosulphate has been used as a dental gold alloy, but this alloy has received complaint of itching in the mouth, loss of taste, and burning sensation in the oral mucosa by a healthy 34 years old woman [3].

15.3 Risks And Toxicity from Oxide Nanoparticle

The body fluid is highly corrosive because of the presence of anions and cations, water, and plasma in the human body system; therefore, nanomaterials can be dissolved easily in the body depending on particle solubility and the surface area where diffusion and dissolution take place. Additionally, surface charges, pH, temperature, ion concentration, chemical structure, and NP roughness can increase the solubility and subsequently toxicity of nanomaterials. For example, more soluble ZnO NPs show stronger toxicity on mammalian cells than less soluble TiO_2 NPs with a lower-solubility rate. The dissolution products of ZnO are Zn^{2+} and $Zn(OH)_2$, which cause severe allergic reactions and illnesses. Many cosmetic and personal care products such as deodorants, soap, toothpaste, shampoo and hair conditioner, sunscreen and cosmetics cream, face powder, lipstick, blush, eye shadow, nail polish, perfume, and aftershave lotion consist of nanomaterials. The TiO_2 particles with size >100 nm are considered biologically inert in both humans and animals systems. Hence, TiO_2 NPs have been widely used in many products, such as white pigment, food colorant, sunscreens, and cosmetic creams. Moreover, the use of TiO_2 NPs is being explored as photocatalyst for sterilizing equipment of environmental microorganisms in the healthcare facility. The addition of NPs in the sunscreens and cosmetics has raised concerns over possible dermal penetration of material, leading to health hazard. For example, TiO_2 and ZnO NPs are used as effective ultraviolet (UV)-blocking agents in sunscreens, and nanoscale liposomes are currently used as delivery vehicles in skincare products. However, some studies have shown that the photogeneration of hydroxyl radicals by TiO_2 and ZnO NPs may lead to oxidative damage to the skin. The generation of free radical may be suppressed by using surface-modified NPs [4,5]. TiO_2 NPs are used in coatings for saltwater vessels to control antifouling and reduce corrosion. TiO_2 is also used in water treatment to kill bacteria and degrade volatile organic compounds in the presence of light. TiO_2 particles under UV light irradiation suppress tumor growth in cultured human bladder cancer cells via reactive oxygen species. The effect of TiO_2 NPs on ocean organisms has not been studied well. In aqueous media, TiO_2 leads to the production of reactive oxygen species, which can cause substantial damage to DNA. Cerium oxide (CeO_2) NPs are added to diesel as a catalyst to reduce toxic exhaust emission gases and particulate emission from diesel vehicles. The use of CeO2 catalyst reduces fuel consumption up to 10% and hence, particulate emissions. However, its toxic effects to human is not reported yet [2,6].

15.4 Challenges

Due to high-surface area to volume ratio, NPs particularly below 10 nm have tendency to agglomerate and are prone to grain growth during synthesis process and/or during sintering, respectively. For the fabrication of nanocomposites, there is challenge to get excellent distribution and dispersion of fibers or particles in the matrix. Compared to bacteria of micrometer size, removal of viruses with size of few tens of nanometers would need membranes having much smaller pores for an efficient virus removal. The smaller pore size, however, would significantly reduce the water flux. Moreover, the separation of photocatalytic powders from treated water prior to discharge is a time-consuming task and limits the application of TiO_2 photocatalysis in water treatment. TiO_2 has low efficiency under natural lighting. In view of above, substantial efforts are required for easier catalyst recovery and recycling. The properties of CNTs-filled polymer nanocomposites are affected by several factors that include the CNTs synthesis and purification process, the geometrical and structural properties of the CNTs, their orientation and distribution in the matrix, the dispersion process, and the fabrication process. Due to their high-aspect ratio and surface area, there are challenges to disperse CNTs uniformly in polymeric matrices. Additionally, the present cost of CNTs limits their large-scale applications. Nevertheless, the functionalization (using coupling agent or surfactant) processes lead to excellent distribution of CNTs or other fillers in the matrices, which may give better improvement in mechanical and thermal properties. However, other properties including electrical conductivity may be deteriorated. Moreover, excessive amount of coupling agent introduces pores in the final nanocomposites which may deteriorate electrical and mechanical properties of the nanocomposites.

References

1. Wijnhoven S. W. P., Peijnenburg W. J. G. M., Herberts C. A. 2009. Nano-silver a review of available data and knowledge gaps in human and environmental risk assessment. *Nanotoxicology* 3(2): 109–138.
2. Borma P. J. A., Berubec D. 2008. A tale of opportunities, uncertainties, and risks. *Nanotoday* 3: 56–59.
3. Panyala N. R., Peña-Méndez E. M., Havel J. 2009. Gold and nano-gold in medicine: Overview, toxicology and perspectives. *J. Appl. Biomed.* 7: 75–91.
4. Maynard A. D. 2006. Nanotechnology: Assessing the risks. *Nanotoday* 1: 22–33.
5. Gwinn M. R., Vallyathan V. 2006. Nanoparticles: Health effects—Pros and cons. *Environ. Health Perspect.* 114: 1818–1825.
6. Buzea C., Pacheco I. I., Robbie K. 2007. Nanomaterials and nanoparticles: Sources and toxicity. *Biointerphases* 2(4): MR17–71.

Index

A

AAMs, *see* Anodized alumina membranes
Abalone Nacre, 61–63
Abalone shell, 61, 62
Abrasion resistance of polymer–clay
    nanocomposites, 196
Absorption of electrons, 254
Acid washing, 114
Active matrix OLEDs (AMOLEDs), 87
Active sensing part, 146
Active smart textiles, 295
Aerogels, 100
Aerosol-assisted CVD, 91
AFM, *see* Atomic force microscopy
Age-hardened aluminum alloys, 70
Aging, 100
Ag NPs, *see* Silver nanoparticles
Air Mass (AM), 287
ALD, *see* Atomic layer deposition
Alkali metal-molten salt-assisted combustion
    method, 102
Alloys, 269, 281
    age-hardened aluminum, 70
    Mg–Ni, 77
    SPD-processed nanostructured TiNi, 79
    superalloys, 73
Alumina (Al_2O_3), 78
    alumina-supported catalysts, 304
    membranes, 138
    NPs, 78
    powder, 12
    suspension, 279
Alumina trihydrate (ATH), 190
Aluminum (Al), 21, 278
    atoms, 184
    cryomilling of aluminum powder, 75
    cylinder's inner wall, 111
    decrease in melting point, 21
    radiators, 277
    sacrificial masks, 300
AM, *see* Air Mass
Amino-terminated hyperbranched polymer
    (HBP-NH$_2$), 290
AMOLEDs, *see* Active matrix OLEDs
Amorphous silver NPs, 92
Anapore filters, 297
Ancient history, nanotechnology in, 66

age-hardened aluminum alloys, 70
    Damascus Sword, 66–68
    Lycurgus cup shining, 68
    Maya Blue Paint, 70
    particle size of Ag *vs.* color, 70
    stained glass, 68–70
Anodic Tafel slope (β_a), 268
Anodized alumina membranes (AAMs), 138
Antibacterial coating, 289–290
Antimicrobial nanocoatings, 295
Antiviral property, 289
Aragonite platelets, 62
Arc discharge
    arc-discharge-grown MWNTs, 108
    method, 108–110
    reactor, 108
Armchair SWNTs, 117
Artifact-free bulk nanocrystalline
    Cu sample, 37
Artificial atoms, 5, 243
Artificial implants, 129
Aspalathus linearis' natural extract, 98
Aspect ratio, 192
ATH, *see* Alumina trihydrate
Atomic force microscopy (AFM), 85, 233,
    260–262
Atomic layer deposition (ALD), 102
    cross-sectional SEM image of
        Al_2O_3 film, 104
    ZnS film formation using, 103
Attrition, 73
Aurous ion (Au^{1+}), 313
Autocorrelation function,
    approximation of, 249
Automotive applications, 277–278
Automotive sector, 300
    applications, 302–303
    automotive industry in Europe, 300–301
    diesel particulate filter, 302
    fuel cell, 301–302
    solar energy, 301
    vehicle radiator, 302
Azadirachta indica extract, 96

B

Backscattered electrons (BSEs), 250, 251
Ball-to-powder ratio (BPR), 75, 76

Ball milling, 27, 74, 77, 78
Bandgap, 44, 98, 142, 243, 286
Barbule, 55, 56
Barium titanate (BaTiO₃), 32, 184
Beetles, 59
Benzyl benzoate, 152
BET
    method, 233, 245–246
    plot, 246
β-sheet nanocrystals, 59
B–H loop, 172, 173
Bimetallic NP synthesis, 98
BIMSM, *see* Brominated isobutylene-co-para-
        methyl styrene elastomer
Biodegradable polymeric nanofibers, 296
Bioinspired nanostructures, 53
Biological
    method, 95
    synthesis, 94–95
Biomedical applications, 228–229, 278–279
Biomedical field
    cancer treatment, 296
    cosmetics, 296–297
    drug delivery, 296
    medical prostheses, 296
    wound dressing, 296
Blocking temperature, 174
Biomedical-grade iron NPs, 77
Blueshift, 44–45, 242
BNNFs, *see* h-BN nanoflakes
Body fluid, 313
Bohr radius, 1, 43
Bonampak archeological site, 70
Bone, 64, 65
Boron-rich phases, 270
Bottom-up approaches, 2, 3, 73, 88, 151; *see also*
        Top-down approach(es)
    ALD, 102–104
    colloidal or wet chemical route, 91–93
    combustion method, 101–102
    CVD, 90–91
    green chemistry route, 94–99
    MBE, 89–90
    PVD, 88
    reverse micelle method, 93–94
    sol-gel method, 99–101
    stabilization of Au nanoparticles, 92
BPR, *see* Ball-to-powder ratio
Bragg's angles, 235
Brakes, 278
Brillouin zone, 119
Brominated isobutylene-co-para-methyl
        styrene elastomer (BIMSM), 303

BSEs, *see* Backscattered electrons
Bulk nanomaterials, *see* 3-D nanomaterials

C

Caffeine, 94
Calcium carbonate (CaCO₃), 61
Camphor sulfonic acid (CSA), 226
Cancellous bone, 64
Cancer
    therapeutics, 278
    treatment, 296
Capillary forces, 61
Capsicum annuum (*C. annuum*), 95
Capture silk, 57
Carbide, 276
Carbonaceous impurities, 113
Carbon monoxide (CO), 48, 111, 303
Carbon nanofibers (CNFs), 5; *see also*
        Nanofibers
Carbon nanoparticles (CNPs), 113
Carbon nanotubes (CNTs), 5, 34, 67, 107, 151,
        184, 276, 295, 310; *see also* Nanotubes
    applications, 124–130
    arc discharge method, 108–110
    artificial implants, 129
    characterization techniques to analyzing
        purity, 112
    CNTs-filled polymer nanocomposites, 198
    coatings, 130
    current–voltage characteristics, 128
    CVD, 110–112
    electrical brush contacts, 128–129
    electrical properties, 119–120
    energy storage devices, 125–126
    field emission emitters, 126–128
    hydrogen storage, 125
    laser ablation, 110
    mechanical properties, 120, 121
    microelectronics, 128
    nanocomposites, 124–125
    plot of modulus *vs.* SWNTs rope diameter, 121
    powder, 298
    purification, 113–116
    Raman spectrum of CVD–MWNT, 118
    RBM mode, 119
    sensors and probes, 129
    sponge, 129–130
    structures, 116–119
    SWNTs, 117, 118
    synthesis, 107–112
    temperature-dependent thermal
        conductivity, 122

TEP, 123
thermal properties, 121–123
water filters, 129–130
Catalysts, 137, 303–305
Cathodic Tafel slope, 268
Cation exchange capacity (CEC), 185
CBE, *see* Chemical beam epitaxy
CCD, *see* Charge-coupled device
CdSe-QDs, 286
CEC, *see* Cation exchange capacity
Cementite (Fe_3C), 66
Centrifugation, 115
Cephalopods, 61
Cerium oxide (CeO_2), 314
Chameleon (*Chamaeleonidae*), 61
Characterization techniques to analyzing
    purity, 112
Charge-coupled device (CCD), 241, 253
Charge exchange capacity, 184
Chemical beam epitaxy (CBE), 135
Chemical capping, 93
Chemical exfoliation, 152
Chemical oxidation method, 113–114
Chemical sensitivity, 46
    ethanol sensing characteristics, 48
    instantaneous alloying, 47
    effect of particle size, 49
    semiconductor gas sensors, 48
Chemical vapor deposition (CVD), 90–91, 107,
    110–112, 135, 151, 153–154
Chemisorption, 283
Chenopodium album (*C. album*), 97
Chinese porcelain, 69
Chiral angle, 116
Chiral SWNTs, 117
Chiral vector, 116
Chlorauric acid ($HAuCl_4$), 92
Chlorophyll, *see* Natural solar cell
Cinnamomum camphora (*C. camphora*), 95
Circumferential vector, 116
CNFs, *see* Carbon nanofibers
CNPs, *see* Carbon nanoparticles
CNTs, *see* Carbon nanotubes
Coarse-grained polycrystalline Cu sample, 37
Coatings, 130
Cobalt (Co), 174–175, 269
Coefficient of thermal expansion (CTE), 9,
    27–28, 124, 141, 159, 183, 190, 223
Coercivity (H_c), 172
Cold combustion, 294
Colloidal
    chemical route, 91–93
    dispersion, 73

gold, 2, 278
silver proteins, 312
systems, 69
Combustion method, 101–102
Combustion synthesis (CS), 101
Compact bone, 64
Constant current mode, 260
Constant height mode, 260
Construction industry, 292–293
Contact printing, 83
Conventional
    materials, 47
    PSAs, 61
    water treatment methods, 129
Coolants, 279–281
Coordination number, 24–25
Copper (Cu), 98
    nanoparticles, 69
    NPs, 293
Copper oxide (CuO), 98
Corrosion, 267
    behavior of nanomaterials
    characteristics, 268
    corrosion resistance of nanocomposite
        coating, 269–273
    corrosion resistance of nanocrystalline
        metals, 268–269
    of materials, 267
Corrosion current (i_{corr}), 268
Corrosion potential (E_{corr}), 268
Corrosion rate (CR), 267
Cosmetics, 229, 296–297
CR, *see* Corrosion rate
Creep, 39–41
Cryogenic-TEM (Cryo-TEM), 255
Cryomilling, 75
Cryosurgery, 279
Crystal anisotropy, 41
CS, *see* Combustion synthesis
CSA, *see* Camphor sulfonic acid
CTE, *see* Coefficient of thermal expansion
Cubical particles
    division of parent cube with edge length, 12
    increased numbers of cubes and surface
        atoms, 15
    size effect on surface area, 11–14
    size effect on surface atoms, 14–15
    variation in specific surface area of
        spherical alumina particles, 12
Cubic $BaTiO_3$ (c-$BaTiO_3$)/PC
    nanocomposites, 209
CuO, *see* Copper oxide
Cuprous oxide nanoparticles, 69

Curie temperature, 25–27, 42
Cuttlefish, 56–57
CVD, *see* Chemical vapor deposition
Cyclic forging, 68

D

Damascene steel, 66
Damascus blades, 66
Damascus Sword, 66–68
Dampers, 279
Damping property, 124
Dark-field image, 254
D-band, *see* Disorder band
DCM, *see* Dichloromethane
Debye temperatures, 42
Deep UV lithography (DUVL), 84
Defect band, 118
Deflection mode, 261
Dendrimers, 299
Dendritic polymers, 299
Dentin, 65–66
Dermal exposure, 310–311
Detection of nucleic acids (DNA), 146
Dextran, 177
Diamond, 38, 300
Dichloromethane (DCM), 208
Dichlorosilane (SiH_2Cl_2), 91
Dielectric constant
    of $BaTiO_3$ ceramic, 208
    insulating material, 49–50
Dielectric materials, 199
Diesel electric generators, coolant for, 279
Diesel particulate filter, 302
Differential scanning calorimetry
        (DSC), 188, 263
Diffraction
    contrast, 254
    mode, 253
Dimensional stability, *see* Coefficient of
        thermal expansion (CTE)
N,N-Dimethylacetamide (DMA), 152
N,N-Dimethylformamide (DMF), 152
2,4-Dinitrotulene (DNT), 229
Dip-pen nanolithography (DPNL), 81, 85–86
Direct method, 233
Direct NIL process, 86
Dislocation motion within grains, 39
Disorder band (D-band), 112, 118
Dispersion, 16, 124, 314
    chemical, 102
    of clay, 185
    CNT, 198

colloidal, 73
of exfoliated clay sheets, 192
of gold nanoparticles, 2
GO precursor, 153, 165
of graphite powder, 152
uniform, 199, 200, 212
DLS, *see* Dynamic light scattering
DMA, *see* *N,N*-Dimethylacetamide
DMF, *see* *N,N*-Dimethylformamide
DNA, *see* Detection of nucleic acids
DNT, *see* 2,4-Dinitrotulene
Double walled carbon nanotubes
        (DWNTs), 119
DPNL, *see* Dip-pen nanolithography
Drug delivery, 179–180, 296
Dry
    adhesion mechanism in gecko lizards, 60
    grinding, 75
    ice reduction method, 154
DSC, *see* Differential scanning calorimetry
DSSCs, *see* Dye-sensitized solar cells
DUVL, *see* Deep UV lithography
DWNTs, *see* Double walled carbon nanotubes
Dye-sensitized solar cells (DSSCs), 66, 285
Dynamic light scattering (DLS), 233, 247
Dynamic seal, 281

E

EBL, *see* Electron-beam lithography
EBSD, *see* Electron backscatter diffraction
ECAP method, *see* Equal channel angular
        pressing method
EDS, *see* Energy-dispersive spectroscopy
EDX, *see* Energy-dispersive x-ray
Effective heat of combustion (EHC), 188
EG, *see* Expandable graphite
EHC, *see* Effective heat of combustion
EIS, *see* Electrochemical impedance
        spectroscopy
Elastic modulus, 33–34
Electrical brush contacts, 128–129
Electrical conductivity of materials, 29
Electrochemical deposition, 138–140
Electrochemical exfoliation method, 152
Electrochemical impedance spectroscopy
        (EIS), 268
Electroless deposition, 140
Electrolytic thinning, 258
Electromagnetic interference (EMI),
        124, 167, 197
Electron-beam lithography (EBL), 81, 84–85
Electron backscatter diffraction (EBSD), 252

Electronic(s), 145
  conductivity, 31
  excitation of semiconductor crystal, 43
Electron microscopy, 250; *see also* Scanning
      probe microscopy
  scanning electron microscopy, 251–253
  transmission electron microscopy, 253–258
Electropolishing, 258
Electrorheological fluids (ER fluids), 279
Electrospinning, 217
Electrospun
  nanofibers, 299
  polymer nanofibers, 228
Electrostatic discharge (ESD), 167, 199
EMI, *see* Electromagnetic interference
Energy-dispersive spectroscopy (EDS), 112, 233
Energy-dispersive x-ray (EDX), 252
Energy storage devices, 125–126
Engineered nanomaterials, 7
Enthalpy of fusion, 21–22
Environmental friendly green processes, 95
Environmental SEM (ESEM), 252
Equal channel angular pressing method
      (ECAP method), 37, 78, 80
Equiaxed nanoparticles or nanocrystals, 6
ER fluids, *see* Electrorheological fluids
Escherichia coli (*E. coli*), 63, 290
ESD, *see* Electrostatic discharge
ESEM, *see* Environmental SEM
Etching, 73
Ethylene-vinyl acetate (EVA), 190, 210
Euphorbia jatropha (*E. jatropha*), 99
Euplectella, 59–60
EUVL, *see* Extreme UV lithography
EVA, *see* Ethylene-vinyl acetate
Exfoliated nanocomposite, 185
Expandable graphite (EG), 155
Extreme UV lithography (EUVL), 84

F

Fabrication of polymer/clay
      nanocomposites, 185
  melt-mixing method, 186–188
  in situ polymerization method, 185
  solution method, 185–186
Face centered cubic symmetry (FCC
      symmetry), 98
Famille rose, see Chinese porcelain
Fatigue, 39–41
FCC symmetry, *see* Face centered cubic
      symmetry
Fe(CO)$_5$, 111, 112

Fe$_2$O$_3$ NPs, 277, 292
Fe$_3$O$_4$ NPs, *see* Magnetite NPs
Fermi surface of graphene, 160
Ferrofluids, 277, 278
FETs, *see* Field effect transistors
Few layer graphene (FLG), 151, 157
FGO, *see* Functionalized graphite oxides
FGS, *see* Functionalized graphene sheets
FIBL, *see* Focused-ion beam lithography
FIB milling, *see* Focused ion beam milling
Field effect transistors (FETs), 107, 128, 145
Field emission emitters, 126–128
Field enhancement factor, 127
Fingerprint of molecule, 244
Flammability resistance, 189
FLG, *see* Few layer graphene
"Floating-and-cleaning" capability, 130
Fluorescence spectroscopy, 142
f-MMT, *see* Functionalized MMT
Focused-ion beam lithography (FIBL), 81
Focused ion beam milling (FIB milling),
      257, 258
Fourier transform infrared spectroscopy
      (FTIR), 233, 244
Fuel cell, 301–302
Fullerene (C$_{60}$), 5, 114
Full width at half maximum (FWHM), 236
Functionalized graphene sheets (FGS), 162
Functionalized graphite oxides (FGO), 155, 165
Functionalized MMT (f-MMT), 189
Functional sensors, 229
FWHM, *see* Full width at half maximum

G

Gadolinium (Gd), 179
γ-butyrolactone (GBL), 152
Gas-phase oxidation, 114
Gas adsorption method, 245
Gas barrier properties, 166
Gas sensor, 47
GBL, *see* γ-butyrolactone
GBs, *see* Grain boundaries
Gecko, 60–61
General Motors (GM), 302
Geothermal power extraction, coolant to, 280
Gerris remigis, see Water strider
Giant magnetoresistance (GMR), 291–292
Glass sponges, 59–60
GM, *see* General Motors
GMR, *see* Giant magnetoresistance
GNS, *see* Graphene nanosheets
GO, *see* Graphene oxide

Gold NPs, 92, 96, 313
Gold ruby glass, 2
GP zones, 70
Grain boundaries (GBs), 4, 7, 36
    density of states for charge carriers, 9
    size effect on, 7–9
    two-dimensional model of nanocrystalline
        material, 8
    volume fraction, 8
Graphene, 151, 299
    applications, 167
    chemical exfoliation, 152
    chemical vapor deposition, 153–154
    dry ice reduction method, 154
    electrical properties, 160–161
    mechanical exfoliation, 151–152
    mechanical properties, 156–158
    properties of graphene/polymer
        nanocomposites, 155, 161–166
    reduction of graphene oxide, 152–153
    structural properties, 155–156
    synthesis, 151
    thermal properties, 159–160
Graphene/CNT and graphene/polymer
        nanocomposites, 167
Graphene composites
    electrical properties, 165–166
    mechanical properties, 161–162
    thermal properties, 163–165
Graphene nanosheets (GNS), 299
Graphene oxide (GO), 152, 240, 299
    reduction, 152–153
    thermal conductivities, 159
Graphene/polymer nanocomposites, 167
    electrical properties of graphene
        composites, 165–166
    gas barrier properties, 166
    mechanical properties of graphene
        composites, 161–162
    properties, 161
    thermal properties of graphene composites,
        163–165
Green chemistry route, 94
    biological synthesis, 94–95
    extracts of plants, metal salts, synthesized
        nanoparticle, and sizes, 96
    factors affecting size and morphology of
        NPs, 99
    metallic NPs synthesis, 96–98
    noble metals, 94
    NPs synthesis, 96
    oxide NPs synthesis, 98–99
Green synthesis, 94

Greigite (Fe_3S_4), 63
Growth chamber, 89

H

Hall–Petch equation, 34, 35
Hardness, 34–36
$HAuCl_4$, *see* Chlorauric acid
h-BN, *see* Hexagonal boron nitride
h-BN nanoflakes (BNNFs), 205
$HBP-NH_2$, *see* Amino-terminated
        hyperbranched polymer
HCs, *see* Hydrocarbons
HDT, *see* Heat deflection temperature
Heat, 287–288
    capacity, 25
Heat deflection temperature (HDT), 189
Heat release rate (HRR), 164, 184, 188
Heavy metals removal, 180
Helium-saturated powder, 245
Hexagonal boron nitride (h-BN), 184, 203
HF acid, *see* Hydrofluoric acid
HFC, *see* Hydrogen fuel cell
Hibiscus rosa sinensis (*H. rosa sinensis*), 95
High-energy ball milling, 74–77
Highly ordered pyrolytic graphite (HOPG), 151
High-performance loudspeakers, coolant in, 280
High-pressure carbon monoxide (HiPCO),
        107, 111
High-pressure torsion (HPT), 78
High-resolution transmission electron
        microscopy (HRTEM), 66, 112, 255
High speeds of vials, 77
High-temperature
    annealing, 115
    oxidation in air, 113–114
HiPCO, *see* High-pressure carbon monoxide
HIV, *see* Human immunodeficiency virus
HNO_3, 114
HOPG, *see* Highly ordered pyrolytic graphite
HPT, *see* High-pressure torsion
HRR, *see* Heat release rate
HRTEM, *see* High-resolution transmission
        electron microscopy
H_2S, 103, 303
Human immunodeficiency virus (HIV), 6
Hummers and Offeman method, 152
Hydrocarbons (HCs), 111, 303
Hydrofluoric acid (HF acid), 137
Hydrogen
    energy, 301
    storage, 125, 281–284
Hydrogen fuel cell (HFC), 302

Hydrophilic coatings, 293–294
Hydrophobic coatings, 293
Hyperthermia, 178–179
Hysteresis *M–H* loop, 171–173

I

IC, *see* Integrated circuit
ICPs, *see* Inherently conducting polymers
ICSD, *see* Inorganic Crystal Structure Database
IGC, *see* Inert gas condensation
Image mode, 253
Imogolite, 5–6
Impact toughness, 36
Incident-photon-to-charge carrier efficiency
        (IPCE), 287
Indirect method, 233
Indium phosphide (InP), 237
Inefficient heat dissipation, 147
Inert gas condensation (IGC), 37, 88, 89
Ingestion exposure, 311–312
Inhalation exposure, 312
Inherently conducting polymers (ICPs), 295
Inorganic Crystal Structure Database
        (ICSD), 234
InP, *see* Indium phosphide
In situ consolidated nanocrystalline Cu
        sample, 37
In situ polymerization method, 185
Instantaneous alloying, 47
Integrated circuit (IC), 82, 147
Intelligent textiles, 295
Intercalated nanocomposite, 185
Intertube scattering, 122
Ion milling, 257
IPCE, *see* Incident-photon-to-charge carrier
        efficiency
Iron oxide magnetic NPs, 278

K

Kubo gap, 44

L

Laser ablation, 110
Laser-assisted CVD, 91
Lattice constant, 31–32
Length scale and calculations, 11
    size effect on surface area of cubical
        particles, 11–14
    size effect on surface atoms of cubical
        particles, 14–15

size effect on surface atoms of spherical
        particles, 15–19
Light-sensitive molecules, 66
Light scattering method, 247–250
Limiting oxygen index (LOI), 188
Linear-response conductance of bundle, 119
Liquid nitrogen temperature (LNT), 37
Liquid-phase oxidation method, 114
Lithography, 73, 81
    DPNL, 85–86
    EBL, 84–85
    NIL, 86–87
    photolithography, 82–84
    scanning probe lithography, 85
    soft lithography, 87
    X-ray lithography, 84
LM, *see* Longitudinal mode
LNT, *see* Liquid nitrogen temperature
Load chamber, 89
Lodestone, 171
LOI, *see* Limiting oxygen index
Longitudinal mode (LM), 45
Lotus effect, 54
Lotus leaf, 53–54
Low-pressure CVD, 91
Low-voltage scanning electron microscopy
        (LVSEM), 253
Lubrication, 280
LVSEM, *see* Low-voltage scanning electron
        microscopy
Lycurgus cup, 2
    Roman Lycurgus cup, 68
    shining, 68

M

MA, *see* Maleic anhydride; Mechanical
        alloying
Machining, coolant for, 279
Magnesium (Mg), 154
Magnetic
    cell separation, 179
    information storage devices, 147
    nanocomposites, 269
    organic/inorganic hybrids, 98
    properties, 41–43
    refrigeration, 180
Magnetic nanomaterials, 171
    hysteresis *M–H* loop, 171–173
    effect of particle size on *M–H* loop, 173–177
Magnetic nanoparticles, 171, 310
    applications, 178
    drug delivery, 179–180

Magnetic nanoparticles (*Continued*)
    hyperthermia, 178–179
    magnetic cell separation, 179
    magnetic refrigeration, 180
    medical diagnostics, 179
    oil removal, 180
    removal of heavy metals, 180
    surface modification, 177–178
Magnetic resonance imaging (MRI), 178, 278
Magnetite NPs (Fe_3O_4 NPs), 63, 77, 171
Magnetorheological fluids (MR fluids), 279
Magnetospirillum magnetotacticum (*M. magnetotacticum*), 63
Magnetotactic bacteria (MTB), 63–64
Maleic anhydride (MA), 185
MA-modified PP oligomers (PP-MA), 187
Mass loss rate (MLR), 188
Maya Blue Paint, 70
MBE, *see* Molecular beam epitaxy
MDC, *see* MOFs (zinc)-derived carbon
Mechanical alloying (MA), 73
    applications, 77–78
    high-energy ball milling, 74–77
    effect of milling time, 76
    nonequilibrium phases, 73
    pros and cons, 78
Mechanical exfoliation, 151–152
Mechanical milling (MM), 74
Mechanical polishing, 257
Medical diagnostics, 179
Medical prostheses, 228, 296
Melt-mixing method, 186–188
Melting point, 21, 140–141
    depression or decrease in, 24–25
    and enthalpy of fusion of Al nanoparticles
        vs. reciprocal of grain size, 22
    as function of Al cluster size, 22
    normalized melting point *vs.* particle size
        for Au nanoparticles, 23
    size-dependent melting behavior, 23
Mesolayers, 63
Mesoporous coatings, 178
Metal alkoxides, 99
Metallic corrosion, 267
Metallic nanoparticles, 5
    risks and toxicity from, 312–313
    synthesis, 96–98
Metallic NWs, 135, 143
Metallic silver, 312
Metalorganic CVD (MOCVD), 91
Metal organic frameworks (MOFs), 282, 284
Metal oxides, 88, 93, 276, 298
N-Methyl-2-pyrrolidone (NMP), 152

Mg–Ni alloys, 77
M–H loop, 173
    effect of particle size on, 173–177
MIC, *see* Microbially induced galvanic corrosion
Microbially induced galvanic corrosion (MIC), 271
Microelectronics, 128
Microporous coatings, 178
Microscale TGA (μ-TGA), 263
Microwave-assisted purification, 115
Milling, 73
MLR, *see* Mass loss rate
MM, *see* Mechanical milling
MMT clay, *see* Montmorillonite clay
MOCVD, *see* Metalorganic CVD
Modified CVD methods, 91
Modified RoM (M-RoM), 212
MOFs, *see* Metal organic frameworks
MOFs (zinc)-derived carbon (MDC), 283
Molecular beam epitaxy (MBE), 89–90
Monosized metallic NPs, 93
Montmorillonite clay (MMT clay), 6, 184–185
Moore's law, 3, 4
Morpho butterfly, 55
Morpho didius (*M. didius*), 55
Mother-of-pearl, 61
Moth eyes, antireflective nanostructures of, 64
Mott–Vanie exciton, 43
Movement of dislocations, 34
MR fluids, *see* Magnetorheological fluids
MRI, *see* Magnetic resonance imaging
M-RoM, *see* Modified RoM
MTB, *see* Magnetotactic bacteria
Multidomains, 173
Multiwalled carbon nanotubes (MWNTs), 66, 107, 196, 225
    MWNT-based electrical brush contacts, 128–129
MWNTs, *see* Multiwalled carbon nanotubes

N

Nanobarium titanate ($BaTiO_3$), 184
Nanobeam electron diffraction (NBD), 256
Nano-Care, 294
Nanoceramic-based coatings, 301
Nanocomposites, 124–125, 147–148, 183
    corrosion resistance nanocomposite coating of, 269–273
    fabrication, 314
    fibers for structural applications, 226–228
Nano cryosurgery, 279

Nanocrystalline
  corrosion resistance of nanocrystalline
      metals, 268–269
  materials, 7, 8
  Ni, 269
Nanocrystalline Cu (nc-Cu), 37
Nano-Cu specimens, 25
Nanodiamond, 300
Nano-Dry, 294–295
Nanofibers
  electrical properties, 225–226
  magnetic properties, 222–223
  mechanical properties, 219–222
  parameter affecting diameter of,
      218–219
  properties, 219
  thermal properties, 223–225
Nanofilters, 228
Nanofluids, 275
  applications, 277
  automotive applications, 277–278
  biomedical applications, 278–279
  coolants, 279–281
  dynamic seal, 281
  stabilization, 275–276
  synthesis, 276–277
  two-phase systems, 275
Nanogenerators, 228
Nanographene platelets (NGPs), 303
Nanoimprint lithography (NIL), 82, 86–87
Nanolithography, 73
Nanomanipulation, 73
Nanomaterials, 1, 2, 309
  0-D, 5
  1-D, 5–6
  2-D, 6
  3-D, 6–7
  antibacterial coating, 289–290
  applications, 275
  automotive sector, 300–303
  biomedical field, 296–297
  catalysts, 303–305
  characterization, 233
  classification, 4–7
  construction industry, 292–293
  electron microscopy, 250–258
  fabrication of nanocomposites, 314
  GMR, 291–292
  hydrogen storage, 281–284
  light scattering method, 247–250
  logarithmical length scale showing size, 7
  nanodiamond, 300
  nanofluids, 275–281

nanoparticle risks and toxicity, 309–312
  nanopore filters, 297
  nanotextiles, 294–295
  optical spectroscopy, 238–245
  risks and toxicity from metallic
      nanoparticle, 312–313
  risks and toxicity from oxide nanoparticle,
      313–314
  risks, toxicity, and challenges, 309
  scanning probe microscopy, 258–262
  self-cleaning coatings, 293–294
  SET, 292
  solar energy, 284–289
  surface area analysis, 245–246
  thermal analyzer, 263–264
  water treatment, 297–300
  x-ray diffraction, 233–238
  XPS, 262–263
  zeta potential, 264
Nanomedicine, 296
Nanometer (nm), 1
Nanometer-sized nanoparticles, 275
Nano-organism, 6
Nanoparticles (NPs), 7, 13, 73, 275, 309
  dermal exposure, 310–311
  factors affecting size and morphology, 99
  ingestion exposure, 311–312
  inhalation exposure, 312
  risks and toxicity, 309
  synthesis, 96
Nano-Pd specimens, 25
Nano-Pel, 294
Nanopore filters, 297
Nano-Pt-CeO$_x$ particles, 302
Nanosilver, 313
Nanosized water droplets, 93
Nanosize effect, 300
Nanostructured magnetic materials, 42
Nanostructured materials, 1, 282
Nanostructures, 1, 285
Nanotechnology, 2, 301
  miniaturization of electronic devices, 3
  in past, 2–4
  in twenty-first century, 4
Nanotextiles
  intelligent textiles, 295
  textile fabrics, 294–295
Nano-touch, 295
Nanotubes, 6, 125; *see also* Carbon nanotubes
      (CNTs)
  actuators, 126
  field-emitting surfaces, 126
Nanotweezers, 129

Nanowires (NWs), 6, 135, 296
    applications, 145
    electrical properties, 144–145
    electronics, 145
    magnetic information storage devices, 147
    nanocomposites, 147–148
    optical properties, 142–144
    properties, 140
    sensors, 146–147
    solar cell, 147
    synthesis method for, 135
    TEGs, 145–146
    template method, 138–140
    thermal properties, 140–142
    VLS method, 135–137
Natural nanocomposite, 64
Natural solar cell, 66, 286
Nature, nanotechnology of, 53
    Abalone Nacre, 61–63
    antireflective nanostructures of
        moth eyes, 64
    beetles, 59
    biological features of DNA, cells, and
        membranes, 65
    chameleon, 61
    cuttlefish, 56–57
    dentin, 65–66
    gecko, 60–61
    glass sponges, 59–60
    lotus leaf, 53–54
    Morpho butterfly, 55
    MTB, 63–64
    natural nanocomposite, 64
    natural solar cell, 66
    peacock feather, 55–56
    spider silks, 57–59
    water strider, 57
NBC warfare, *see* Nuclear, biological, and
        chemical warfare
NBD, *see* Nanobeam electron diffraction
nc-Cu, *see* Nanocrystalline Cu
Nd-rich phases, *see* Neodymium-rich phases
Neodymium-rich phases (Nd-rich phases), 270
Nervin, 313
NGPs, *see* Nanographene platelets
Nickel (Ni), 255
NIL, *see* Nanoimprint lithography
Nitric acid treatment, 202
Nitride, 276
Nitrogen oxides, 303
NMP, *see* N-Methyl-2-pyrrolidone
NMR, *see* Nuclear magnetic resonance
Noble metals, 94

Nonequilibrium phases, 73
Nonmetallic nanotubes-based sensors, 129
Nontoxic and skin-friendly antibacterial
        agent, 290
Nonwoven nanofibrous membranes, 296
NPs, *see* Nanoparticles
Nuclear, biological, and chemical warfare
        (NBC warfare), 229
Nuclear magnetic resonance (NMR), 263, 278
Nuclepore filters, 297
NVP, *see* N-Vinyl-2-pyrrolidone
NWs, *see* Nanowires

O

ODS, *see* Oxide dispersion strengthened
Oil removal, 180
OLEDs, *see* Organic light emitting diodes
One-dimensional anisotropic structures (1-D
        anisotropic structures), 135
1-D nanomaterials, 5–6
Optical properties, 43–45
Optical spectroscopy, 238
    FTIR, 244–245
    PL spectroscopy, 243–244
    Raman spectroscopy, 239–240
    UV–Vis spectroscopy, 240–243
Organic-phosphate-FGO, 155
Organic light emitting diodes (OLEDs), 87, 128
Oxidation reaction, 267
Oxide-ion conductors, 29
Oxide dispersion strengthened (ODS), 73
    materials, 77
    superalloys, 73
Oxide nanoparticles (Oxide NPs), 2
    risks and toxicity from, 313–314
    synthesis, 98–99, 101

P

PACVD, *see* Plasma-assisted CVD
PAEK, *see* Poly(aryletherketone)
Palladium (Pd), 98, 303
    NPs, 96
Palygorskite crystals, 70
PAN, *see* Polyacrylonitrile
PANI, *see* Polyaniline
Particle size effect on properties of
        nanomaterials, 21, 31
Particle size effect on properties of
        nanomaterials, 21, 31
    Arrhenius plots, 30
    chemical sensitivity, 46–49

dielectric constant, 49–50
electrical properties, 29–31
electrical conductivity of polycrystalline
    cerium oxide, 30
evolution of bandgap and density
    of states, 44
fundamental properties of metals, 42
lattice constant, 31–32
magnetic properties, 41–43
mechanical properties, 33–41
optical absorption spectra, 46
optical properties, 43–45
phase transformation, 33
resistivities of nanocrystalline Ni compared
    to normal crystalline Ni, 29
specific saturation magnetization and
    coercivity, 41
SPR, 45
TEM images of $BaTiO_3$ nanoparticles, 50
temperature of monoclinic-tetragonal
    transformation, 33
thermal properties, 21–28
variation of particle sizes, 50
wear resistance, 45–46
Parts per million (ppm), 2
Passive smart textiles, 295
PBA, *see* 1-Pyrenebutyric acid
PBI, *see* Polybenzimidazole
PC, *see* Polycarbonate
PCA, *see* Process control agent
PCBs, *see* Printed circuit boards
PCL, *see* Poly(ε-caprolactone)
PCMs, *see* Phase change materials
PCS, *see* Photon correlation spectroscopy
PDMS, *see* Polydimethylsiloxane
PE, *see* Polyethylene
Peacock feather, 55–56
Peak, 245
PECVD, *see* Plasma-enhanced CVD
PEG, *see* Poly(ethylene glycol)s
PEM fuel cell, *see* Proton exchange membrane
    fuel cell
PEN, *see* Poly(ethylene naphthalate)
PEO, *see* Poly(ethylene oxide)
Percolation threshold, 197, 198
Permanent magnets, 173
Permeability of polymer–clay nanocomposites,
    191–193
Permittivity, *see* Dielectric constant
PET, *see* Polyethylene terephthalate
PGC, *see* Phenyl glycidol chlorophosphate
Phase change materials (PCMs), 295
Phase transformation, 33

Phenyl glycidol chlorophosphate (PGC), 155
Phonons, 288
Phosphorus (P), 269
Photocatalysis, 293
Photoelectrochemical cells, 147, 285–287
Photoelectrons, 262
Photolithography, 82–84
Photoluminescence peaks (PL peaks), 239
Photoluminescence spectroscopy (PL
    spectroscopy), 142, 233, 243–244
Photomultiplier tube (PMT), 241
Photon correlation spectroscopy (PCS), 247
Photovoltaic cell, *see* Solar cell
Photovoltaic cells, *see* Photoelectrochemical
    cells
Photovoltaics, 285
pH role, 267
Physical method, 114–116
Physical vapor deposition (PVD), 88
Physisorption, 283
π orbital, 117
Planetary ball mill, 74, 75
Plant-based biosynthesis methods, 95
Plant extracts, 95, 96
Plasma-assisted CVD (PACVD), 111
Plasma-enhanced CVD (PECVD), 91, 111
Plastic membranes, 138
Platinum (Pt), 303
    NPs, 96
PLD method, *see* Pulsed laser deposition
    method
PL peaks, *see* Photoluminescence peaks
PL spectroscopy, *see* Photoluminescence
    spectroscopy
PMMA, *see* Poly(methyl methacrylate)
PMT, *see* Photomultiplier tube
Poly(aryletherketone) (PAEK), 207
Poly(ethylene glycol)s (PEG), 177
Poly(ethylene naphthalate) (PEN), 223
Poly(ethylene oxide) (PEO), 177, 219, 225
Poly(methyl methacrylate) (PMMA), 185, 201,
    225, 226
Poly(tetrafluoroethylene) (PTFE), 295
Poly(trimethyl hexamethylene
    terephthalamide) (PA 6[3]T), 220
Poly(vinyl alcohol) (PVA), 162
Poly(vinyl chloride) (PVC), 228
Poly(vinylpyrrolidone) (PVP), 177
Poly(ε-caprolactone) (PCL), 221
Polyacrylonitrile (PAN), 227
Polyaniline (PANI), 226, 271, 272, 284, 295
Polybenzimidazole (PBI), 227
Polycarbonate (PC), 153

Polydimethylsiloxane (PDMS), 86, 87
Polyethylene (PE), 78, 185, 223
Polyethylene terephthalate (PET), 138, 223
Polyimide–clay hybrid, 192
Polymer/BaTiO$_3$ nanocomposites, 208
    dielectric constant of BaTiO$_3$ ceramic, 208
    electrical properties, 209–211
    mechanical properties, 212–213
    preparation of PC/BaTiO$_3$ nanocomposites,
        208–209
    thermal properties, 211–212
Polymer/BN nanocomposites, 203
    mechanical properties, 205
    thermal properties, 205–207
    wear resistance, 207–208
Polymer/BN nanocomposites, 205–207
Polymer–clay nanocomposites, 184
    abrasion resistance, 196
    characterization, 188
    fabrication, 185–188
    mechanical properties, 193–196
    permeability, 191–193
    properties, 189
    structure of montmorillonite, 184–185
    thermal properties, 189–191
    wear resistance, 196
Polymer/CNT nanocomposites, 196
    electrical properties, 196–199
    mechanical properties, 199–201
    thermal conductivity, 202–203, 204
Polymeric nanofibers, 217, 299–300
    applications, 226
    biomedical application, 228–229
    functional sensors, 229
    nanocomposite fibers for structural
        applications, 226–228
    nanofilters, 228
    nanogenerators, 228
    parameter affecting diameter of nanofibers,
        218–219
    properties of nanofibers, 219–226
    protective fabric for military, 229
    synthesis, 217–218
Polymer nanocomposites, 183
    polymer/BaTiO$_3$ nanocomposites, 208–213
    polymer/BN nanocomposites, 203–208
    polymer–clay nanocomposites, 184–196
    polymer/CNT nanocomposites, 196–203, 204
    processing, properties, and
        applications of, 183
Polymer(s), 183
    matrix composites, 183
    membranes, 138

    nanofibers, 229, 296–297
    stabilizers, 93
Polypropylene (PP), 185
Polypyrrole (PPy), 225, 295
Polysaccharides, 177
Polystyrene (PS), 161
Polyvinyl alcohol (PVA), 92, 185, 200
Polyvinylidene difluoride (PVDF), 194
Powder
    contamination level, 78
    suspension method, 257
    XRD method, 233
PP-MA, *see* MA-modified PP oligomers
PP, *see* Polypropylene
ppm, *see* Parts per million
PPy, *see* Polypyrrole
Pre-thinning, 257
Pressure-sensitive adhesives (PSAs), 61
Pressure injection technique, 140
Pressurized water reactor (PWR), 279
Printed circuit boards (PCBs), 191
Probes, 129
Process control agent (PCA), 77, 185
Projection printing, 83
Protective fabric for military, 229
Protein nanofibers, 296
Proton exchange membrane fuel cell (PEM fuel
    cell), 284
Proximity printing, 83
PS, *see* Polystyrene
PSAs, *see* Pressure-sensitive adhesives
Pt, *see* Platinum
PTFE, *see* Poly(tetrafluoroethylene)
Pulsed laser deposition method
    (PLD method), 29
Purification of CNTs, 113
    chemical oxidation method, 113–114
    EDS, 116
    HRTEM, 115
    physical method, 114–116
Purple of Cassius, 69
PVA, *see* Poly(vinyl alcohol); Polyvinyl alcohol
PVC, *see* Poly(vinyl chloride)
PVD, *see* Physical vapor deposition
PVDF, *see* Polyvinylidene difluoride
PVP, *see* Poly(vinylpyrrolidone)
PWR, *see* Pressurized water reactor
1-Pyrenebutyric acid (PBA), 205

Q

Quantum dot-sensitized solar cell
    (QD-SSC), 286

Quantum dots (QDs), 1, 5, 242
  solar cells, 285
Quantum size effect, 142
  confinement effect, 44

R

Radial breathing modes (RBM), 118
Radiator, 277–278
Radio frequency signals (RF signals), 278
Raman effect, 239
Raman scattering, 239
Raman shift, 239
Raman spectra of graphene, 155
Raman spectroscopy, 112, 118, 233,
    239–240
Rayleigh scattering, 239
Rayleigh's equation, 83
RBM, *see* Radial breathing modes
Redshift, 45, 144
Reduced GO (rGO), 153
Reducing reagents, 93
Reduction reaction, 267
Reflection high-energy electron diffraction
    (RHEED), 89
Remanence (M_r), 172
Remediation process, 297
Renewable energy, 285
Resolution limit (R), 83
Respiratory system, 312
Reverse micelle method, 93–94
RF signals, *see* Radio frequency signals
rGO, *see* Reduced GO
RHEED, *see* Reflection high-energy electron
    diffraction
Rhodium (Rh), 303
Risks
  from metallic nanoparticle, 312–313
  from oxide nanoparticle, 313–314
RoM, *see* Rule of mixtures
"Roman Lycurgus cup", 68, 69
Rosa rugosa (*R. rugosa*), 97
Royal Mail Ship Titanic sinking, 36
Ruby-red glasses, 69
Rule of mixtures (RoM), 190, 212

S

SAD, *see* Selected-area diffraction
Saturation magnetization (M_s), 172
SBFET, *see* Schottky barrier field effect
    transistor
SBR, *see* Styrene-butadiene rubber

Scanning electron microscopy (SEM), 112,
    233; *see also* Transmission electron
    microscopy
  basic principle, 251–252
  specimen preparation for, 252–253
Scanning probe lithography, 85
Scanning probe microscopy (SPM), 233, 258; *see
    also* Electron microscopy
  atomic force microscopy, 260–262
  basic principle of STM, 258–260
Scanning-TEM (STEM), 256
Scanning tunneling microscopy (STM), 85, 233
  basic principle, 258–260
School of learning nature, 53
Schottky barrier field effect transistor
    (SBFET), 146
SCS method, *see* Solution combustion synthesis
    method
SD, *see* Single domain
Secondary electrons (SEs), 250, 251
Secondary ion mass spectroscopy (SIMS), 262
Selected-area diffraction (SAD), 253, 256
Self-cleaning coatings, 293
  hydrophilic coatings, 293–294
  hydrophobic coating, 293
Self-cleaning effect, 54
Self-propagating high-temperature synthesis
    (SHS), 101, 102
SEM, *see* Scanning electron microscopy
Semiconductor gas sensors, 48
Sensors, 129, 146–147
Separation of NPs, 94
SEs, *see* Secondary electrons
SET, *see* Single-electron transistor
Setae, 61
Severe plastic deformation (SPD), 36, 78
  ECAP, 80
  general tendency of change in strength and
      ductility during, 82
  effect of increasing diameter to grain size
      ratio, 81
  medical implants, 80
  SPD-processed nanostructured TiNi alloy, 79
  techniques, 79
Severe plastic torsion straining (SPTS), 42
Shape anisotropy, 41
SHS, *see* Self-propagating high-temperature
    synthesis
SiC, *see* Silicon carbide
σ–π rehybridization, 117
SiH_2Cl_2, *see* Dichlorosilane
Silane (SiH_4), 91
Silica-coated nanoparticles, 177

Silica nanoparticles (SiO₂ NPs), 292
Silicon-based sol-gel process, 100
Silicon (Si), 135
Silicon carbide (SiC), 154
Silicon tetrachloride (SiCl₄), 91
Silk, 57, 58
Silver (Ag), 98
Silver chloride (AgCl), 312–313
Silver nanoparticles (Ag NPs), 96, 229, 289, 312; *see also* Oxide nanoparticles (Oxide NPs)
Silver nitrate (AgNO₃), 312–313
SIMS, *see* Secondary ion mass spectroscopy
Single-composition nanoparticles, 7
Single-electron transistor (SET), 275, 292
Single-junction photovoltaic cell, 286
Single domain (SD), 63
Single wall nanotubes (SWNTs), 107
SiNW-FETs, *see* Si NW field effect transistors
Si NW field effect transistors (SiNW-FETs), 146
SiO₂ NPs, *see* Silica nanoparticles
Skin, 310
SLS, *see* Static light scattering
Smart textiles, *see* Intelligent textiles
Sodium cholate (NaC), 152
Soft lithography, 87
Sol-gel method, 99–101
    hydrolysis and condensation reactions, 100
    synthesis of oxide nanoparticles, 101
"Sol", 99
Solar cells, *see* Photoelectrochemical cells
Solar energy, 284, 301
    photoelectrochemical cells, 285–287
    thermoelectric devices, 287–289
Solar water heating (SWH), 279–280
Solution combustion synthesis method (SCS method), 101, 102
Solution method, 185–186
Spatial resolution of SEM, 252
Spatulae, 60
SPD, *see* Severe plastic deformation
Specimen preparation
    for SEM, 252–253
    for TEM, 256–258
Spectroscopy-based techniques, 233
Spherical metallic nanoparticles, 143
Spherical particles; *see also* Cubical particles
    calculated percentages of atoms in bulk and on particle surface, 17
    diamond unit cell of silicon, 19
    size effect on surface atoms, 15–19
Spider capture silk, 57
Spider dragline silks, 57
Spider silks, 57

superhydrophobic water strider leg and biomimetic materials, 58
    tensile strength, 59
SPM, *see* Scanning probe microscopy
SPR, *see* Surface plasmon resonance
SPTS, *see* Severe plastic torsion straining
Sputtered films, 88
SS, *see* Stainless steel
Stabilizing agent, 94
Stained glass, 68–70
    windows, 2
Stainless steel (SS), 76
Staphylococcus aureus (*S. aureus*), 290
Static light scattering (SLS), 247
Stearic hindrance, 93
STEM, *see* Scanning-TEM
STM, *see* Scanning tunneling microscopy
STM lithography (STML), 85
Strength, 34–36
Strengthening factor, 34
Strong reduction reaction, 93
Styrene-butadiene rubber (SBR), 303
Superhydrophilic coating, 294
Superhydrophobic properties, 293
Superparamagnetic nanoparticles, 180
Superparamagnetic particles, 174
Surface
    anisotropy, 41
    area analysis, 233
    area analysis, 245–246
    charging, 252
    modification of magnetic nanoparticles, 177–178
    relaxation, 32
Surface plasmon resonance (SPR), 4, 45, 69, 243
Surfactant, *see* Process control agent (PCA)
SWH, *see* Solar water heating
SWNTs, *see* Single wall nanotubes
Synthesis of nanomaterials, 73
    bottom-up approaches, 88–104
    top-down approaches, 73–87
Synthesized CNTs, 113

T

Tapping mode, 261
TEGs, *see* Thermoelectric generators
TEM, *see* Transmission electron microscopy
Template method, 138
    electrochemical deposition, 138–140
    electroless deposition, 140
    pressure injection technique, 140
Tensile strength of single nylon-6 nanofiber, 219

TEP, *see* Thermoelectric power
Tetragonal BaTiO$_3$ (t-BaTiO$_3$)/PC
    nanocomposites, 209
Textile fabrics, 294–295
TFTs, *see* Thin-film transistors
TGA, *see* Thermogravimetric analysis
Thermal conductivity, 141–142, 159
    polymer/CNT nanocomposites,
        202–203, 204
Thermal CVD method, 110–111
Thermal cycling, 68
Thermal exfoliation method, 153
Thermal expansion, 159–160
Thermal interface materials (TIMs), 145, 147
Thermal stability, 160
Thermoelectric devices, 287–289
Thermoelectric generators (TEGs), 145–146
Thermoelectric power (TEP), 123
Thermogravimetric analysis (TGA), 112, 160,
    188, 263
Thermoplastic polyolefin (TPO), 191
Thin-film
    interference, 56
    recording media, 42
Thin-film transistors (TFTs), 128
THR, *see* Total heat release
Three-dimensional structure (3-D structure),
    129–130
3-D nanomaterials, 6–7
Threshold field values, 127
TiAl, *see* Titanium aluminum
TIMs, *see* Thermal interface materials
TiN, *see* Titanium nitride
Tin oxide (SnO$_2$), 98
Tissue template, 229
Titanium aluminum (TiAl), 34
Titanium nitride (TiN), 255
Tittanium dioxide (TiO$_2$)
    nanocrystals, 286
    nanomaterials, 298, 314
    nanopowders, 298
TM, *see* Transverse mode
TOP, *see* Trioctylphosphine
Top-down approach(es), 2, 73, 151; *see also*
    Bottom-up approaches
    lithography, 81–87
    MA, 73–78
    SPD, 78–81
Total heat release (THR), 165
Toughness, 36
    ECAP method, 37
    fracture toughness of diamond–SiC
        nanocomposites, 39
    impact toughness of nanostructured
        and coarse-grained Ti *vs.* testing
        temperature, 36
    tensile engineering stress–strain curves for
        three Cu samples, 37
    tensile stress–strain curve, 38
Toxicity
    from metallic nanoparticle, 312–313
    from oxide nanoparticle, 313–314
Toyota, 302
TPO, *see* Thermoplastic polyolefin
Transducer part, 146
Transfer chamber, 89
Transistors, 128
Transition metal catalysts, 111
Translational vector, 116
Transmission electron microscopy (TEM), 233;
        see also Scanning electron microscopy
    basic principle, 253–256
    specimen preparation for, 256–258
Transverse mode (TM), 45
Trioctylphosphine (TOP), 287
Trisodium citrate (Na$_3$C$_6$H$_5$O$_7$), 92
Two-band model, 123
2-D nanomaterials, 6

U

UFG metals, *see* Ultrafine-grained metals
UHV chamber, *see* Ultrahigh-vacuum chamber
Ultrafine-grained metals (UFG metals), 79
"Ultrafine grain size" materials, 8
Ultrahigh-vacuum chamber (UHV chamber),
    88, 89
Ultramicrotomy, 257
Ultraphobic properties, 293
Ultraviolet light (UV), 82
    blocking agents, 314
    light, 82, 286
Ultraviolet photoelectron spectroscopy
    (UPS), 125
Umklapp scattering, 122
Unit vectors, 116
UPS, *see* Ultraviolet photoelectron spectroscopy
UV, *see* Ultraviolet light
UV–visible spectroscopy (UV–Vis
    spectroscopy), 233, 240–243

V

Vapor–liquid–solid (VLS), 135–137
Vehicle radiator, 302
Venus Flower Basket, *see* Glass sponges

Vertical CNT-FETs, 128
Vibrational spectroscopy, 239
Vibratory ball mill, 75
Vickers hardness, 35
N-Vinyl-2-pyrrolidone (NVP), 152
VLS, *see* Vapor–liquid–solid

W

Wastewater, 180, 297
Water filters, 129–130
Water-in-oil microemulsion, *see* Reverse
        micelle method
Water lily, *see* Lotus leaf
Water strider (*Gerris remigis*), 57
Water treatment, 297
    CNT powder, 298
    dendrimers, 299
    graphene, 299
    polymeric nanofibers, 299–300
    TiO$_2$ Nanomaterials, 298
    zerovalent metals, 298
Wavelength-dispersive spectroscopy
        (WDS), 233
WC–Co nanocomposites, 39
WDS, *see* Wavelength-dispersive spectroscopy
Wear resistance, 45–46
    polymer/BN nanocomposites, 207–208
    of polymer–clay nanocomposites, 196

Wet chemical route, 91–93
Wet grinding, 75
Williamson–Hall equation, 236
Wound dressing, 229, 296

X

X-ray diffraction (XRD), 155, 187, 233–238
X-ray lithography, 84
X-ray photoelectron spectroscopy (XPS), 112,
        262–263

Y

Young modulus
    of nanostructured Cu, 43
    of single nylon-6 nanofiber, 219
Yttria (Y$_2$O$_3$), 33

Z

Zero current potential (ZCP), 269
0-D nanomaterials, 5
Zerovalent metals, 298
Zeta potential (ζ potential), 264
Zinc oxide (ZnO), 290
    NPs, 99
Zirconia, 33
ZnCl$_2$, 103

T - #0176 - 111024 - C88 - 254/178/16 - PB - 9780367572785 - Gloss Lamination